Matter

" The Building Blocks of The Elements "

Edited by Paul F. Kisak

Contents

Chapter 1

Matter

This article is about the concept in the physical sciences. For other uses, see Matter (disambiguation).

Before the 20th century, the term **matter** included **ordinary matter** composed of atoms and excluded other energy phenomena such as light or sound. This concept of matter may be generalized from atoms to include any objects having mass even when at rest, but this is ill-defined because an object's mass can arise from its (possibly massless) constituents' motion and interaction energies. Thus, matter does not have a universal definition, nor is it a fundamental concept in physics today. Matter is also used loosely as a general term for the substance that makes up all observable physical objects.[1][2]

All the objects from everyday life that we can bump into, touch or squeeze are composed of atoms. This atomic matter is in turn made up of interacting subatomic particles—usually a nucleus of protons and neutrons, and a cloud of orbiting electrons.[3][4] Typically, science considers these composite particles matter because they have both rest mass and volume. By contrast, massless particles, such as photons, are not considered matter, because they have neither rest mass nor volume. However, not all particles with rest mass have a classical volume, since fundamental particles such as quarks and leptons (sometimes equated with matter) are considered "point particles" with no effective size or volume. Nevertheless, quarks and leptons together make up "ordinary matter", and their interactions contribute to the effective volume of the composite particles that make up ordinary matter.

Matter commonly exists in four *states* (or *phases*): solid, liquid and gas, and plasma. However, advances in experimental techniques have revealed other previously theoretical phases, such as Bose–Einstein condensates and fermionic condensates. A focus on an elementary-particle view of matter also leads to new phases of matter, such as the quark–gluon plasma.[5] For much of the history of the natural sciences people have contemplated the exact nature of matter. The idea that matter was built of discrete building blocks, the so-called *particulate theory of matter*, was first put forward by the Greek philosophers Leucippus (~490 BC) and Democritus (~470–380 BC).[6]

Matter should not be confused with mass, as the two are not quite the same in modern physics.[7] For example, mass is a conserved quantity, which means that its value is unchanging through time, within closed systems. However, matter is *not* conserved in such systems, although this is not obvious in ordinary conditions on Earth, where matter is approximately conserved. Still, special relativity shows that matter may disappear by conversion into energy, even inside closed systems, and it can also be created from energy, within such systems. However, because *mass* (like energy) can neither be created nor destroyed, the quantity of mass and the quantity of energy remain the same during a transformation of matter (which represents a certain amount of energy) into non-material (i.e., non-matter) energy. This is also true in the reverse transformation of energy into matter.

Different fields of science use the term matter in different, and sometimes incompatible, ways. Some of these ways are based on loose historical meanings, from a time when there was no reason to distinguish mass and matter. As such, there is no single universally agreed scientific meaning of the word "matter". Scientifically, the term "mass" is well-defined, but "matter" is not. Sometimes in the field of physics "matter" is simply equated with particles that exhibit rest mass (i.e., that cannot travel at the speed of light), such as quarks and leptons. However, in both physics and chemistry, matter exhibits both wave-like and particle-like properties, the so-called wave–particle duality.[8][9][10]

1.1   Definition

1.1.1   Common definition

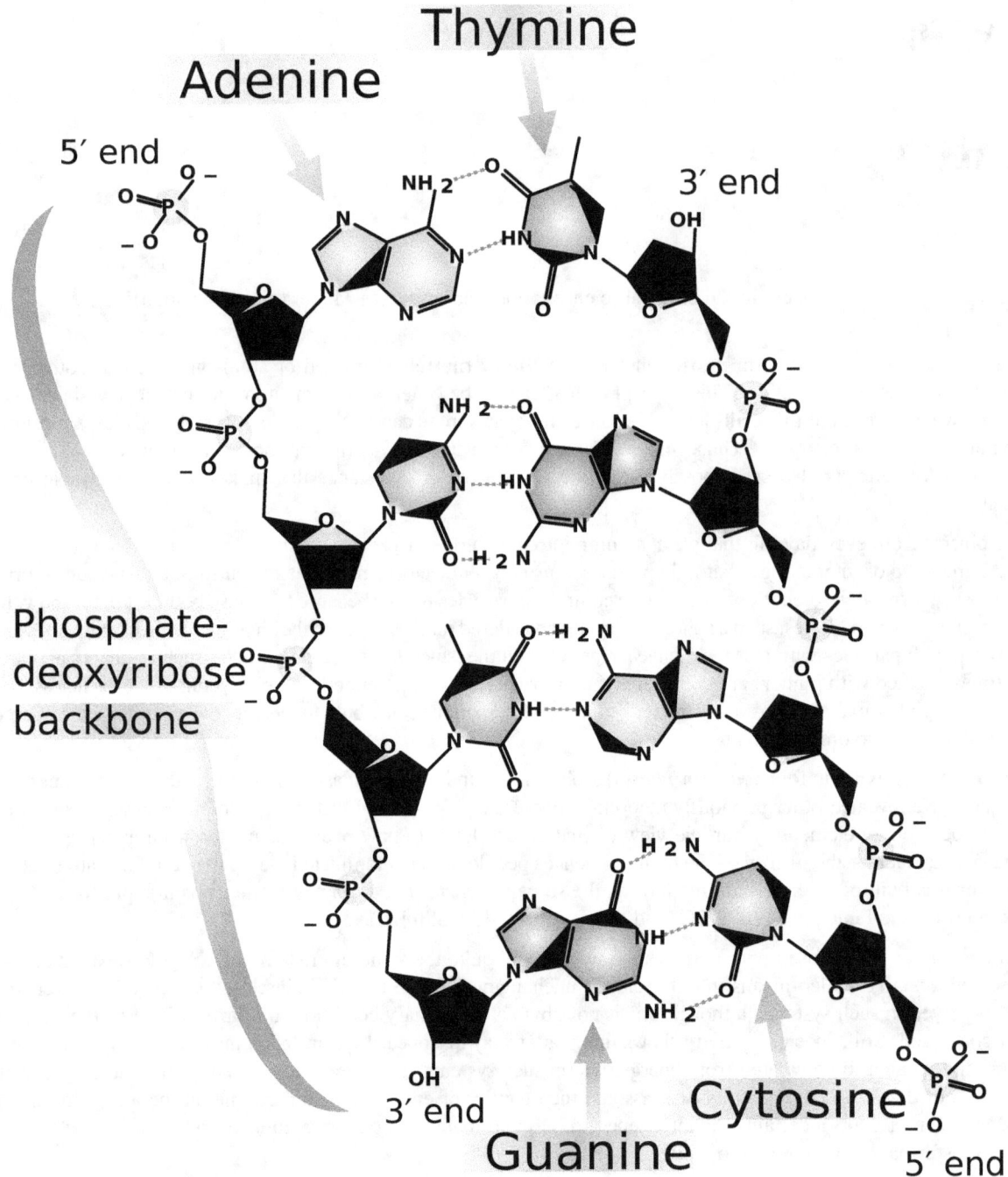

The DNA molecule is an example of matter *under the "atoms and molecules" definition.*

The common definition of matter is *anything that has mass and volume (occupies space).*[11][12] For example, a car would be said to be made of matter, as it occupies space, and has mass.

The observation that matter occupies space goes back to antiquity. However, an explanation for why matter occupies space is recent, and is argued to be a result of the phenomenon described in the Pauli exclusion principle.[13][14] Two particular examples where the exclusion principle clearly relates matter to the occupation of space are white dwarf stars and neutron stars, discussed further below.

1.1.2 Relativity

Main article: Mass–energy equivalence

In the context of relativity, mass is not an additive quantity, in the sense that one can add the rest masses of particles in a system to get the total rest mass of the system.[1] Thus, in relativity usually a more general view is that it is not the sum of rest masses, but the energy–momentum tensor that quantifies the amount of matter. This tensor gives the rest mass for the entire system. "Matter" therefore is sometimes considered as anything that contributes to the energy–momentum of a system, that is, anything that is not purely gravity.[15][16] This view is commonly held in fields that deal with general relativity such as cosmology. In this view, light and other massless particles and fields are part of matter.

The reason for this is that in this definition, electromagnetic radiation (such as light) as well as the energy of electromagnetic fields contributes to the mass of systems, and therefore appears to add matter to them. For example, light radiation (or thermal radiation) trapped inside a box would contribute to the mass of the box, as would any kind of energy inside the box, including the kinetic energy of particles held by the box. Nevertheless, isolated individual particles of light (photons) and the isolated kinetic energy of massive particles, are normally not considered to be *matter*.

A difference between matter and mass therefore may seem to arise when single particles are examined. In such cases, the mass of single photons is zero. For particles with rest mass, such as leptons and quarks, isolation of the particle in a frame where it is not moving, removes its kinetic energy.

A source of definition difficulty in relativity arises from two definitions of mass in common use, one of which is formally equivalent to total energy (and is thus observer dependent), and the other of which is referred to as rest mass or invariant mass and is independent of the observer. Only "rest mass" is loosely equated with matter (since it can be weighed). Invariant mass is usually applied in physics to unbound systems of particles. However, energies which contribute to the "invariant mass" may be weighed also in special circumstances, such as when a system that has invariant mass is confined and has no net momentum (as in the box example above). Thus, a photon with no mass may (confusingly) still add mass to a system in which it is trapped. The same is true of the kinetic energy of particles, which by definition is not part of their rest mass, but which does add rest mass to systems in which these particles reside (an example is the mass added by the motion of gas molecules of a bottle of gas, or by the thermal energy of any hot object).

Since such mass (kinetic energies of particles, the energy of trapped electromagnetic radiation and stored potential energy of repulsive fields) is measured as part of the mass of ordinary *matter* in complex systems, the "matter" status of "massless particles" and fields of force becomes unclear in such systems. These problems contribute to the lack of a rigorous definition of matter in science, although mass is easier to define as the total stress–energy above (this is also what is weighed on a scale, and what is the source of gravity).

1.1.3 Atoms definition

A definition of "matter" based on its physical and chemical structure is: *matter is made up of atoms*.[17] As an example, deoxyribonucleic acid molecules (DNA) are matter under this definition because they are made of atoms. This definition can extend to include charged atoms and molecules, so as to include plasmas (gases of ions) and electrolytes (ionic solutions), which are not obviously included in the atoms definition. Alternatively, one can adopt the *protons, neutrons, and electrons* definition.

1.1.4 Protons, neutrons and electrons definition

A definition of "matter" more fine-scale than the atoms and molecules definition is: *matter is made up of what atoms and molecules are made of*, meaning anything made of positively charged protons, neutral neutrons, and negatively charged electrons.[18] This definition goes beyond atoms and molecules, however, to include substances made from these building blocks that are *not* simply atoms or molecules, for example white dwarf matter—typically, carbon and oxygen nuclei in a sea of degenerate electrons. At a microscopic level, the constituent "particles" of matter such as protons, neutrons, and electrons obey the laws of quantum mechanics and exhibit wave–particle duality. At an even deeper level, protons and neutrons are made up of quarks and the force fields (gluons) that bind them together (see Quarks and leptons definition below).

1.1.5   Quarks and leptons definition

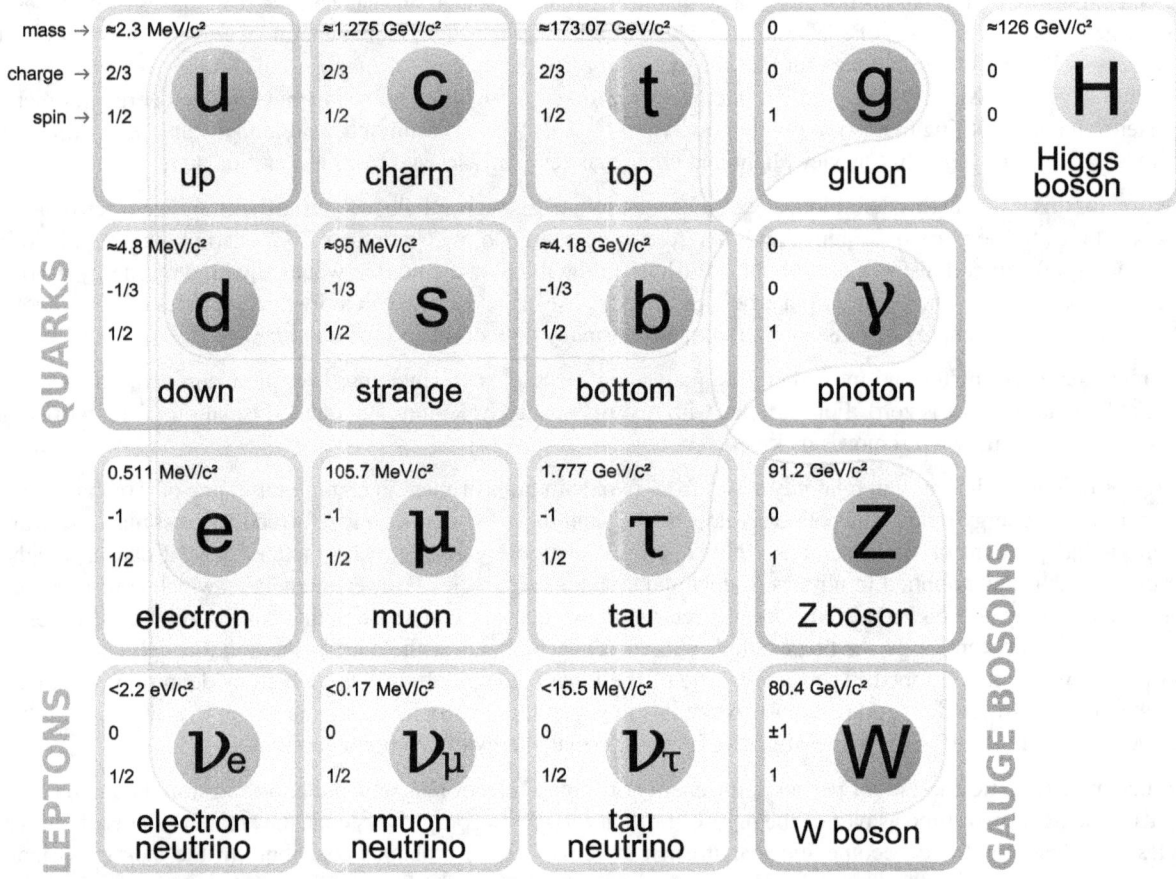

Under the "quarks and leptons" definition, the elementary and composite particles made of the quarks (in purple) and leptons (in green) would be matter—while the gauge bosons (in red) would not be matter. However, interaction energy inherent to composite particles (for example, gluons involved in neutrons and protons) contribute to the mass of ordinary matter.

As seen in the above discussion, many early definitions of what can be called *ordinary matter* were based upon its structure or *building blocks*. On the scale of elementary particles, a definition that follows this tradition can be stated as: *ordinary matter is everything that is composed of elementary fermions, namely quarks and leptons.*[19][20] The connection between these formulations follows.

Leptons (the most famous being the electron), and quarks (of which baryons, such as protons and neutrons, are made) combine to form atoms, which in turn form molecules. Because atoms and molecules are said to be matter, it is natural to phrase the definition as: *ordinary matter is anything that is made of the same things that atoms and molecules are made of.* (However, notice that one also can make from these building blocks matter that is *not* atoms or molecules.) Then, because electrons are leptons, and protons, and neutrons are made of quarks, this definition in turn leads to the definition of matter as being *quarks and leptons*, which are the two types of elementary fermions. Carithers and Grannis state: *Ordinary matter is composed entirely of first-generation particles, namely the [up] and [down] quarks, plus the electron and its neutrino.*[20] (Higher generations particles quickly decay into first-generation particles, and thus are not commonly encountered.[21])

This definition of ordinary matter is more subtle than it first appears. All the particles that make up ordinary matter (leptons and quarks) are elementary fermions, while all the force carriers are elementary bosons.[22] The W and Z bosons that mediate the weak force are not made of quarks or leptons, and so are not ordinary matter, even if they have mass.[23] In other words, mass is not something that is exclusive to ordinary matter.

The quark–lepton definition of ordinary matter, however, identifies not only the elementary building blocks of matter, but also includes composites made from the constituents (atoms and molecules, for example). Such composites contain an interaction energy that holds the constituents together, and may constitute the bulk of the mass of the composite. As an example, to a great extent, the mass of an atom is simply the sum of the masses of its constituent protons, neutrons and electrons. However, digging deeper, the protons and neutrons are made up of quarks bound together by gluon fields (see dynamics of quantum chromodynamics) and these gluons fields contribute significantly to the mass of hadrons.[24] In other words, most of what composes the "mass" of ordinary matter is due to the binding energy of quarks within protons and neutrons.[25] For example, the sum of the mass of the three quarks in a nucleon is approximately 12.5 MeV/c^2, which is low compared to the mass of a nucleon (approximately 938 MeV/c^2).[21][26] The bottom line is that most of the mass of everyday objects comes from the interaction energy of its elementary components.

1.1.6 Smaller building blocks issue

The Standard Model groups matter particles into three generations, where each generation consists of two quarks and two leptons. The first generation is the *up* and *down* quarks, the *electron* and the *electron neutrino*; the second includes the *charm* and *strange* quarks, the *muon* and the *muon neutrino*; the third generation consists of the *top* and *bottom* quarks and the *tau* and *tau neutrino*.[27] The most natural explanation for this would be that quarks and leptons of higher generations are excited states of the first generations. If this turns out to be the case, it would imply that quarks and leptons are composite particles, rather than elementary particles.[28]

1.2 Structure

In particle physics, fermions are particles that obey Fermi–Dirac statistics. Fermions can be elementary, like the electron— or composite, like the proton and neutron. In the Standard Model, there are two types of elementary fermions: quarks and leptons, which are discussed next.

1.2.1 Quarks

Main article: Quark

Quarks are particles of spin-½, implying that they are fermions. They carry an electric charge of −⅓ e (down-type quarks) or +⅔ e (up-type quarks). For comparison, an electron has a charge of −1 e. They also carry colour charge, which is the equivalent of the electric charge for the strong interaction. Quarks also undergo radioactive decay, meaning that they are subject to the weak interaction. Quarks are massive particles, and therefore are also subject to gravity.

Baryonic matter

Main article: Baryon

Baryons are strongly interacting fermions, and so are subject to Fermi–Dirac statistics. Amongst the baryons are the protons and neutrons, which occur in atomic nuclei, but many other unstable baryons exist as well. The term baryon usually refers to triquarks—particles made of three quarks. "Exotic" baryons made of four quarks and one antiquark are known as the pentaquarks, but their existence is not generally accepted.

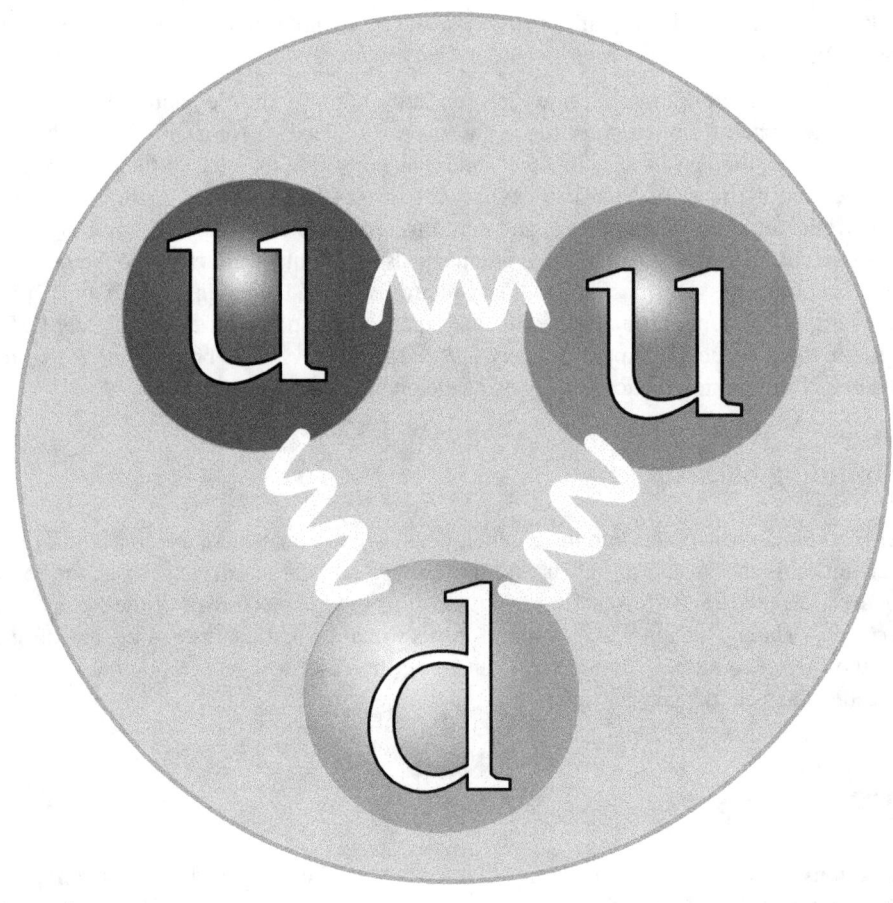

Quark structure of a proton: 2 up quarks and 1 down quark.

Baryonic matter is the part of the universe that is made of baryons (including all atoms). This part of the universe does not include dark energy, dark matter, black holes or various forms of degenerate matter, such as compose white dwarf stars and neutron stars. Microwave light seen by Wilkinson Microwave Anisotropy Probe (WMAP), suggests that only about 4.6% of that part of the universe within range of the best telescopes (that is, matter that may be visible because light could reach us from it), is made of baryonic matter. About 23% is dark matter, and about 72% is dark energy.[30]

Degenerate matter

Main article: Degenerate matter

In physics, **degenerate matter** refers to the ground state of a gas of fermions at a temperature near absolute zero.[31] The Pauli exclusion principle requires that only two fermions can occupy a quantum state, one spin-up and the other spin-down.

A comparison between the white dwarf IK Pegasi B (center), its A-class companion IK Pegasi A (left) and the Sun (right). This white dwarf has a surface temperature of 35,500 K.

Hence, at zero temperature, the fermions fill up sufficient levels to accommodate all the available fermions—and in the case of many fermions, the maximum kinetic energy (called the *Fermi energy*) and the pressure of the gas becomes very large, and depends on the number of fermions rather than the temperature, unlike normal states of matter.

Degenerate matter is thought to occur during the evolution of heavy stars.[32] The demonstration by Subrahmanyan Chandrasekhar that white dwarf stars have a maximum allowed mass because of the exclusion principle caused a revolution in the theory of star evolution.[33]

Degenerate matter includes the part of the universe that is made up of neutron stars and white dwarfs.

Strange matter

Main article: Strange matter

Strange matter is a particular form of quark matter, usually thought of as a *liquid* of up, down, and strange quarks. It is contrasted with nuclear matter, which is a liquid of neutrons and protons (which themselves are built out of up and down quarks), and with non-strange quark matter, which is a quark liquid that contains only up and down quarks. At high enough density, strange matter is expected to be color superconducting. Strange matter is hypothesized to occur in the core of neutron stars, or, more speculatively, as isolated droplets that may vary in size from femtometers (strangelets) to kilometers (quark stars).

Two meanings of the term "strange matter"   In particle physics and astrophysics, the term is used in two ways, one broader and the other more specific.

1. The broader meaning is just quark matter that contains three flavors of quarks: up, down, and strange. In this definition, there is a critical pressure and an associated critical density, and when nuclear matter (made of protons and neutrons) is compressed beyond this density, the protons and neutrons dissociate into quarks, yielding quark matter (probably strange matter).

2. The narrower meaning is quark matter that is *more stable than nuclear matter*. The idea that this could happen is the "strange matter hypothesis" of Bodmer[34] and Witten.[35] In this definition, the critical pressure is zero: the true ground state of matter is *always* quark matter. The nuclei that we see in the matter around us, which are droplets of nuclear matter, are actually metastable, and given enough time (or the right external stimulus) would decay into droplets of strange matter, i.e. strangelets.

1.2.2 Leptons

Main article: Lepton

Leptons are particles of spin-$\frac{1}{2}$, meaning that they are fermions. They carry an electric charge of −1 e (charged leptons) or 0 e (neutrinos). Unlike quarks, leptons do not carry colour charge, meaning that they do not experience the strong interaction. Leptons also undergo radioactive decay, meaning that they are subject to the weak interaction. Leptons are massive particles, therefore are subject to gravity.

1.3 Phases

Main article: Phase (matter)
See also: Phase diagram and State of matter

In bulk, matter can exist in several different forms, or states of aggregation, known as *phases*,[39] depending on ambient pressure, temperature and volume.[40] A phase is a form of matter that has a relatively uniform chemical composition and physical properties (such as density, specific heat, refractive index, and so forth). These phases include the three familiar ones (solids, liquids, and gases), as well as more exotic states of matter (such as plasmas, superfluids, supersolids, Bose–Einstein condensates, ...). A *fluid* may be a liquid, gas or plasma. There are also paramagnetic and ferromagnetic phases of magnetic materials. As conditions change, matter may change from one phase into another. These phenomena are called phase transitions, and are studied in the field of thermodynamics. In nanomaterials, the vastly increased ratio of surface area to volume results in matter that can exhibit properties entirely different from those of bulk material, and not well described by any bulk phase (see nanomaterials for more details).

Phases are sometimes called *states of matter*, but this term can lead to confusion with thermodynamic states. For example, two gases maintained at different pressures are in different *thermodynamic states* (different pressures), but in the same *phase* (both are gases).

1.4 Antimatter

Main article: Antimatter

In particle physics and quantum chemistry, **antimatter** is matter that is composed of the antiparticles of those that constitute ordinary matter. If a particle and its antiparticle come into contact with each other, the two annihilate; that is, they may both be converted into other particles with equal energy in accordance with Einstein's equation $E = mc^2$. These new particles may be high-energy photons (gamma rays) or other particle–antiparticle pairs. The resulting particles are endowed with an amount of kinetic energy equal to the difference between the rest mass of the products of the annihilation and the rest mass of the original particle–antiparticle pair, which is often quite large.

Antimatter is not found naturally on Earth, except very briefly and in vanishingly small quantities (as the result of radioactive decay, lightning or cosmic rays). This is because antimatter that came to exist on Earth outside the confines

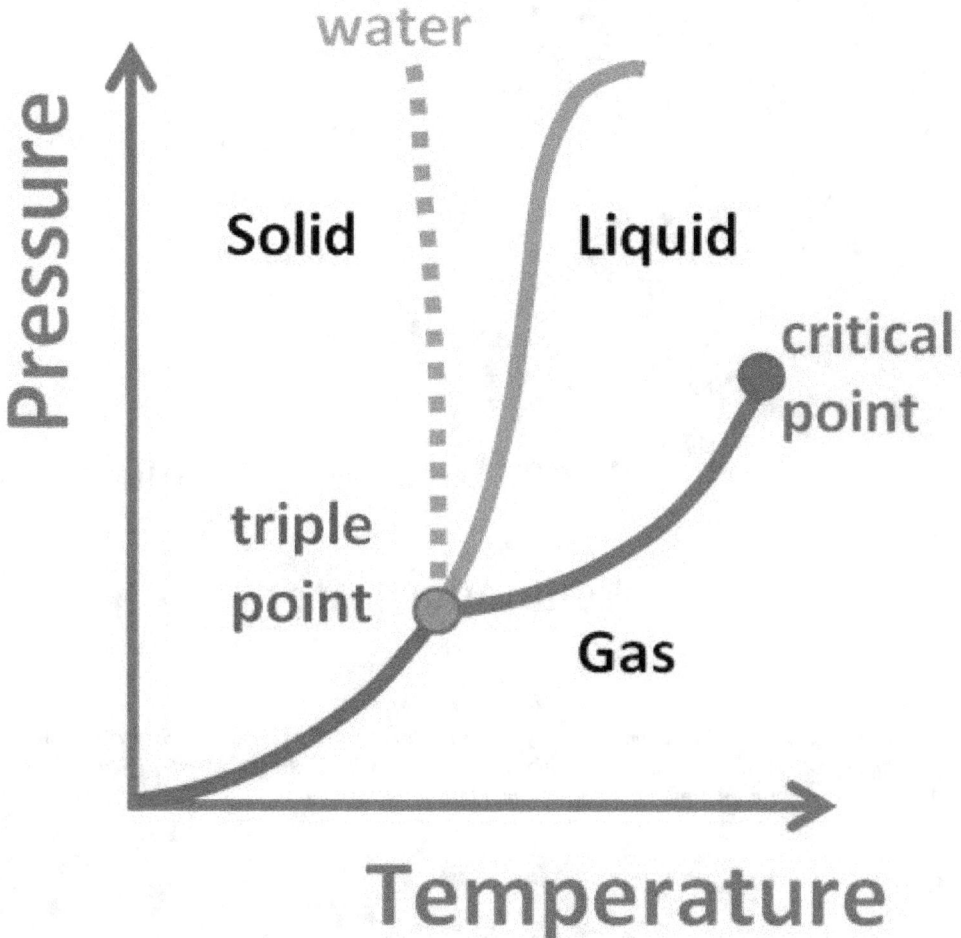

Phase diagram for a typical substance at a fixed volume. Vertical axis is Pressure, horizontal axis is Temperature. The green line marks the freezing point (above the green line is solid, below it is liquid) and the blue line the boiling point (above it is liquid and below it is gas). So, for example, at higher T, a higher P is necessary to maintain the substance in liquid phase. At the triple point the three phases; liquid, gas and solid; can coexist. Above the critical point there is no detectable difference between the phases. The dotted line shows the anomalous behavior of water: ice melts at constant temperature with increasing pressure.[38]

of a suitable physics laboratory would almost instantly meet the ordinary matter that Earth is made of, and be annihilated. Antiparticles and some stable antimatter (such as antihydrogen) can be made in tiny amounts, but not in enough quantity to do more than test a few of its theoretical properties.

There is considerable speculation both in science and science fiction as to why the observable universe is apparently almost entirely matter, and whether other places are almost entirely antimatter instead. In the early universe, it is thought that matter and antimatter were equally represented, and the disappearance of antimatter requires an asymmetry in physical laws called the charge parity (or CP symmetry) violation. CP symmetry violation can be obtained from the Standard Model,[41] but at this time the apparent asymmetry of matter and antimatter in the visible universe is one of the great unsolved problems in physics. Possible processes by which it came about are explored in more detail under baryogenesis.

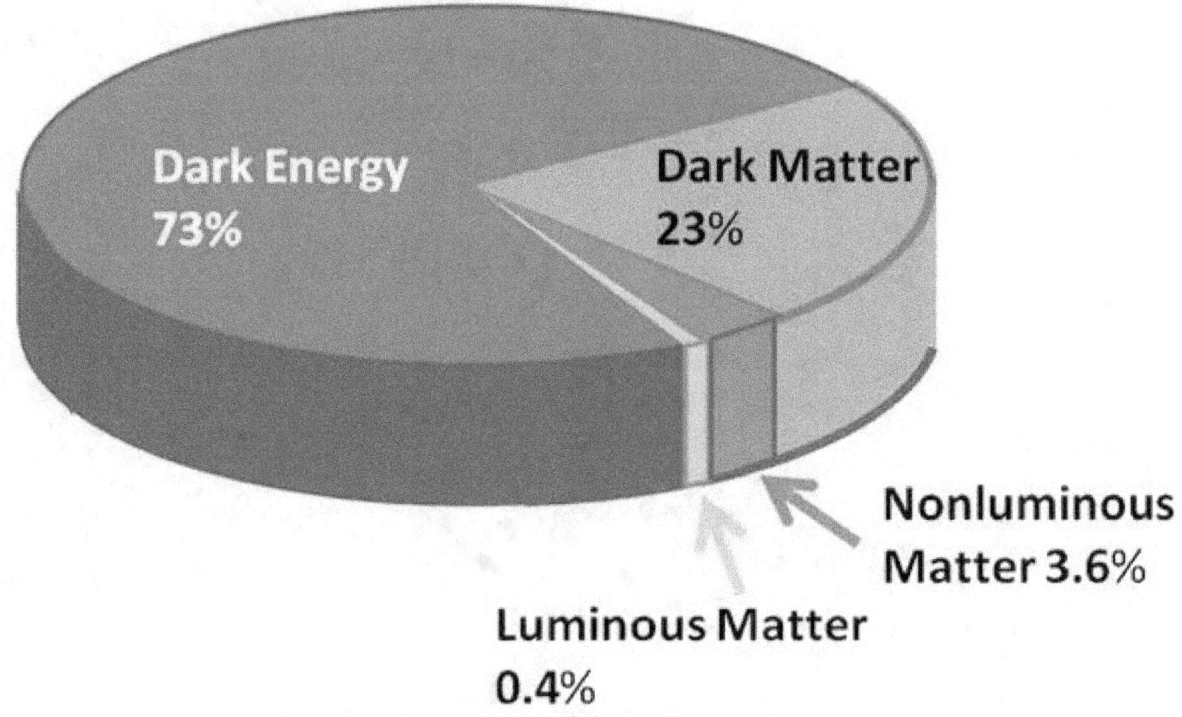

Pie chart showing the fractions of energy in the universe contributed by different sources. Ordinary matter is divided into luminous matter (the stars and luminous gases and 0.005% radiation) and nonluminous matter (intergalactic gas and about 0.1% neutrinos and 0.04% supermassive black holes). Ordinary matter is uncommon. Modeled after Ostriker and Steinhardt.[42] For more information, see NASA.

1.5   Other types

Ordinary matter, in the quarks and leptons definition, constitutes about 4% of the energy of the observable universe. The remaining energy is theorized to be due to exotic forms, of which 23% is dark matter[43][44] and 73% is dark energy.[45][46]

1.5.1   Dark matter

Main articles: Dark matter, Lambda-CDM model and WIMPs
See also: Galaxy formation and evolution and Dark matter halo

In astrophysics and cosmology, **dark matter** is matter of unknown composition that does not emit or reflect enough electromagnetic radiation to be observed directly, but whose presence can be inferred from gravitational effects on visible matter.[50][51] Observational evidence of the early universe and the big bang theory require that this matter have energy and mass, but is not composed of either elementary fermions (as above) OR gauge bosons. The commonly accepted view is that most of the dark matter is non-baryonic in nature.[50] As such, it is composed of particles as yet unobserved in the laboratory. Perhaps they are supersymmetric particles,[52] which are not Standard Model particles, but relics formed at very high energies in the early phase of the universe and still floating about.[50]

1.5.2   Dark energy

Main article: Dark energy
See also: Big bang § Dark energy

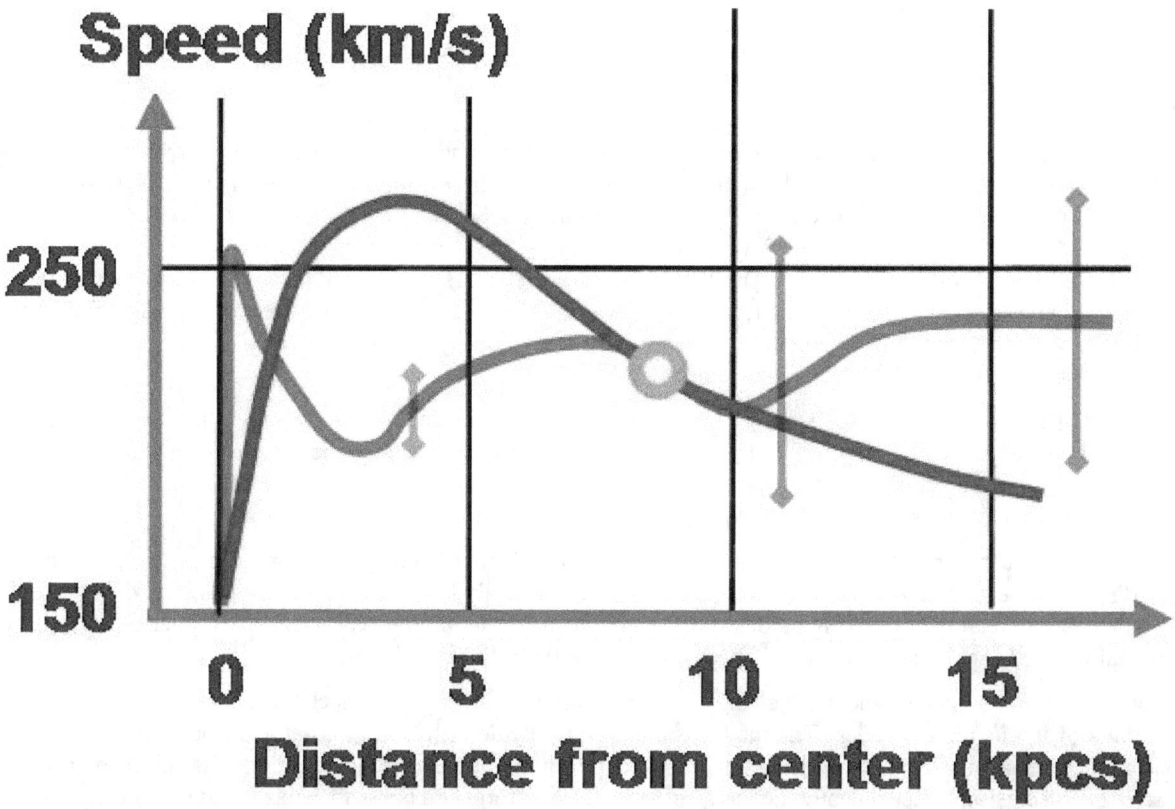

Galaxy rotation curve for the Milky Way. Vertical axis is speed of rotation about the galactic center. Horizontal axis is distance from the galactic center. The sun is marked with a yellow ball. The observed curve of speed of rotation is blue. The predicted curve based upon stellar mass and gas in the Milky Way is red. The difference is due to dark matter or perhaps a modification of the law of gravity.[47][48][49] Scatter in observations is indicated roughly by gray bars.

In cosmology, **dark energy** is the name given to the antigravitating influence that is accelerating the rate of expansion of the universe. It is known not to be composed of known particles like protons, neutrons or electrons, nor of the particles of dark matter, because these all gravitate.[53][54]

Fully 70% of the matter density in the universe appears to be in the form of dark energy. Twenty-six percent is dark matter. Only 4% is ordinary matter. So less than 1 part in 20 is made out of matter we have observed experimentally or described in the standard model of particle physics. Of the other 96%, apart from the properties just mentioned, we know absolutely nothing.

— Lee Smolin: *The Trouble with Physics*, p. 16

1.5.3 Exotic matter

Main article: Exotic matter

Exotic matter is a hypothetical concept of particle physics. It covers any material that violates one or more classical conditions or is not made of known baryonic particles. Such materials would possess qualities like negative mass or being repelled rather than attracted by gravity.

1.6   Historical development

1.6.1   Origins

The pre-Socratics were among the first recorded speculators about the underlying nature of the visible world. Thales (c. 624 BC–c. 546 BC) regarded water as the fundamental material of the world. Anaximander (c. 610 BC–c. 546 BC) posited that the basic material was wholly characterless or limitless: the Infinite (*apeiron*). Anaximenes (flourished 585 BC, d. 528 BC) posited that the basic stuff was *pneuma* or air. Heraclitus (c. 535–c. 475 BC) seems to say the basic element is fire, though perhaps he means that all is change. Empedocles (c. 490–430 BC) spoke of four elements of which everything was made: earth, water, air, and fire.[55] Meanwhile, Parmenides argued that change does not exist, and Democritus argued that everything is composed of minuscule, inert bodies of all shapes called atoms, a philosophy called atomism. All of these notions had deep philosophical problems.[56]

Aristotle (384 BC – 322 BC) was the first to put the conception on a sound philosophical basis, which he did in his natural philosophy, especially in *Physics* book I.[57] He adopted as reasonable suppositions the four Empedoclean elements, but added a fifth, aether. Nevertheless, these elements are not basic in Aristotle's mind. Rather they, like everything else in the visible world, are composed of the basic *principles* matter and form.

The word Aristotle uses for matter, ὕλη (*hyle* or *hule*), can be literally translated as wood or timber, that is, "raw material" for building.[58] Indeed, Aristotle's conception of matter is intrinsically linked to something being made or composed. In other words, in contrast to the early modern conception of matter as simply occupying space, matter for Aristotle is definitionally linked to process or change: matter is what underlies a change of substance.

For example, a horse eats grass: the horse changes the grass into itself; the grass as such does not persist in the horse, but some aspect of it—its matter—does. The matter is not specifically described (e.g., as atoms), but consists of whatever persists in the change of substance from grass to horse. Matter in this understanding does not exist independently (i.e., as a substance), but exists interdependently (i.e., as a "principle") with form and only insofar as it underlies change. It can be helpful to conceive of the relationship of matter and form as very similar to that between parts and whole. For Aristotle, matter as such can only *receive* actuality from form; it has no activity or actuality in itself, similar to the way that parts as such only have their existence *in* a whole (otherwise they would be independent wholes).

1.6.2   Early modernity

René Descartes (1596–1650) originated the modern conception of matter. He was primarily a geometer. Instead of, like Aristotle, deducing the existence of matter from the physical reality of change, Descartes arbitrarily postulated matter to be an abstract, mathematical substance that occupies space:

> So, extension in length, breadth, and depth, constitutes the nature of bodily substance; and thought constitutes the nature of thinking substance. And everything else attributable to body presupposes extension, and is only a mode of extended
> — René Descartes, *Principles of Philosophy*[59]

For Descartes, matter has only the property of extension, so its only activity aside from locomotion is to exclude other bodies:[60] this is the mechanical philosophy. Descartes makes an absolute distinction between mind, which he defines as unextended, thinking substance, and matter, which he defines as unthinking, extended substance.[61] They are independent things. In contrast, Aristotle defines matter and the formal/forming principle as complementary *principles* that together compose one independent thing (substance). In short, Aristotle defines matter (roughly speaking) as what things are actually made of (with a *potential* independent existence), but Descartes elevates matter to an actual independent thing in itself.

The continuity and difference between Descartes' and Aristotle's conceptions is noteworthy. In both conceptions, matter is passive or inert. In the respective conceptions matter has different relationships to intelligence. For Aristotle, matter and intelligence (form) exist together in an interdependent relationship, whereas for Descartes, matter and intelligence (mind) are definitionally opposed, independent substances.[62]

Descartes' justification for restricting the inherent qualities of matter to extension is its permanence, but his real criterion is not permanence (which equally applied to color and resistance), but his desire to use geometry to explain all material properties.[63] Like Descartes, Hobbes, Boyle, and Locke argued that the inherent properties of bodies were limited to extension, and that so-called secondary qualities, like color, were only products of human perception.[64]

Isaac Newton (1643–1727) inherited Descartes' mechanical conception of matter. In the third of his "Rules of Reasoning in Philosophy", Newton lists the universal qualities of matter as "extension, hardness, impenetrability, mobility, and inertia".[65] Similarly in *Optics* he conjectures that God created matter as "solid, massy, hard, impenetrable, movable particles", which were "...even so very hard as never to wear or break in pieces".[66] The "primary" properties of matter were amenable to mathematical description, unlike "secondary" qualities such as color or taste. Like Descartes, Newton rejected the essential nature of secondary qualities.[67]

Newton developed Descartes' notion of matter by restoring to matter intrinsic properties in addition to extension (at least on a limited basis), such as mass. Newton's use of gravitational force, which worked "at a distance", effectively repudiated Descartes' mechanics, in which interactions happened exclusively by contact.[68]

Though Newton's gravity would seem to be a *power* of bodies, Newton himself did not admit it to be an *essential* property of matter. Carrying the logic forward more consistently, Joseph Priestley argued that corporeal properties transcend contact mechanics: chemical properties require the *capacity* for attraction.[68] He argued matter has other inherent powers besides the so-called primary qualities of Descartes, et al.[69]

Since Priestley's time, there has been a massive expansion in knowledge of the constituents of the material world (viz., molecules, atoms, subatomic particles), but there has been no further development in the *definition* of matter. Rather the question has been set aside. Noam Chomsky summarizes the situation that has prevailed since that time:

> What is the concept of body that finally emerged?[...] The answer is that there is no clear and definite conception of body.[...] Rather, the material world is whatever we discover it to be, with whatever properties it must be assumed to have for the purposes of explanatory theory. Any intelligible theory that offers genuine explanations and that can be assimilated to the core notions of physics becomes part of the theory of the material world, part of our account of body. If we have such a theory in some domain, we seek to assimilate it to the core notions of physics, perhaps modifying these notions as we carry out this enterprise.
> — Noam Chomsky, '**Language and problems of knowledge: the Managua lectures***, p. 144*[68]

So matter is whatever physics studies and the object of study of physics is matter: there is no independent general definition of matter, apart from its fitting into the methodology of measurement and controlled experimentation. In sum, the boundaries between what constitutes matter and everything else remains as vague as the demarcation problem of delimiting science from everything else.[70]

1.6.3 Late nineteenth and early twentieth centuries

In the 19th century, following the development of the periodic table, and of atomic theory, atoms were seen as being the fundamental constituents of matter; atoms formed molecules and compounds.[71]

The common definition in terms of occupying space and having mass is in contrast with most physical and chemical definitions of matter, which rely instead upon its structure and upon attributes not necessarily related to volume and mass. At the turn of the nineteenth century, the knowledge of matter began a rapid evolution.

Aspects of the Newtonian view still held sway. James Clerk Maxwell discussed matter in his work *Matter and Motion*.[72] He carefully separates "matter" from space and time, and defines it in terms of the object referred to in Newton's first law of motion.

However, the Newtonian picture was not the whole story. In the 19th century, the term "matter" was actively discussed by a host of scientists and philosophers, and a brief outline can be found in Levere.[73] A textbook discussion from 1870 suggests matter is what is made up of atoms:[74]

> Three divisions of matter are recognized in science: masses, molecules and atoms.
> A Mass of matter is any portion of matter appreciable by the senses.

A Molecule is the smallest particle of matter into which a body can be divided without losing its identity.
An Atom is a still smaller particle produced by division of a molecule.

Rather than simply having the attributes of mass and occupying space, matter was held to have chemical and electrical properties. The famous physicist J. J. Thomson wrote about the "constitution of matter" and was concerned with the possible connection between matter and electrical charge.[75]

1.6.4   Later developments

There is an entire literature concerning the "structure of matter", ranging from the "electrical structure" in the early 20th century,[76] to the more recent "quark structure of matter", introduced today with the remark: *Understanding the quark structure of matter has been one of the most important advances in contemporary physics.*[77] In this connection, physicists speak of *matter fields*, and speak of particles as "quantum excitations of a mode of the matter field".[8][9] And here is a quote from de Sabbata and Gasperini: "With the word "matter" we denote, in this context, the sources of the interactions, that is spinor fields (like quarks and leptons), which are believed to be the fundamental components of matter, or scalar fields, like the Higgs particles, which are used to introduced mass in a gauge theory (and that, however, could be composed of more fundamental fermion fields)."[78]

The modern conception of matter has been refined many times in history, in light of the improvement in knowledge of just *what* the basic building blocks are, and in how they interact.

In the late 19th century with the discovery of the electron, and in the early 20th century, with the discovery of the atomic nucleus, and the birth of particle physics, matter was seen as made up of electrons, protons and neutrons interacting to form atoms. Today, we know that even protons and neutrons are not indivisible, they can be divided into quarks, while electrons are part of a particle family called leptons. Both quarks and leptons are elementary particles, and are currently seen as being the fundamental constituents of matter.[79]

These quarks and leptons interact through four fundamental forces: gravity, electromagnetism, weak interactions, and strong interactions. The Standard Model of particle physics is currently the best explanation for all of physics, but despite decades of efforts, gravity cannot yet be accounted for at the quantum level; it is only described by classical physics (see quantum gravity and graviton).[80] Interactions between quarks and leptons are the result of an exchange of force-carrying particles (such as photons) between quarks and leptons.[81] The force-carrying particles are not themselves building blocks. As one consequence, mass and energy (which cannot be created or destroyed) cannot always be related to matter (which can be created out of non-matter particles such as photons, or even out of pure energy, such as kinetic energy). Force carriers are usually not considered matter: the carriers of the electric force (photons) possess energy (see Planck relation) and the carriers of the weak force (W and Z bosons) are massive, but neither are considered matter either.[82] However, while these particles are not considered matter, they do contribute to the total mass of atoms, subatomic particles, and all systems that contain them.[83][84]

1.6.5   Summary

The term "matter" is used throughout physics in a bewildering variety of contexts: for example, one refers to "condensed matter physics",[85] "elementary matter",[86] "partonic" matter, "dark" matter, "anti"-matter, "strange" matter, and "nuclear" matter. In discussions of matter and antimatter, normal matter has been referred to by Alfvén as *koinomatter* (Gk. *common matter*).[87] It is fair to say that in physics, there is no broad consensus as to a general definition of matter, and the term "matter" usually is used in conjunction with a specifying modifier.

1.7 See also

1.8 References

[1] R. Penrose (1991). "The mass of the classical vacuum". In S. Saunders, H.R. Brown. *The Philosophy of Vacuum*. Oxford University Press. p. 21. ISBN 0-19-824449-5.

[2] "Matter (physics)". *McGraw-Hill's Access Science: Encyclopedia of Science and Technology Online*. Retrieved 2009-05-24.

[3] P. Davies (1992). *The New Physics: A Synthesis*. Cambridge University Press. p. 1. ISBN 0-521-43831-4.

[4] G. 't Hooft (1997). *In search of the ultimate building blocks*. Cambridge University Press. p. 6. ISBN 0-521-57883-3.

[5] "RHIC Scientists Serve Up "Perfect" Liquid" (Press release). Brookhaven National Laboratory. 18 April 2005. Retrieved 2009-09-15.

[6] J. Olmsted; G.M. Williams (1996). *Chemistry: The Molecular Science* (2nd ed.). Jones & Bartlett. p. 40. ISBN 0-8151-8450-6.

[7] J. Mongillo (2007). *Nanotechnology 101*. Greenwood Publishing. p. 30. ISBN 0-313-33880-9.

[8] P.C.W. Davies (1979). *The Forces of Nature*. Cambridge University Press. p. 116. ISBN 0-521-22523-X.

[9] S. Weinberg (1998). *The Quantum Theory of Fields*. Cambridge University Press. p. 2. ISBN 0-521-55002-5.

[10] M. Masujima (2008). *Path Integral Quantization and Stochastic Quantization*. Springer. p. 103. ISBN 3-540-87850-5.

[11] S.M. Walker; A. King (2005). *What is Matter?*. Lerner Publications. p. 7. ISBN 0-8225-5131-4.

[12] J.Kenkel; P.B. Kelter; D.S. Hage (2000). *Chemistry: An Industry-based Introduction with CD-ROM*. CRC Press. p. 2. ISBN 1-56670-303-4. All basic science textbooks define *matter* as simply the collective aggregate of all material substances that occupy space and have mass or weight.

[13] K.A. Peacock (2008). *The Quantum Revolution: A Historical Perspective*. Greenwood Publishing Group. p. 47. ISBN 0-313-33448-X.

[14] M.H. Krieger (1998). *Constitutions of Matter: Mathematically Modeling the Most Everyday of Physical Phenomena*. University of Chicago Press. p. 22. ISBN 0-226-45305-7.

[15] S.M. Caroll (2004). *Spacetime and Geometry*. Addison Wesley. pp. 163–164. ISBN 0-8053-8732-3.

[16] P. Davies (1992). *The New Physics: A Synthesis*. Cambridge University Press. p. 499. ISBN 0-521-43831-4. **Matter fields**: the fields whose quanta describe the elementary particles that make up the material content of the Universe (as opposed to the gravitons and their supersymmetric partners).

[17] G. F. Barker (1870). "Divisions of matter". *A text-book of elementary chemistry: theoretical and inorganic*. John F Morton & Co. p. 2. ISBN 978-1-4460-2206-1.

[18] M. de Podesta (2002). *Understanding the Properties of Matter* (2nd ed.). CRC Press. p. 8. ISBN 0-415-25788-3.

[19] B. Povh; K. Rith; C. Scholz; F. Zetsche; M. Lavelle (2004). "Part I: Analysis: The building blocks of matter". *Particles and Nuclei: An Introduction to the Physical Concepts* (4th ed.). Springer. ISBN 3-540-20168-8.

[20] B. Carithers, P. Grannis (1995). "Discovery of the Top Quark" (PDF). *Beam Line* (SLAC National Accelerator Laboratory) **25** (3): 4–16.

[21] D. Green (2005). *High PT physics at hadron colliders*. Cambridge University Press. p. 23. ISBN 0-521-83509-7.

[22] L. Smolin (2007). *The Trouble with Physics: The Rise of String Theory, the Fall of a Science, and What Comes Next*. Mariner Books. p. 67. ISBN 0-618-91868-X.

[23] The W boson mass is 80.398 GeV; see Figure 1 in C. Amsler *et al.* (Particle Data Group) (2008). "Review of Particle Physics: The Mass and Width of the W Boson" (PDF). *Physics Letters B* **667**: 1. Bibcode:2008PhLB..667....1P. doi:10.1016/j.physlet

[24] I.J.R. Aitchison; A.J.G. Hey (2004). *Gauge Theories in Particle Physics*. CRC Press. p. 48. ISBN 0-7503-0864-8.

[25] B. Povh; K. Rith; C. Scholz; F. Zetsche; M. Lavelle (2004). *Particles and Nuclei: An Introduction to the Physical Concepts.* Springer. p. 103. ISBN 3-540-20168-8.

[26] T. Hatsuda (2008). "Quark–gluon plasma and QCD". In H. Akai. *Condensed matter theories* **21**. Nova Publishers. p. 296. ISBN 1-60021-501-7.

[27] K.W Staley (2004). "Origins of the Third Generation of Matter". *The Evidence for the Top Quark.* Cambridge University Press. p. 8. ISBN 0-521-82710-8.

[28] Y. Ne'eman; Y. Kirsh (1996). *The Particle Hunters* (2nd ed.). Cambridge University Press. p. 276. ISBN 0-521-47686-0. [T]he most natural explanation to the existence of higher generations of quarks and leptons is that they correspond to excited states of the first generation, and experience suggests that excited systems must be composite

[29] C. Amsler *et al.* (Particle Data Group) (2008). "Reviews of Particle Physics: Quarks" (PDF). *Physics Letters B* **667**: 1. Bibcode:2008PhLB..667....1P. doi:10.1016/j.physletb.2008.07.018.

[30] "Five Year Results on the Oldest Light in the Universe". NASA. 2008. Retrieved 2008-05-02.

[31] H.S. Goldberg; M.D. Scadron (1987). *Physics of Stellar Evolution and Cosmology.* Taylor & Francis. p. 202. ISBN 0-677-05540-4.

[32] H.S. Goldberg; M.D. Scadron (1987). *Physics of Stellar Evolution and Cosmology.* Taylor & Francis. p. 233. ISBN 0-677-05540-4.

[33] J.-P. Luminet; A. Bullough; A. King (1992). *Black Holes.* Cambridge University Press. p. 75. ISBN 0-521-40906-3.

[34] A. Bodmer (1971). "Collapsed Nuclei". *Physical Review D* **4** (6): 1601. Bibcode:1971PhRvD...4.1601B. doi:10.1103/Phys.

[35] E. Witten (1984). "Cosmic Separation of Phases". *Physical Review D* **30** (2): 272. Bibcode:1984PhRvD..30..272W. doi:10.

[36] C. Amsler *et al.* (Particle Data Group) (2008). "Review of Particle Physics: Leptons" (PDF). *Physics Letters B* **667**: 1. Bibcode:2008PhLB..667....1P. doi:10.1016/j.physletb.2008.07.018.

[37] C. Amsler *et al.* (Particle Data Group) (2008). "Review of Particle Physics: Neutrinos Properties" (PDF). *Physics Letters B* **667**: 1. Bibcode:2008PhLB..667....1P. doi:10.1016/j.physletb.2008.07.018.

[38] S. R. Logan (1998). *Physical Chemistry for the Biomedical Sciences.* CRC Press. pp. 110–111. ISBN 0-7484-0710-3.

[39] P.J. Collings (2002). "Chapter 1: States of Matter". *Liquid Crystals: Nature's Delicate Phase of Matter.* Princeton University Press. ISBN 0-691-08672-9.

[40] D.H. Trevena (1975). "Chapter 1.2: Changes of phase". *The Liquid Phase.* Taylor & Francis. ISBN 978-0-85109-031-3.

[41] National Research Council (US) (2006). *Revealing the hidden nature of space and time.* National Academies Press. p. 46. ISBN 0-309-10194-8.

[42] J.P. Ostriker; P.J. Steinhardt (2003). "New Light on Dark Matter". *Science* **300** (5627): 1909–13. arXiv:astro-ph/0306402. Bibcode:2003Sci...300.1909O. doi:10.1126/science.1085976. PMID 12817140.

[43] K. Pretzl (2004). "Dark Matter, Massive Neutrinos and Susy Particles". *Structure and Dynamics of Elementary Matter.* Walter Greiner. p. 289. ISBN 1-4020-2446-0.

[44] K. Freeman; G. McNamara (2006). "What can the matter be?". *In Search of Dark Matter.* Birkhäuser Verlag. p. 105. ISBN 0-387-27616-5.

[45] J.C. Wheeler (2007). *Cosmic Catastrophes: Exploding Stars, Black Holes, and Mapping the Universe.* Cambridge University Press. p. 282. ISBN 0-521-85714-7.

[46] J. Gribbin (2007). *The Origins of the Future: Ten Questions for the Next Ten Years.* Yale University Press. p. 151. ISBN 0-300-12596-8.

[47] P. Schneider (2006). *Extragalactic Astronomy and Cosmology.* Springer. p. 4, Fig. 1.4. ISBN 3-540-33174-3.

[48] T. Koupelis; K.F. Kuhn (2007). *In Quest of the Universe.* Jones & Bartlett Publishers. p. 492; Fig. 16.13. ISBN 0-7637-4387-9.

[49] M. H. Jones; R. J. Lambourne; D. J. Adams (2004). *An Introduction to Galaxies and Cosmology*. Cambridge University Press. p. 21; Fig. 1.13. ISBN 0-521-54623-0.

[50] D. Majumdar (2007). "Dark matter — possible candidates and direct detection". arXiv:hep-ph/0703310 [hep-ph].

[51] K.A. Olive (2003). "Theoretical Advanced Study Institute lectures on dark matter". arXiv:astro-ph/0301505 [astro-ph].

[52] K.A. Olive (2009). "Colliders and Cosmology".*European Physical Journal C***59**(2): 269–295.arXiv:0806.1208.Bibc. doi:10.1140/epjc/s10052-008-0738-8.

[53] J.C. Wheeler (2007). *Cosmic Catastrophes*. Cambridge University Press. p. 282. ISBN 0-521-85714-7.

[54] L. Smolin (2007). *The Trouble with Physics*. Mariner Books. p. 16. ISBN 0-618-91868-X.

[55] S. Toulmin; J. Goodfield (1962). *The Architecture of Matter*. University of Chicago Press. pp. 48–54.

[56] Discussed by Aristotle in *Physics*, esp. book I, but also later; as well as *Metaphysics* I–II.

[57] For a good explanation and elaboration, see R.J. Connell (1966). *Matter and Becoming*. Priory Press.

[58] H. G. Liddell; R. Scott; J. M. Whiton (1891). *A lexicon abridged from Liddell & Scott's Greek–English lexicon*. Harper and Brothers. p. 725.

[59] R. Descartes (1644). "The Principles of Human Knowledge". *Principles of Philosophy I*. p. 53.

[60] though even this property seems to be non-essential (René Descartes, *Principles of Philosophy* II [1644], "On the Principles of Material Things", no. 4.)

[61] R. Descartes (1644). "The Principles of Human Knowledge". *Principles of Philosophy I*. pp. 8, 54, 63.

[62] D.L. Schindler (1986). "The Problem of Mechanism". In D.L. Schindler. *Beyond Mechanism*. University Press of America.

[63] E.A. Burtt, *Metaphysical Foundations of Modern Science* (Garden City, New York: Doubleday and Company, 1954), 117–118.

[64] J.E. McGuire and P.M. Heimann, "The Rejection of Newton's Concept of Matter in the Eighteenth Century", *The Concept of Matter in Modern Philosophy* ed. Ernan McMullin (Notre Dame: University of Notre Dame Press, 1978), 104–118 (105).

[65] Isaac Newton, *Mathematical Principles of* Natural Philosophy, *trans. A. Motte, revised by F. Cajori (Berkeley: University of California Press, 1934), pp. 398–400. Further analyzed by Maurice A. Finocchiaro, "Newton's Third Rule of Philosophizing: A Role for Logic in Historiography",* Isis *65:1 (Mar. 1974), pp. 66–73.*

[66] Isaac Newton, *Optics*, Book III, pt. 1, query 31.

[67] McGuire and Heimann, 104.

[68] N. Chomsky (1988). *Language and problems of knowledge: the Managua lectures* (2nd ed.). MIT Press. p. 144. ISBN 0-262-53070-8.

[69] McGuire and Heimann, 113.

[70] Nevertheless, it remains true that the mathematization regarded as requisite for a modern physical theory carries its own implicit notion of matter, which is very like Descartes', despite the demonstrated vacuity of the latter's notions.

[71] M. Wenham (2005). *Understanding Primary Science: Ideas, Concepts and Explanations* (2nd ed.). Paul Chapman Educational Publishing. p. 115. ISBN 1-4129-0163-4.

[72] J.C. Maxwell (1876). *Matter and Motion*. Society for Promoting Christian Knowledge. p. 18. ISBN 0-486-66895-9.

[73] T.H. Levere (1993). "Introduction". *Affinity and Matter: Elements of Chemical Philosophy, 1800–1865*. Taylor & Francis. ISBN 2-88124-583-8.

[74] G.F. Barker (1870). "Introduction". *A Text Book of Elementary Chemistry: Theoretical and Inorganic*. John P. Morton and Company. p. 2.

[75] J. J. Thomson (1909). "Preface". *Electricity and Matter*. A. Constable.

[76] O.W. Richardson (1914). "Chapter 1". *The Electron Theory of Matter*. The University Press.

[77] M. Jacob (1992). *The Quark Structure of Matter*. World Scientific. ISBN 981-02-3687-5.

[78] V. de Sabbata; M. Gasperini (1985). *Introduction to Gravitation*. World Scientific. p. 293. ISBN 9971-5-0049-3.

[79] The history of the concept of matter is a history of the fundamental *length scales* used to define matter. Different building blocks apply depending upon whether one defines matter on an atomic or elementary particle level. One may use a definition that matter is atoms, or that matter is hadrons, or that matter is leptons and quarks depending upon the scale at which one wishes to define matter. B. Povh; K. Rith; C. Scholz; F. Zetsche; M. Lavelle (2004). "Fundamental constituents of matter". *Particles and Nuclei: An Introduction to the Physical Concepts* (4th ed.). Springer. ISBN 3-540-20168-8.

[80] J. Allday (2001). *Quarks, Leptons and the Big Bang*. CRC Press. p. 12. ISBN 0-7503-0806-0.

[81] B.A. Schumm (2004). *Deep Down Things: The Breathtaking Beauty of Particle Physics*. Johns Hopkins University Press. p. 57. ISBN 0-8018-7971-X.

[82] See for example, M. Jibu; K. Yasue (1995). *Quantum Brain Dynamics and Consciousness*. John Benjamins Publishing Company. p. 62. ISBN 1-55619-183-9., B. Martin (2009). *Nuclear and Particle Physics* (2nd ed.). John Wiley & Sons. p. 125. ISBN 0-470-74275-5. and K. W. Plaxco; M. Gross (2006). *Astrobiology: A Brief Introduction*. Johns Hopkins University Press. p. 23. ISBN 0-8018-8367-9.

[83] P. A. Tipler; R. A. Llewellyn (2002). *Modern Physics*. Macmillan. pp. 89–91, 94–95. ISBN 0-7167-4345-0.

[84] P. Schmüser; H. Spitzer (2002). "Particles". In L. Bergmann et al. *Constituents of Matter: Atoms, Molecules, Nuclei*. CRC Press. pp. 773 *ff*. ISBN 0-8493-1202-7.

[85] P. M. Chaikin; T. C. Lubensky (2000). *Principles of Condensed Matter Physics*. Cambridge University Press. p. xvii. ISBN 0-521-79450-1.

[86] W. Greiner (2003). W. Greiner, M.G. Itkis, G. Reinhardt, M.C. Güçlü, ed. *Structure and Dynamics of Elementary Matter*. Springer. p. xii. ISBN 1-4020-2445-2.

[87] P. Sukys (1999). *Lifting the Scientific Veil: Science Appreciation for the Nonscientist*. Rowman & Littlefield. p. 87. ISBN 0-8476-9600-6.

1.9   Further reading

- Lillian Hoddeson; Michael Riordan, eds. (1997). *The Rise of the Standard Model*. Cambridge University Press. ISBN 0-521-57816-7.

- Timothy Paul Smith (2004). "The search for quarks in ordinary matter". *Hidden Worlds*. Princeton University Press. p. 1. ISBN 0-691-05773-7.

- Harald Fritzsch (2005). *Elementary Particles: Building blocks of matter*. World Scientific. p. 1. ISBN 981-256-141-2.

- Bertrand Russell (1992). "The philosophy of matter". *A Critical Exposition of the Philosophy of Leibniz* (Reprint of 1937 2nd ed.). Routledge. p. 88. ISBN 0-415-08296-X.

- Stephen Toulmin and June Goodfield, *The Architecture of Matter* (Chicago: University of Chicago Press, 1962).

- Richard J. Connell, *Matter and Becoming* (Chicago: The Priory Press, 1966).

- Ernan McMullin, *The Concept of Matter in Greek and Medieval Philosophy* (Notre Dame, Indiana: Univ. of Notre Dame Press, 1965).

- Ernan McMullin, *The Concept of Matter in Modern Philosophy* (Notre Dame, Indiana: University of Notre Dame Press, 1978).

1.10 External links

- Visionlearning Module on Matter

- Matter in the universe How much Matter is in the Universe?

- NASA on superfluid core of neutron star

- Matter and Energy: A False Dichotomy – Conversations About Science with Theoretical Physicist Matt Strassler

Chapter 2

State of matter

Not to be confused with Phase (matter).

In physics, a **state of matter** is one of the distinct forms that matter takes on. Four states of matter are observable in everyday life: solid, liquid, gas, and plasma. Many other states are known, such as Bose–Einstein condensates and neutron-degenerate matter, but these only occur in extreme situations such as ultra cold or ultra dense matter. Other states, such as quark–gluon plasmas, are believed to be possible but remain theoretical for now. For a complete list of all exotic states of matter, see the list of states of matter.

Historically, the distinction is made based on qualitative differences in properties. Matter in the solid state maintains a fixed volume and shape, with component particles (atoms, molecules or ions) close together and fixed into place. Matter in the liquid state maintains a fixed volume, but has a variable shape that adapts to fit its container. Its particles are still close together but move freely. Matter in the gaseous state has both variable volume and shape, adapting both to fit its container. Its particles are neither close together nor fixed in place. Matter in the plasma state has variable volume and shape, but as well as neutral atoms, it contains a significant number of ions and electrons, both of which can move around freely. Plasma is the most common form of visible matter in the universe.[1]

The term phase is sometimes used as a synonym for state of matter, but a system can contain several immiscible phases of the same state of matter (see Phase (matter) for more discussion of the difference between the two terms).

2.1 The four fundamental states

2.1.1 Solid

Main article: Solid

In a solid the particles (ions, atoms or molecules) are closely packed together. The forces between particles are strong so that the particles cannot move freely but can only vibrate. As a result, a solid has a stable, definite shape, and a definite volume. Solids can only change their shape by force, as when broken or cut.

In crystalline solids, the particles (atoms, molecules, or ions) are packed in a regularly ordered, repeating pattern. There are various different crystal structures, and the same substance can have more than one structure (or solid phase). For example, iron has a body-centred cubic structure at temperatures below 912 °C, and a face-centred cubic structure between 912 and 1394 °C. Ice has fifteen known crystal structures, or fifteen solid phases, which exist at various temperatures and pressures.[2]

Glasses and other non-crystalline, amorphous solids without long-range order are not thermal equilibrium ground states; therefore they are described below as nonclassical states of matter.

The four fundamental states of matter. Clockwise from top left, they are solid, liquid, plasma, and gas, represented by an ice sculpture, a drop of water, electrical arcing from a tesla coil, and the air around clouds, respectively.

A crystalline solid: atomic resolution image of strontium titanate. Brighter atoms are Sr and darker ones are Ti.

Solids can be transformed into liquids by melting, and liquids can be transformed into solids by freezing. Solids can also change directly into gases through the process of sublimation, and gases can likewise change directly into solids through deposition.

2.1.2 Liquid

Structure of a classical monatomic liquid. Atoms have many nearest neighbors in contact, yet no long-range order is present.

Main article: Liquid

A liquid is a nearly incompressible fluid that conforms to the shape of its container but retains a (nearly) constant volume independent of pressure. The volume is definite if the temperature and pressure are constant. When a solid is heated above its melting point, it becomes liquid, given that the pressure is higher than the triple point of the substance. Intermolecular (or interatomic or interionic) forces are still important, but the molecules have enough energy to move relative to each other and the structure is mobile. This means that the shape of a liquid is not definite but is determined by its container. The volume is usually greater than that of the corresponding solid, the best known exception being water, H_2O. The highest temperature at which a given liquid can exist is its critical temperature.[3]

2.1.3 Gas

Main article: Gas

A gas is a compressible fluid. Not only will a gas conform to the shape of its container but it will also expand to fill the container.

In a gas, the molecules have enough kinetic energy so that the effect of intermolecular forces is small (or zero for an ideal gas), and the typical distance between neighboring molecules is much greater than the molecular size. A gas has no definite shape or volume, but occupies the entire container in which it is confined. A liquid may be converted to a gas by heating at constant pressure to the boiling point, or else by reducing the pressure at constant temperature.

At temperatures below its critical temperature, a gas is also called a vapor, and can be liquefied by compression alone

The spaces between gas molecules are very big. Gas molecules have very weak or no bonds at all. The molecules in "gas" can move freely and fast.

without cooling. A vapor can exist in equilibrium with a liquid (or solid), in which case the gas pressure equals the vapor pressure of the liquid (or solid).

A supercritical fluid (SCF) is a gas whose temperature and pressure are above the critical temperature and critical pressure respectively. In this state, the distinction between liquid and gas disappears. A supercritical fluid has the physical properties of a gas, but its high density confers solvent properties in some cases, which leads to useful applications. For example, supercritical carbon dioxide is used to extract caffeine in the manufacture of decaffeinated coffee.[4]

2.1.4 Plasma

Main article: Plasma (physics)

Like a gas, plasma does not have definite shape or volume. Unlike gases, plasmas are electrically conductive, produce magnetic fields and electric currents, and respond strongly to electromagnetic forces. Positively charged nuclei swim in a "sea" of freely-moving disassociated electrons, similar to the way such charges exist in conductive metal. In fact it is this electron "sea" that allows matter in the plasma state to conduct electricity.

The plasma state is often misunderstood, but it is actually quite common on Earth, and the majority of people observe it on a regular basis without even realizing it. Lightning, electric sparks, fluorescent lights, neon lights, plasma televisions, some types of flame and the stars are all examples of illuminated matter in the plasma state.

A gas is usually converted to a plasma in one of two ways, either from a huge voltage difference between two points, or by exposing it to extremely high temperatures.

Heating matter to high temperatures causes electrons to leave the atoms, resulting in the presence of free electrons. At very high temperatures, such as those present in stars, it is assumed that essentially all electrons are "free", and that a very high-energy plasma is essentially bare nuclei swimming in a sea of electrons.

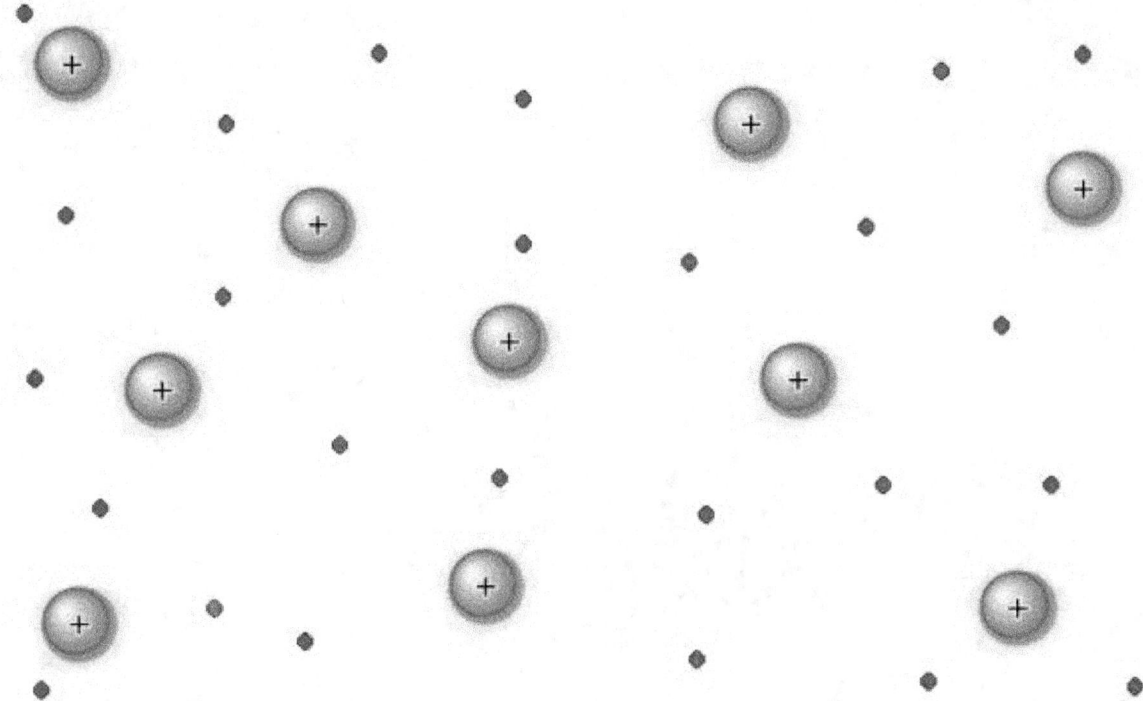

In a plasma, electrons are ripped away from their nuclei, forming an electron "sea". This gives it the ability to conduct electricity.

2.2   Phase transitions

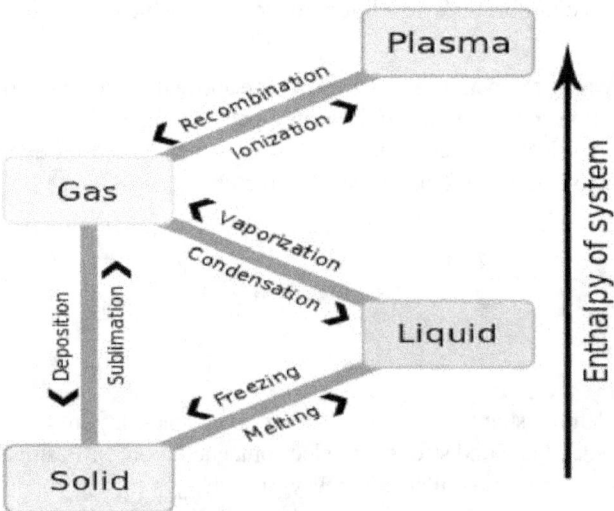

This diagram illustrates transitions between the four fundamental states of matter.

A state of matter is also characterized by phase transitions. A phase transition indicates a change in structure and can be recognized by an abrupt change in properties. A distinct state of matter can be defined as any set of states distinguished from any other set of states by a phase transition. Water can be said to have several distinct solid states.[5] The appearance of superconductivity is associated with a phase transition, so there are superconductive states. Likewise, ferromagnetic states are demarcated by phase transitions and have distinctive properties. When the change of state occurs in stages the intermediate steps are called mesophases. Such phases have been exploited by the introduction of liquid crystal technology. [6][7]

The state or *phase* of a given set of matter can change depending on pressure and temperature conditions, transitioning to other phases as these conditions change to favor their existence; for example, solid transitions to liquid with an increase in temperature. Near absolute zero, a substance exists as a solid. As heat is added to this substance it melts into a liquid at its melting point, boils into a gas at its boiling point, and if heated high enough would enter a plasma state in which the electrons are so energized that they leave their parent atoms.

Forms of matter that are not composed of molecules and are organized by different forces can also be considered different states of matter. Superfluids (like Fermionic condensate) and the quark–gluon plasma are examples.

In a chemical equation, the state of matter of the chemicals may be shown as (s) for solid, (l) for liquid, and (g) for gas. An aqueous solution is denoted (aq). Matter in the plasma state is seldom used (if at all) in chemical equations, so there is no standard symbol to denote it. In the rare equations that plasma is used in plasma is symbolized as (p).

2.3 Non-classical states

2.3.1 Glass

Main article: Glass

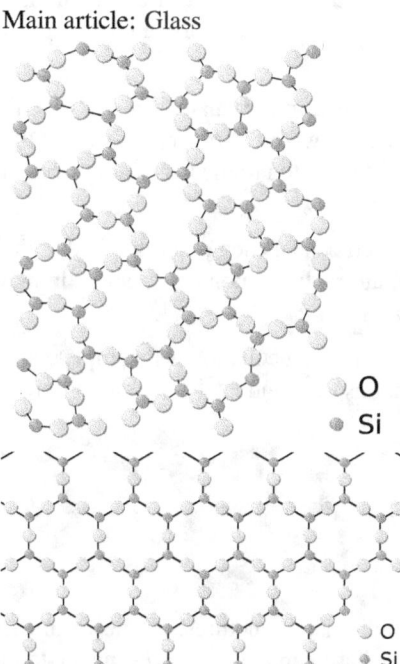

Schematic representation of a random-network glassy form (left) and ordered crystalline lattice (right) of identical chemical composition.

Glass is a non-crystalline or amorphous solid material that exhibits a glass transition when heated towards the liquid state. Glasses can be made of quite different classes of materials: inorganic networks (such as window glass, made of silicate plus additives), metallic alloys, ionic melts, aqueous solutions, molecular liquids, and polymers. Thermodynamically, a glass is in a metastable state with respect to its crystalline counterpart. The conversion rate, however, is practically zero.

2.3.2 Crystals with some degree of disorder

A plastic crystal is a molecular solid with long-range positional order but with constituent molecules retaining rotational freedom; in an orientational glass this degree of freedom is frozen in a quenched disordered state.

Similarly, in a spin glass magnetic disorder is frozen.

2.3.3   Liquid crystal states

Main article: Liquid crystal

Liquid crystal states have properties intermediate between mobile liquids and ordered solids. Generally, they are able to flow like a liquid, but exhibiting long-range order. For example, the nematic phase consists of long rod-like molecules such as para-azoxyanisole, which is nematic in the temperature range 118–136 °C.[8] In this state the molecules flow as in a liquid, but they all point in the same direction (within each domain) and cannot rotate freely.

Other types of liquid crystals are described in the main article on these states. Several types have technological importance, for example, in liquid crystal displays.

2.3.4   Magnetically ordered

Transition metal atoms often have magnetic moments due to the net spin of electrons that remain unpaired and do not form chemical bonds. In some solids the magnetic moments on different atoms are ordered and can form a ferromagnet, an antiferromagnet or a ferrimagnet.

In a ferromagnet—for instance, solid iron—the magnetic moment on each atom is aligned in the same direction (within a magnetic domain). If the domains are also aligned, the solid is a permanent magnet, which is magnetic even in the absence of an external magnetic field. The magnetization disappears when the magnet is heated to the Curie point, which for iron is 768 °C.

An antiferromagnet has two networks of equal and opposite magnetic moments, which cancel each other out so that the net magnetization is zero. For example, in nickel(II) oxide (NiO), half the nickel atoms have moments aligned in one direction and half in the opposite direction.

In a ferrimagnet, the two networks of magnetic moments are opposite but unequal, so that cancellation is incomplete and there is a non-zero net magnetization. An example is magnetite (Fe_3O_4), which contains Fe^{2+} and Fe^{3+} ions with different magnetic moments.

2.3.5   Microphase-separated

Main article: Copolymer

Copolymers can undergo microphase separation to form a diverse array of periodic nanostructures, as shown in the example of the styrene-butadiene-styrene block copolymer shown at right. Microphase separation can be understood by analogy to the phase separation between oil and water. Due to chemical incompatibility between the blocks, block copolymers undergo a similar phase separation. However, because the blocks are covalently bonded to each other, they cannot demix macroscopically as water and oil can, and so instead the blocks form nanometer-sized structures. Depending on the relative lengths of each block and the overall block topology of the polymer, many morphologies can be obtained, each its own phase of matter.

2.3.6   Quantum spin liquid

Main article: Quantum spin liquid

A disordered state in a system of interacting quantum spins which preserves its disorder to very low temperatures, unlike other disordered states.

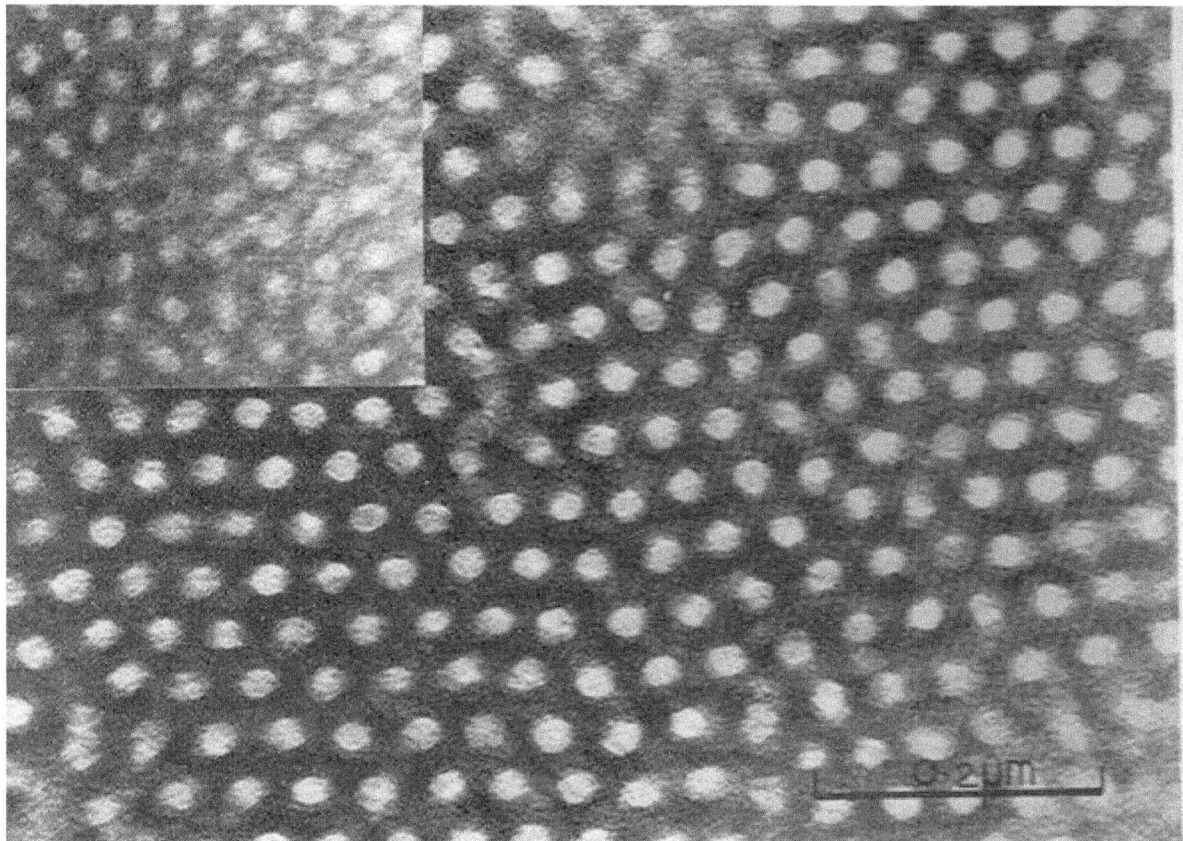

SBS block copolymer in TEM

2.4 Low-temperature states

2.4.1 Superfluid

Main article: Superfluid

Close to absolute zero, some liquids form a second liquid state described as **superfluid** because it has zero viscosity (or infinite fluidity; i.e., flowing without friction). This was discovered in 1937 for helium, which forms a superfluid below the lambda temperature of 2.17 K. In this state it will attempt to "climb" out of its container.[9] It also has infinite thermal conductivity so that no temperature gradient can form in a superfluid. Placing a superfluid in a spinning container will result in quantized vortices.

These properties are explained by the theory that the common isotope helium-4 forms a Bose–Einstein condensate (see next section) in the superfluid state. More recently, Fermionic condensate superfluids have been formed at even lower temperatures by the rare isotope helium-3 and by lithium-6.[10]

2.4.2 Bose–Einstein condensate

Main article: Bose–Einstein condensate

In 1924, Albert Einstein and Satyendra Nath Bose predicted the "Bose–Einstein condensate" (BEC), sometimes referred to as the fifth state of matter. In a BEC, matter stops behaving as independent particles, and collapses into a single quantum state that can be described with a single, uniform wavefunction.

Liquid helium in a superfluid phase creeps up on the walls of the cup in a Rollin film, eventually dripping out from the cup.

In the gas phase, the Bose–Einstein condensate remained an unverified theoretical prediction for many years. In 1995, the research groups of Eric Cornell and Carl Wieman, of JILA at the University of Colorado at Boulder, produced the first such condensate experimentally. A Bose–Einstein condensate is "colder" than a solid. It may occur when atoms have very similar (or the same) quantum levels, at temperatures very close to absolute zero (−273.15 °C).

2.4.3   Fermionic condensate

Main article: Fermionic condensate

A *fermionic condensate* is similar to the Bose–Einstein condensate but composed of fermions. The Pauli exclusion principle prevents fermions from entering the same quantum state, but a pair of fermions can behave as a boson, and multiple such pairs can then enter the same quantum state without restriction.

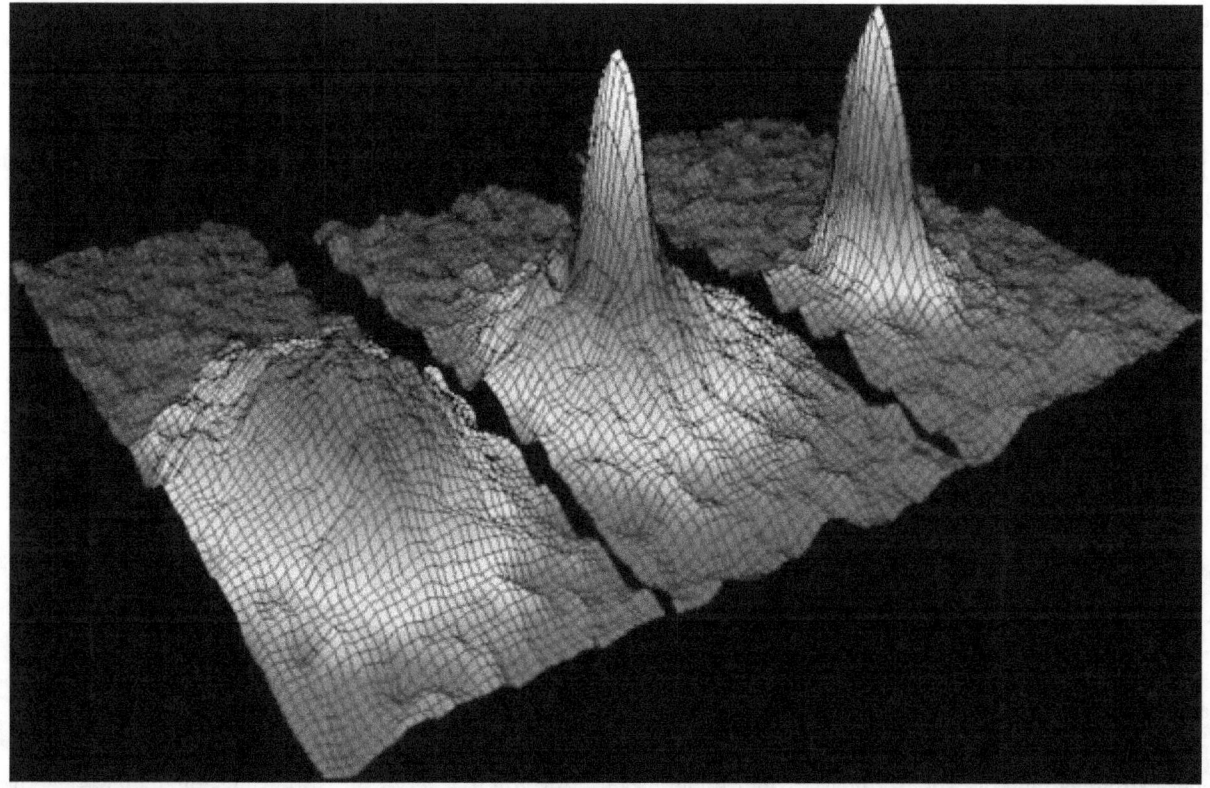

Velocity in a gas of rubidium as it is cooled: the starting material is on the left, and Bose–Einstein condensate is on the right.

2.4.4 Rydberg molecule

One of the metastable states of strongly non-ideal plasma is Rydberg matter, which forms upon condensation of excited atoms. These atoms can also turn into ions and electrons if they reach a certain temperature. In April 2009, *Nature* reported the creation of Rydberg molecules from a Rydberg atom and a ground state atom,[11] confirming that such a state of matter could exist.[12] The experiment was performed using ultracold rubidium atoms.

2.4.5 Quantum Hall state

Main article: Quantum Hall effect

A *quantum Hall state* gives rise to quantized Hall voltage measured in the direction perpendicular to the current flow. A *quantum spin Hall state* is a theoretical phase that may pave the way for the development of electronic devices that dissipate less energy and generate less heat. This is a derivation of the Quantum Hall state of matter.

2.4.6 Strange matter

Main article: Strange matter

Strange matter is a type of quark matter that may exist inside some neutron stars close to the Tolman–Oppenheimer–Volkoff limit (approximately 2–3 solar masses). It may be stable at lower energy states once formed.

2.4.7 Photonic matter

Main article: Photonic matter

In photonic matter, photons behave as if they had mass, and can interact with each other, even forming photonic "molecules". This is in contrast to the usual properties of photons, which have no rest mass, and cannot interact.

2.4.8 Dropleton

Main article: Dropleton

A "quantum fog" of electrons and holes that flow around each other and even ripple like a liquid, rather than existing as discrete pairs.[13]

2.5   High-energy states

2.5.1 Degenerate matter

Main article: Degenerate matter

Under extremely high pressure, ordinary matter undergoes a transition to a series of exotic states of matter collectively known as degenerate matter. In these conditions, the structure of matter is supported by the Pauli exclusion principle. These are of great interest to astrophysicists, because these high-pressure conditions are believed to exist inside stars that have used up their nuclear fusion "fuel", such as the white dwarfs and neutron stars.

Electron-degenerate matter is found inside white dwarf stars. Electrons remain bound to atoms but are able to transfer to adjacent atoms. Neutron-degenerate matter is found in neutron stars. Vast gravitational pressure compresses atoms so strongly that the electrons are forced to combine with protons via inverse beta-decay, resulting in a superdense conglomeration of neutrons. (Normally free neutrons outside an atomic nucleus will decay with a half life of just under 15 minutes, but in a neutron star, as in the nucleus of an atom, other effects stabilize the neutrons.)

2.5.2 Quark–gluon plasma

Main article: Quark–gluon plasma

Quark–gluon plasma is a phase in which quarks become free and able to move independently (rather than being perpetually bound into particles) in a sea of gluons (subatomic particles that transmit the strong force that binds quarks together); this is similar to splitting molecules into atoms. This state may be briefly attainable in particle accelerators, and allows scientists to observe the properties of individual quarks, and not just theorize. See also Strangeness production.

Quark–gluon plasma was discovered at CERN in 2000.

2.5.3 Color-glass condensate

Main article: Color-glass condensate

Color-glass condensate is a type of matter theorized to exist in atomic nuclei traveling near the speed of light. According to Einstein's theory of relativity, a high-energy nucleus appears length contracted, or compressed, along its direction of motion. As a result, the gluons inside the nucleus appear to a stationary observer as a "gluonic wall" traveling near the speed of light. At very high energies, the density of the gluons in this wall is seen to increase greatly. Unlike the quark–gluon plasma produced in the collision of such walls, the color-glass condensate describes the walls themselves, and is an intrinsic property of the particles that can only be observed under high-energy conditions such as those at RHIC and possibly at the Large Hadron Collider as well.

2.6   Very high energy states

The **gravitational singularity** predicted by general relativity to exist at the center of a black hole is *not* a phase of matter; it is not a material object at all (although the mass-energy of matter contributed to its creation) but rather a property of spacetime at a location. It could be argued, of course, that all particles are properties of spacetime at a location,[14] leaving a half-note of controversy on the subject.

2.7   Other proposed states

2.7.1   Supersolid

Main article: Supersolid

A supersolid is a spatially ordered material (that is, a solid or crystal) with superfluid properties. Similar to a superfluid, a supersolid is able to move without friction but retains a rigid shape. Although a supersolid is a solid, it exhibits so many characteristic properties different from other solids that many argue it is another state of matter.[15]

2.7.2   String-net liquid

Main article: String-net liquid

In a string-net liquid, atoms have apparently unstable arrangement, like a liquid, but are still consistent in overall pattern, like a solid. When in a normal solid state, the atoms of matter align themselves in a grid pattern, so that the spin of any electron is the opposite of the spin of all electrons touching it. But in a string-net liquid, atoms are arranged in some pattern that requires some electrons to have neighbors with the same spin. This gives rise to curious properties, as well as supporting some unusual proposals about the fundamental conditions of the universe itself.

2.7.3   Superglass

Main article: Superglass

A superglass is a phase of matter characterized, at the same time, by superfluidity and a frozen amorphous structure.

2.7.4   Dark matter

Main article: Dark matter

While dark matter is estimated to comprise 83% of the mass of matter in the universe, most of its properties remain a mystery due to the fact that it neither absorbs nor emits electromagnetic radiation, and there are many competing theories regarding what dark matter is actually made of. Thus, while it is hypothesized to exist and comprise the vast majority of matter in the universe, almost all of its properties are unknown and a matter of speculation, because it has only been observed through its gravitational effects.[16][17]

2.7.5   Equilibrium gel

Main article: Equilibrium gel

Equilibrium gel is made from a synthetic clay called Laponite. Unlike other gels, it maintains the same consistency throughout its structure and is stable, which means it does not separate into sections of solid mass and those of more liquid mass. Equilibrium gel filtration liquid chromatography is a technique used for the quantitation of ligand binding.[18]

2.8   See also

- Hidden states of matter
- Classical element
- Condensed matter physics
- Cooling curve
- Phase (matter)
- Supercooling
- Superheating

2.9   Notes and references

[1] It is often stated that more than 99% of the material in the visible universe is plasma. See, for instance, D. A. Gurnett; A. Bhattacharjee (2005). *Introduction to Plasma Physics: With Space and Laboratory Applications.* Cambridge, UK: Cambridge University Press. p. 2. ISBN 0-521-36483-3. and K Scherer; H Fichtner; B Heber (2005). *Space Weather: The Physics Behind a Slogan.* Berlin: Springer. p. 138. ISBN 3-540-22907-8.. Essentially, all of the visible light from space comes from stars, which are plasmas with a temperature such that they radiate strongly at visible wavelengths. Most of the ordinary (or baryonic) matter in the universe, however, is found in the intergalactic medium, which is also a plasma, but much hotter, so that it radiates primarily as X-rays. The current scientific consensus is that about 96% of the total energy density in the universe is not plasma or any other form of ordinary matter, but a combination of cold dark matter and dark energy.

[2] M.A. Wahab (2005). *Solid State Physics: Structure and Properties of Materials.* Alpha Science. pp. 1–3. ISBN 1-84265-218-4.

[3] F. White (2003). *Fluid Mechanics.* McGraw-Hill. p. 4. ISBN 0-07-240217-2.

[4] G. Turrell (1997). *Gas Dynamics: Theory and Applications.* John Wiley & Sons. pp. 3–5. ISBN 0-471-97573-7.

[5] M. Chaplin (20 August 2009). "Water phase Diagram". *Water Structure and Science.* Retrieved 23 February 2010.

[6] D.L. Goodstein (1985). *States of Matter.* Dover Phoenix. ISBN 978-0-486-49506-4.

[7] A.P. Sutton (1993). *Electronic Structure of Materials.* Oxford Science Publications. pp. 10–12. ISBN 978-0-19-851754-2.

[8] Shao, Y.; Zerda, T. W. (1998). "Phase Transitions of Liquid Crystal PAA in Confined Geometries". *Journal of Physical Chemistry B* **102** (18): 3387–3394. doi:10.1021/jp9734437.

[9] J.R. Minkel (20 February 2009). "Strange but True: Superfluid Helium Can Climb Walls". *Scientific American.* Retrieved 23 February 2010.

[10] L. Valigra (22 June 2005). "MIT physicists create new form of matter". MIT News. Retrieved 23 February 2010.

[11]V. Bendkowsky et al. (2009). "Observation of Ultralong-Range Rydberg Molecules". *Nature* **458**(7241): 1005–8.. doi:10.1038/nature07945. PMID 19396141.

[12] V. Gill (23 April 2009). "World First for Strange Molecule". BBC News. Retrieved 23 February 2010.

[13] http://www.iflscience.com/physics/new-state-matter-discovered#3Oe9x65kkHViXABt.99

[14] David Chalmers; David Manley; Ryan Wasserman (2009). *Metametaphysics: New Essays on the Foundations of Ontology.* Oxford University Press. pp. 378–. ISBN 978-0-19-954604-6.

[15] G. Murthy et al. (1997). "Superfluids and Supersolids on Frustrated Two-Dimensional Lattices". *Physical Review B* **55** (5): 3104. arXiv:cond-mat/9607217. Bibcode:1997PhRvB..55.3104M. doi:10.1103/PhysRevB.55.3104.

[16] Trimble, Virginia (1987). "Existence and nature of dark matter in the universe". *Annual Review of Astronomy and Astrophysics* **25**: 425–472. Bibcode:1987ARA&A..25..425T. doi:10.1146/annurev.aa.25.090187.002233.

[17] Hinshaw, Gary F. (29 January 2010). "What is the universe made of?". *Universe 101*. NASA website. Retrieved 17 March 2010.

[18] Cartlidge, Edwin (12 January 2012). "New State of Matter Seen in Clay". *Technology*. Science Now website. Retrieved 10 September 2013.

2.10 External links

- 2005-06-22, MIT News: MIT physicists create new form of matter Citat: "... They have become the first to create a new type of matter, a gas of atoms that shows high-temperature superfluidity."

- 2003-10-10, Science Daily: Metallic Phase For Bosons Implies New State Of Matter

- 2004-01-15, ScienceDaily: Probable Discovery Of A New, Supersolid, Phase Of Matter Citat: "...We apparently have observed, for the first time, a solid material with the characteristics of a superfluid...but because all its particles are in the identical quantum state, it remains a solid even though its component particles are continually flowing..."

- 2004-01-29, ScienceDaily: NIST/University Of Colorado Scientists Create New Form Of Matter: A Fermionic Condensate

- Short videos demonstrating of States of Matter, solids, liquids and gases by Prof. J M Murrell, University of Sussex

Chapter 3

Solid

For other uses, see Solid (disambiguation).

Single crystalline form of solid insulin.

Solid is one of the four fundamental states of matter (the others being liquid, gas, and plasma). It is characterized by structural rigidity and resistance to changes of shape or volume. Unlike a liquid, a solid object does not flow to take on the shape of its container, nor does it expand to fill the entire volume available to it like a gas does. The atoms in a solid are tightly bound to each other, either in a regular geometric lattice (crystalline solids, which include metals and ordinary ice) or irregularly (an amorphous solid such as common window glass).

The branch of physics that deals with solids is called solid-state physics, and is the main branch of condensed matter physics (which also includes liquids). Materials science is primarily concerned with the physical and chemical properties of solids. Solid-state chemistry is especially concerned with the synthesis of novel materials, as well as the science of identification and chemical composition.

3.1 Microscopic description

Model of closely packed atoms within a crystalline solid.

The atoms, molecules or ions which make up solids may be arranged in an orderly repeating pattern, or irregularly. Materials whose constituents are arranged in a regular pattern are known as crystals. In some cases, the regular ordering can continue unbroken over a large scale, for example diamonds, where each diamond is a single crystal. Solid objects that are large enough to see and handle are rarely composed of a single crystal, but instead are made of a large number of single crystals, known as crystallites, whose size can vary from a few nanometers to several meters. Such materials are called polycrystalline. Almost all common metals, and many ceramics, are polycrystalline.

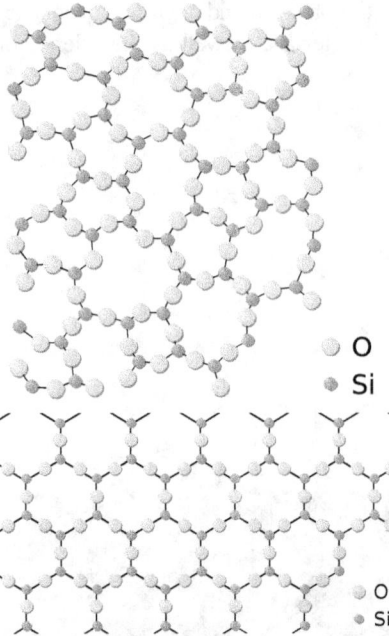

Schematic representation of a random-network glassy form (left) and ordered crystalline lattice (right) of identical chemical composition.

In other materials, there is no long-range order in the position of the atoms. These solids are known as amorphous solids; examples include polystyrene and glass.

Whether a solid is crystalline or amorphous depends on the material involved, and the conditions in which it was formed. Solids which are formed by slow cooling will tend to be crystalline, while solids which are frozen rapidly are more likely to be amorphous. Likewise, the specific crystal structure adopted by a crystalline solid depends on the material involved and on how it was formed.

While many common objects, such as an ice cube or a coin, are chemically identical throughout, many other common materials comprise a number of different substances packed together. For example, a typical rock is an aggregate of several different minerals and mineraloids, with no specific chemical composition. Wood is a natural organic material consisting primarily of cellulose fibers embedded in a matrix of organic lignin. In materials science, composites of more than one constituent material can be designed to have desired properties.

3.2   Classes of solids

Further information: Bonding in solids

The forces between the atoms in a solid can take a variety of forms. For example, a crystal of sodium chloride (common salt) is made up of ionic sodium and chlorine, which are held together by ionic bonds. In diamond or silicon, the atoms share electrons and form covalent bonds. In metals, electrons are shared in metallic bonding. Some solids, particularly most organic compounds, are held together with van der Waals forces resulting from the polarization of the electronic charge cloud on each molecule. The dissimilarities between the types of solid result from the differences between their bonding.

3.2.1   Metals

Main article: Metal

Metals typically are strong, dense, and good conductors of both electricity and heat. The bulk of the elements in the

The pinnacle of New York's Chrysler Building, the world's tallest steel-supported brick building, is clad with stainless steel.

periodic table, those to the left of a diagonal line drawn from boron to polonium, are metals. Mixtures of two or more elements in which the major component is a metal are known as alloys.

People have been using metals for a variety of purposes since prehistoric times. The strength and reliability of metals has led to their widespread use in construction of buildings and other structures, as well as in most vehicles, many appliances and tools, pipes, road signs and railroad tracks. Iron and aluminium are the two most commonly used structural metals, and they are also the most abundant metals in the Earth's crust. Iron is most commonly used in the form of an alloy, steel, which contains up to 2.1% carbon, making it much harder than pure iron.

Because metals are good conductors of electricity, they are valuable in electrical appliances and for carrying an electric current over long distances with little energy loss or dissipation. Thus, electrical power grids rely on metal cables to distribute electricity. Home electrical systems, for example, are wired with copper for its good conducting properties and easy machinability. The high thermal conductivity of most metals also makes them useful for stovetop cooking utensils.

The study of metallic elements and their alloys makes up a significant portion of the fields of solid-state chemistry, physics, materials science and engineering.

Metallic solids are held together by a high density of shared, delocalized electrons, known as "metallic bonding". In a metal, atoms readily lose their outermost ("valence") electrons, forming positive ions. The free electrons are spread over the entire solid, which is held together firmly by electrostatic interactions between the ions and the electron cloud.[1] The large number of free electrons gives metals their high values of electrical and thermal conductivity. The free electrons also prevent transmission of visible light, making metals opaque, shiny and lustrous.

More advanced models of metal properties consider the effect of the positive ions cores on the delocalised electrons. As most metals have crystalline structure, those ions are usually arranged into a periodic lattice. Mathematically, the potential of the ion cores can be treated by various models, the simplest being the nearly free electron model.

3.2.2   Minerals

Main article: Minerals

Minerals are naturally occurring solids formed through various geological processes under high pressures. To be classified as a true mineral, a substance must have a crystal structure with uniform physical properties throughout. Minerals range in composition from pure elements and simple salts to very complex silicates with thousands of known forms. In contrast, a rock sample is a random aggregate of minerals and/or mineraloids, and has no specific chemical composition. The vast majority of the rocks of the Earth's crust consist of quartz (crystalline SiO_2), feldspar, mica, chlorite, kaolin, calcite, epidote, olivine, augite, hornblende, magnetite, hematite, limonite and a few other minerals. Some minerals, like quartz, mica or feldspar are common, while others have been found in only a few locations worldwide. The largest group of minerals by far is the silicates (most rocks are $\geq$95% silicates), which are composed largely of silicon and oxygen, with the addition of ions of aluminium, magnesium, iron, calcium and other metals.

3.2.3   Ceramics

Main article: Ceramic engineering

Ceramic solids are composed of inorganic compounds, usually oxides of chemical elements. They are chemically inert, and often are capable of withstanding chemical erosion that occurs in an acidic or caustic environment. Ceramics generally can withstand high temperatures ranging from 1000 to 1600 °C (1800 to 3000 °F). Exceptions include non-oxide inorganic materials, such as nitrides, borides and carbides.

Traditional ceramic raw materials include clay minerals such as kaolinite, more recent materials include aluminium oxide (alumina). The modern ceramic materials, which are classified as advanced ceramics, include silicon carbide and tungsten carbide. Both are valued for their abrasion resistance, and hence find use in such applications as the wear plates of crushing equipment in mining operations.

Most ceramic materials, such as alumina and its compounds, are formed from fine powders, yielding a fine grained

A collection of various minerals.

polycrystalline microstructure which is filled with light scattering centers comparable to the wavelength of visible light. Thus, they are generally opaque materials, as opposed to transparent materials. Recent nanoscale (e.g. sol-gel) technology has, however, made possible the production of polycrystalline transparent ceramics such as transparent alumina and alumina compounds for such applications as high-power lasers. Advanced ceramics are also used in the medicine, electrical and electronics industries.

Ceramic engineering is the science and technology of creating solid-state ceramic materials, parts and devices. This is done either by the action of heat, or, at lower temperatures, using precipitation reactions from chemical solutions. The term includes the purification of raw materials, the study and production of the chemical compounds concerned, their formation into components, and the study of their structure, composition and properties.

Mechanically speaking, ceramic materials are brittle, hard, strong in compression and weak in shearing and tension. Brittle materials may exhibit significant tensile strength by supporting a static load. Toughness indicates how much energy a material can absorb before mechanical failure, while fracture toughness (denoted KI_c) describes the ability of a material with inherent microstructural flaws to resist fracture via crack growth and propagation. If a material has a large value of fracture toughness, the basic principles of fracture mechanics suggest that it will most likely undergo ductile fracture. Brittle fracture is very characteristic of most ceramic and glass-ceramic materials which typically exhibit low (and inconsistent) values of KI_c.

For an example of applications of ceramics, the extreme hardness of Zirconia is utilized in the manufacture of knife blades, as well as other industrial cutting tools. Ceramics such as alumina, boron carbide and silicon carbide have been used in bulletproof vests to repel large-caliber rifle fire. Silicon nitride parts are used in ceramic ball bearings, where their high hardness makes them wear resistant. In general, ceramics are also chemically resistant and can be used in wet

Si$_3$N$_4$ ceramic bearing parts

environments where steel bearings would be susceptible to oxidation (or rust).

As another example of ceramic applications, in the early 1980s, Toyota researched production of an adiabatic ceramic engine with an operating temperature of over 6000 °F (3300 °C). Ceramic engines do not require a cooling system and hence allow a major weight reduction and therefore greater fuel efficiency. In a conventional metallic engine, much of the energy released from the fuel must be dissipated as waste heat in order to prevent a meltdown of the metallic parts. Work is also being done in developing ceramic parts for gas turbine engines. Turbine engines made with ceramics could operate more efficiently, giving aircraft greater range and payload for a set amount of fuel. However, such engines are not in production because the manufacturing of ceramic parts in the sufficient precision and durability is difficult and costly. Processing methods often result in a wide distribution of microscopic flaws which frequently play a detrimental role in the sintering process, resulting in the proliferation of cracks, and ultimate mechanical failure.

3.2.4   Glass ceramics

Main article: Glass-ceramic

Glass-ceramic materials share many properties with both non-crystalline glasses and crystalline ceramics. They are formed as a glass, and then partially crystallized by heat treatment, producing both amorphous and crystalline phases so that crystalline grains are embedded within a non-crystalline intergranular phase.

A high strength glass-ceramic cooktop with negligible thermal expansion.

Glass-ceramics are used to make cookware (originally known by the brand name CorningWare) and stovetops which have both high resistance to thermal shock and extremely low permeability to liquids. The negative coefficient of thermal expansion of the crystalline ceramic phase can be balanced with the positive coefficient of the glassy phase. At a certain point (~70% crystalline) the glass-ceramic has a net coefficient of thermal expansion close to zero. This type of glass-ceramic exhibits excellent mechanical properties and can sustain repeated and quick temperature changes up to 1000 °C.

Glass ceramics may also occur naturally when lightning strikes the crystalline (e.g. quartz) grains found in most beach sand. In this case, the extreme and immediate heat of the lightning (~2500 °C) creates hollow, branching rootlike structures called fulgurite via fusion.

3.2.5 Organic solids

Main article: Organic chemistry

Organic chemistry studies the structure, properties, composition, reactions, and preparation by synthesis (or other means) of chemical compounds of carbon and hydrogen, which may contain any number of other elements such as nitrogen, oxygen and the halogens: fluorine, chlorine, bromine and iodine. Some organic compounds may also contain the elements phosphorus or sulfur. Examples of organic solids include wood, paraffin wax, naphthalene and a wide variety of polymers and plastics.

Wood

Main article: Wood

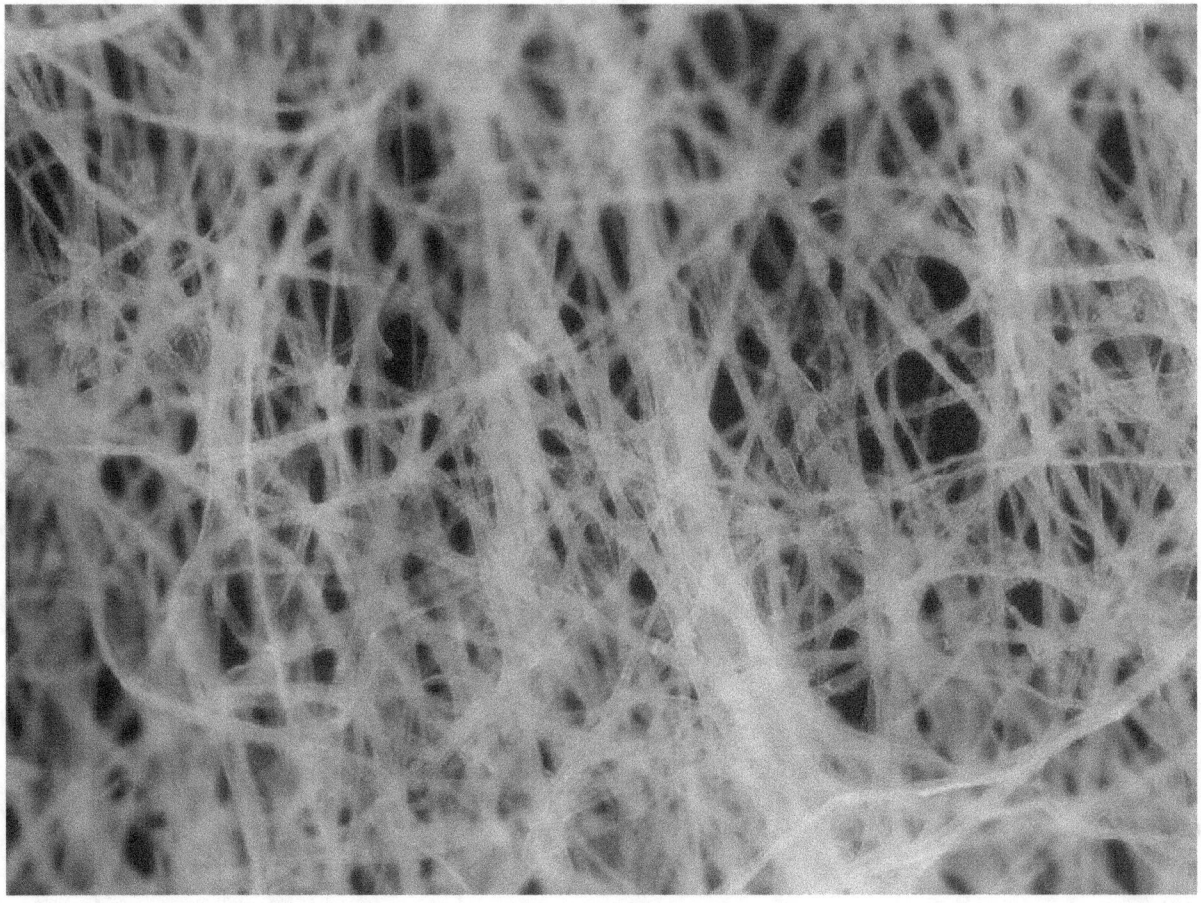

The individual wood pulp fibers in this sample are around 10 μm in diameter.

Wood is a natural organic material consisting primarily of cellulose fibers embedded in a matrix of lignin. Regarding mechanical properties, the fibers are strong in tension, and the lignin matrix resists compression. Thus wood has been an important construction material since humans began building shelters and using boats. Wood to be used for construction work is commonly known as *lumber* or *timber*. In construction, wood is not only a structural material, but is also used to form the mould for concrete.

Wood-based materials are also extensively used for packaging (e.g. cardboard) and paper which are both created from the refined pulp. The chemical pulping processes use a combination of high temperature and alkaline (kraft) or acidic (sulfite) chemicals to break the chemical bonds of the lignin before burning it out.

Polymers

Main article: Polymer

One important property of carbon in organic chemistry is that it can form certain compounds, the individual molecules of which are capable of attaching themselves to one another, thereby forming a chain or a network. The process is called polymerization and the chains or networks polymers, while the source compound is a monomer. Two main groups of polymers exist: those artificially manufactured are referred to as industrial polymers or synthetic polymers (plastics) and those naturally occurring as biopolymers.

Monomers can have various chemical substituents, or functional groups, which can affect the chemical properties of organic compounds, such as solubility and chemical reactivity, as well as the physical properties, such as hardness, density, mechanical or tensile strength, abrasion resistance, heat resistance, transparency, color, etc.. In proteins, these differences

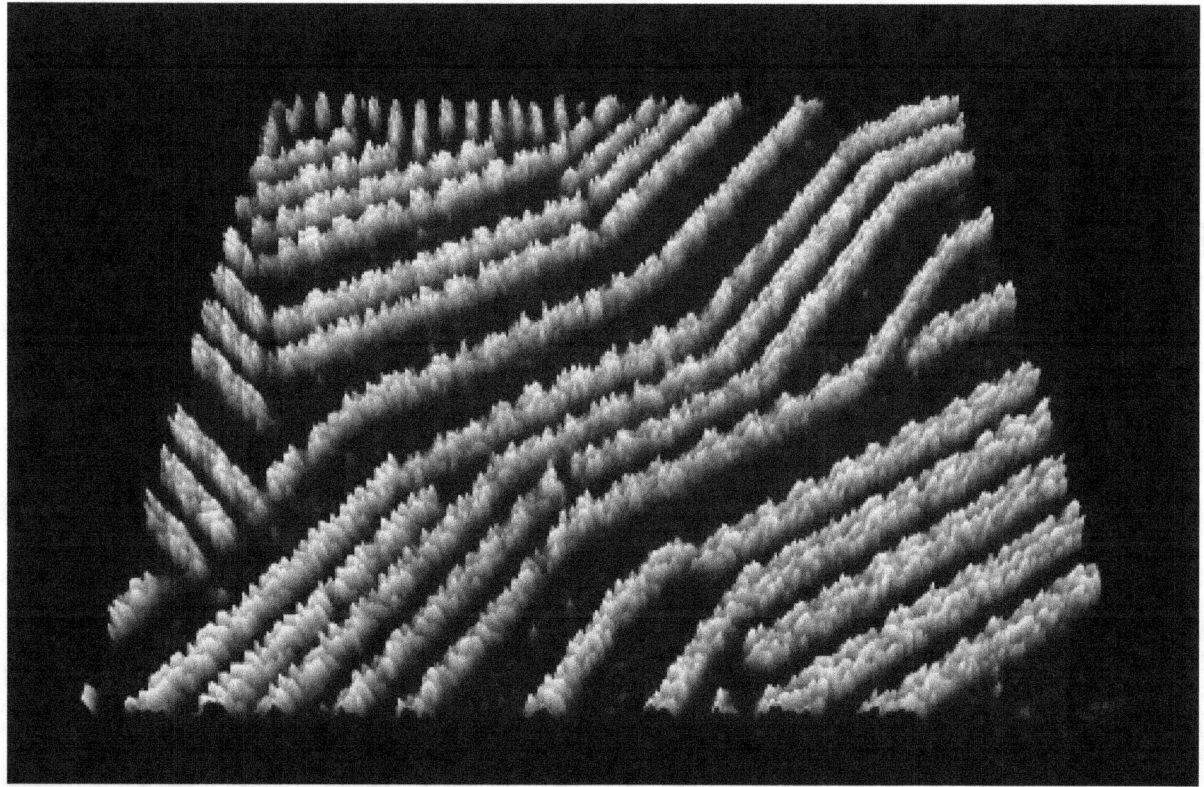

STM image of self-assembled supramolecular chains of the organic semiconductor quinacridone on graphite.

give the polymer the ability to adopt a biologically active conformation in preference to others (see self-assembly).

People have been using natural organic polymers for centuries in the form of waxes and shellac which is classified as a thermoplastic polymer. A plant polymer named cellulose provided the tensile strength for natural fibers and ropes, and by the early 19th century natural rubber was in widespread use. Polymers are the raw materials (the resins) used to make what are commonly called plastics. Plastics are the final product, created after one or more polymers or additives have been added to a resin during processing, which is then shaped into a final form. Polymers which have been around, and which are in current widespread use, include carbon-based polyethylene, polypropylene, polyvinyl chloride, polystyrene, nylons, polyesters, acrylics, polyurethane, and polycarbonates, and silicon-based silicones. Plastics are generally classified as "commodity", "specialty" and "engineering" plastics.

3.2.6 Composite materials

Main article: Composite material

Composite materials contain two or more macroscopic phases, one of which is often ceramic. For example, a continuous matrix, and a dispersed phase of ceramic particles or fibers.

Applications of composite materials range from structural elements such as steel-reinforced concrete, to the thermally insulative tiles which play a key and integral role in NASA's Space Shuttle thermal protection system which is used to protect the surface of the shuttle from the heat of re-entry into the Earth's atmosphere. One example is Reinforced Carbon-Carbon (RCC), the light gray material which withstands reentry temperatures up to 1510 °C (2750 °F) and protects the nose cap and leading edges of Space Shuttle's wings. RCC is a laminated composite material made from graphite rayon cloth and impregnated with a phenolic resin. After curing at high temperature in an autoclave, the laminate is pyrolized to convert the resin to carbon, impregnated with furfural alcohol in a vacuum chamber, and cured/pyrolized to convert the furfural alcohol to carbon. In order to provide oxidation resistance for reuse capability, the outer layers of the RCC

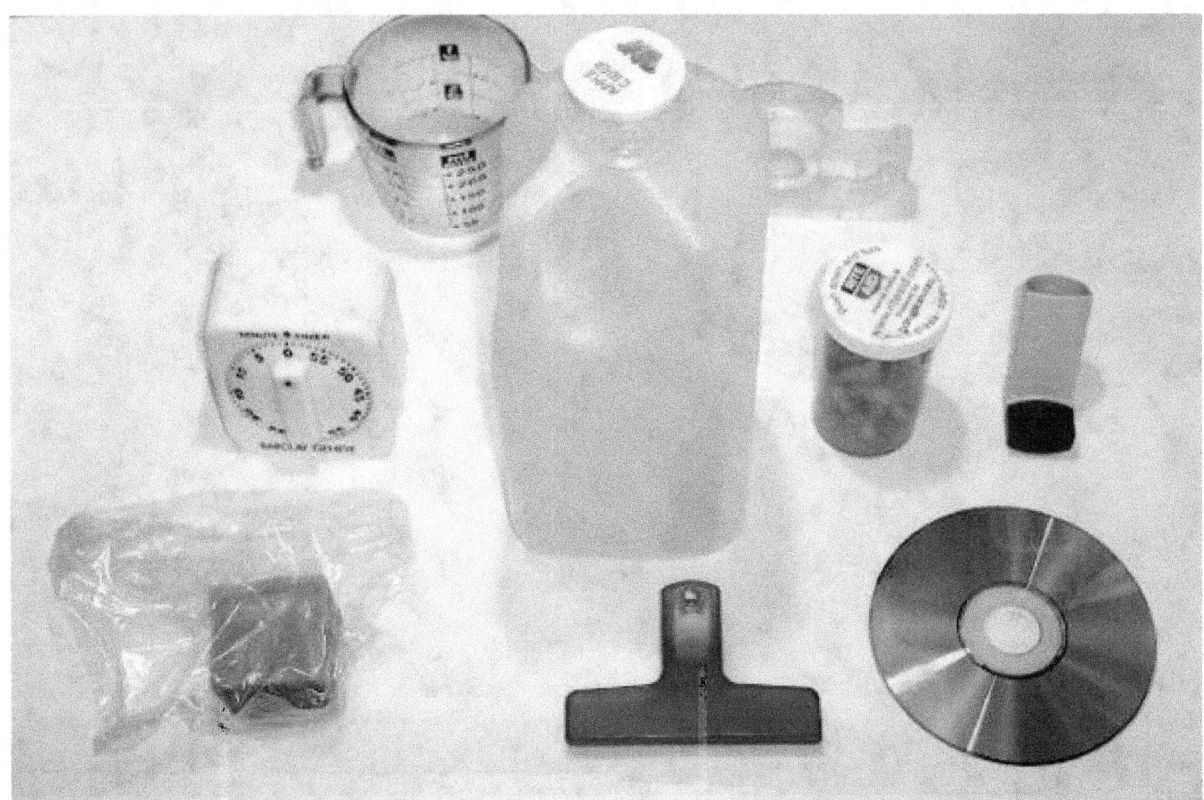

Household items made of various kinds of plastic.

are converted to silicon carbide.

Domestic examples of composites can be seen in the "plastic" casings of television sets, cell-phones and so on. These plastic casings are usually a composite made up of a thermoplastic matrix such as acrylonitrile butadiene styrene (ABS) in which calcium carbonate chalk, talc, glass fibers or carbon fibers have been added for strength, bulk, or electro-static dispersion. These additions may be referred to as reinforcing fibers, or dispersants, depending on their purpose.

Thus, the matrix material surrounds and supports the reinforcement materials by maintaining their relative positions. The reinforcements impart their special mechanical and physical properties to enhance the matrix properties. A synergism produces material properties unavailable from the individual constituent materials, while the wide variety of matrix and strengthening materials provides the designer with the choice of an optimum combination.

3.2.7 Semiconductors

Main article: Semiconductors

Semiconductors are materials that have an electrical resistivity (and conductivity) between that of metallic conductors and non-metallic insulators. They can be found in the periodic table moving diagonally downward right from boron. They separate the electrical conductors (or metals, to the left) from the insulators (to the right).

Devices made from semiconductor materials are the foundation of modern electronics, including radio, computers, telephones, etc. Semiconductor devices include the transistor, solar cells, diodes and integrated circuits. Solar photovoltaic panels are large semiconductor devices that directly convert light into electrical energy.

In a metallic conductor, current is carried by the flow of electrons", but in semiconductors, current can be carried either by electrons or by the positively charged "holes" in the electronic band structure of the material. Common semiconductor materials include silicon, germanium and gallium arsenide.

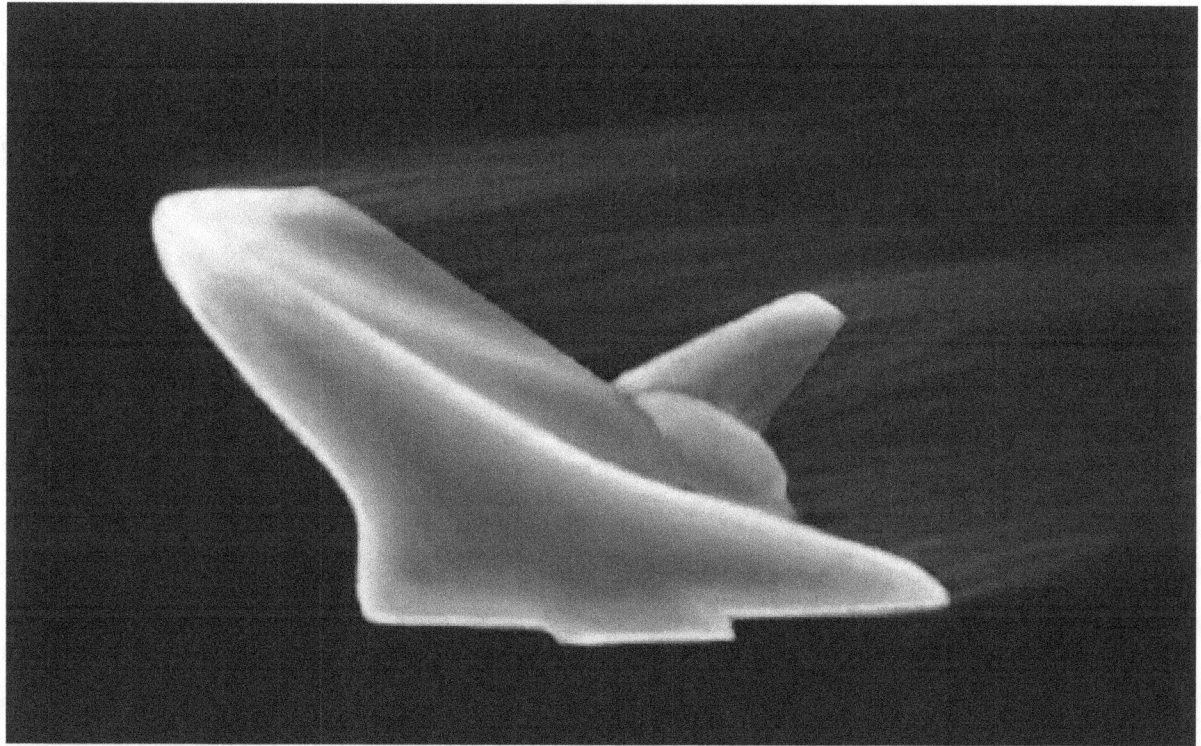

Simulation of the outside of the Space Shuttle as it heats up to over 1500 °C during re-entry

3.2.8 Nanomaterials

Main article: Nanotechnology

Many traditional solids exhibit different properties when they shrink to nanometer sizes. For example, nanoparticles of usually yellow gold and gray silicon are red in color; gold nanoparticles melt at much lower temperatures (~300 °C for 2.5 nm size) than the gold slabs (1064 °C);[2] and metallic nanowires are much stronger than the corresponding bulk metals.[3][4] The high surface area of nanoparticles makes them extremely attractive for certain applications in the field of energy. For example, platinum metals may be provide improvements as automotive fuel catalysts, as well as proton exchange membrane (PEM) fuel cells. Also, ceramic oxides (or cermets) of lanthanum, cerium, manganese and nickel are now being developed as solid oxide fuel cells (SOFC). Lithium, lithium–titanate and tantalum nanoparticles are being applied in lithium ion batteries. Silicon nanoparticles have been shown to dramatically expand the storage capacity of lithium ion batteries during the expansion/contraction cycle. Silicon nanowires cycle without significant degradation and present the potential for use in batteries with greatly expanded storage times. Silicon nanoparticles are also being used in new forms of solar energy cells. Thin film deposition of silicon quantum dots on the polycrystalline silicon substrate of a photovoltaic (solar) cell increases voltage output as much as 60% by fluorescing the incoming light prior to capture. Here again, surface area of the nanoparticles (and thin films) plays a critical role in maximizing the amount of absorbed radiation.

3.2.9 Biomaterials

Main article: Biomaterials

Many natural (or biological) materials are complex composites with remarkable mechanical properties. These complex structures, which have risen from hundreds of million years of evolution, are inspiring materials scientists in the design of novel materials. Their defining characteristics include structural hierarchy, multifunctionality and self-healing capability. Self-organization is also a fundamental feature of many biological materials and the manner by which the structures are assembled from the molecular level up. Thus, self-assembly is emerging as a new strategy in the chemical synthesis of

A cloth of woven carbon fiber filaments, a common element in composite materials

high performance biomaterials.

3.3   Physical properties

Physical properties of elements and compounds which provide conclusive evidence of chemical composition include odor, color, volume, density (mass per unit volume), melting point, boiling point, heat capacity, physical form and shape at room temperature (solid, liquid or gas; cubic, trigonal crystals, etc.), hardness, porosity, index of refraction and many others. This section discusses some physical properties of materials in the solid state.

3.3.1   Mechanical

The mechanical properties of materials describe characteristics such as their strength and resistance to deformation. For example, steel beams are used in construction because of their high strength, meaning that they neither break nor bend significantly under the applied load.

Mechanical properties include elasticity and plasticity, tensile strength, compressive strength, shear strength, fracture toughness, ductility (low in brittle materials), and indentation hardness. Solid mechanics is the study of the behavior of solid matter under external actions such as external forces and temperature changes.

A solid does not exhibit macroscopic flow, as fluids do. Any degree of departure from its original shape is called deformation. The proportion of deformation to original size is called strain. If the applied stress is sufficiently low,

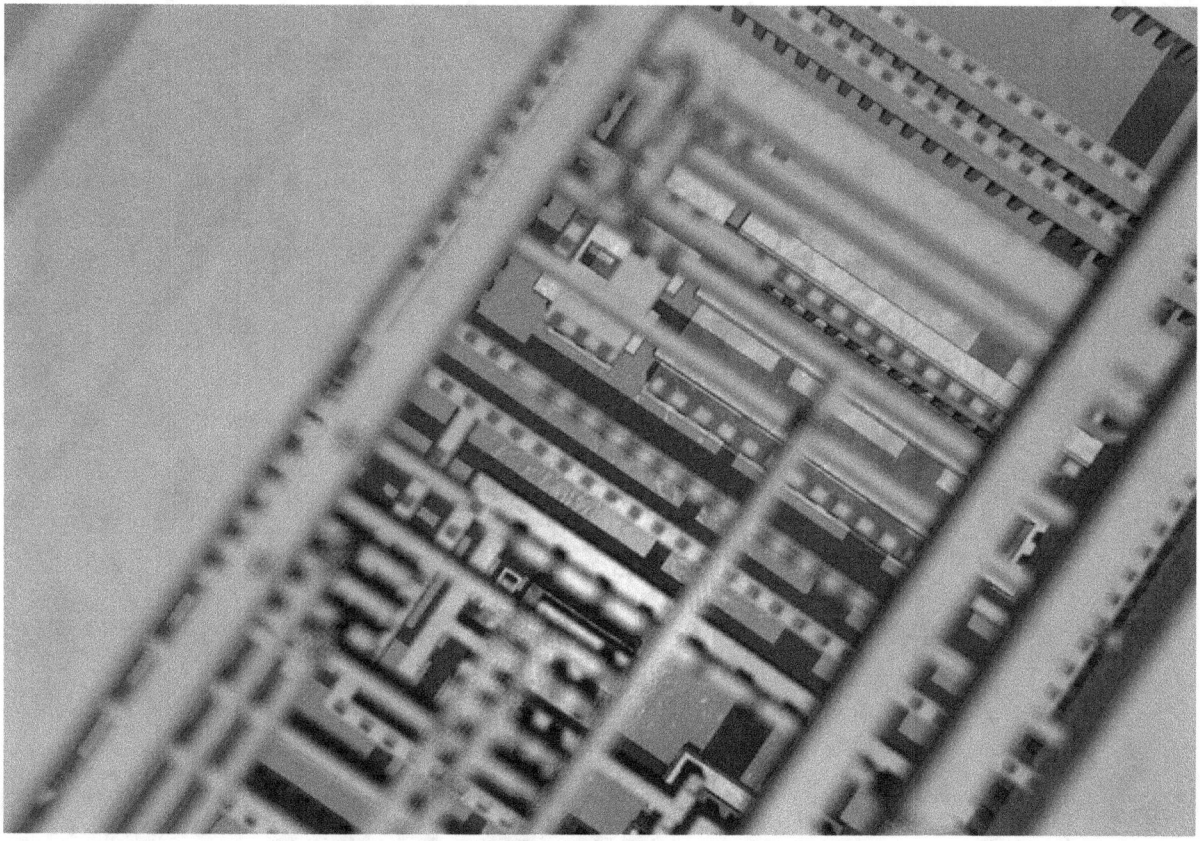

Semiconductor chip on crystalline silicon substrate.

almost all solid materials behave in such a way that the strain is directly proportional to the stress (Hooke's law). The coefficient of the proportion is called the modulus of elasticity or Young's modulus. This region of deformation is known as the linearly elastic region. Three models can describe how a solid responds to an applied stress:

- Elasticity – When an applied stress is removed, the material returns to its undeformed state.

- Viscoelasticity – These are materials that behave elastically, but also have damping. When the applied stress is removed, work has to be done against the damping effects and is converted to heat within the material. This results in a hysteresis loop in the stress–strain curve. This implies that the mechanical response has a time-dependence.

- Plasticity – Materials that behave elastically generally do so when the applied stress is less than a yield value. When the stress is greater than the yield stress, the material behaves plastically and does not return to its previous state. That is, irreversible plastic deformation (or viscous flow) occurs after yield which is permanent.

Many materials become weaker at high temperatures. Materials which retain their strength at high temperatures, called refractory materials, are useful for many purposes. For example, glass-ceramics have become extremely useful for countertop cooking, as they exhibit excellent mechanical properties and can sustain repeated and quick temperature changes up to 1000 °C. In the aerospace industry, high performance materials used in the design of aircraft and/or spacecraft exteriors must have a high resistance to thermal shock. Thus, synthetic fibers spun out of organic polymers and polymer/ceramic/metal composite materials and fiber-reinforced polymers are now being designed with this purpose in mind.

3.3.2 Thermal

Because solids have thermal energy, their atoms vibrate about fixed mean positions within the ordered (or disordered) lattice. The spectrum of lattice vibrations in a crystalline or glassy network provides the foundation for the kinetic theory

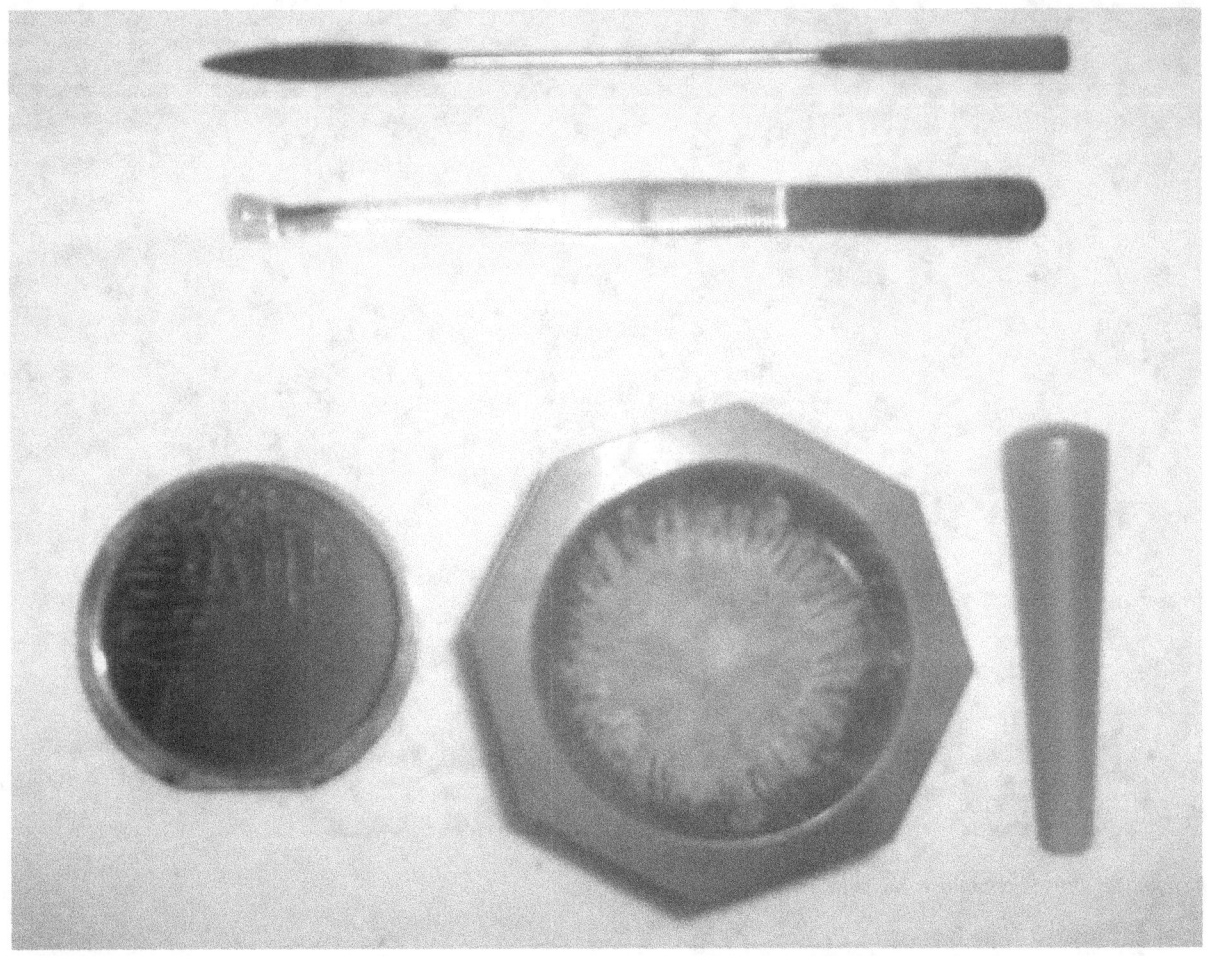

Bulk silicon (left) and silicon nanopowder (right)

of solids. This motion occurs at the atomic level, and thus cannot be observed or detected without highly specialized equipment, such as that used in spectroscopy.

Thermal properties of solids include thermal conductivity, which is the property of a material that indicates its ability to conduct heat. Solids also have a specific heat capacity, which is the capacity of a material to store energy in the form of heat (or thermal lattice vibrations).

3.3.3   Electrical

Electrical properties include conductivity, resistance, impedance and capacitance. Electrical conductors such as metals and alloys are contrasted with electrical insulators such as glasses and ceramics. Semiconductors behave somewhere in between. Whereas conductivity in metals is caused by electrons, both electrons and holes contribute to current in semiconductors. Alternatively, ions support electric current in ionic conductors.

Many materials also exhibit superconductivity at low temperatures; they include metallic elements such as tin and aluminium, various metallic alloys, some heavily doped semiconductors, and certain ceramics. The electrical resistivity of most electrical (metallic) conductors generally decreases gradually as the temperature is lowered, but remains finite. In a superconductor however, the resistance drops abruptly to zero when the material is cooled below its critical temperature. An electric current flowing in a loop of superconducting wire can persist indefinitely with no power source.

A dielectric, or electrical insulator, is a substance that is highly resistant to the flow of electric current. A dielectric, such as plastic, tends to concentrate an applied electric field within itself which property is used in capacitors. A capacitor is

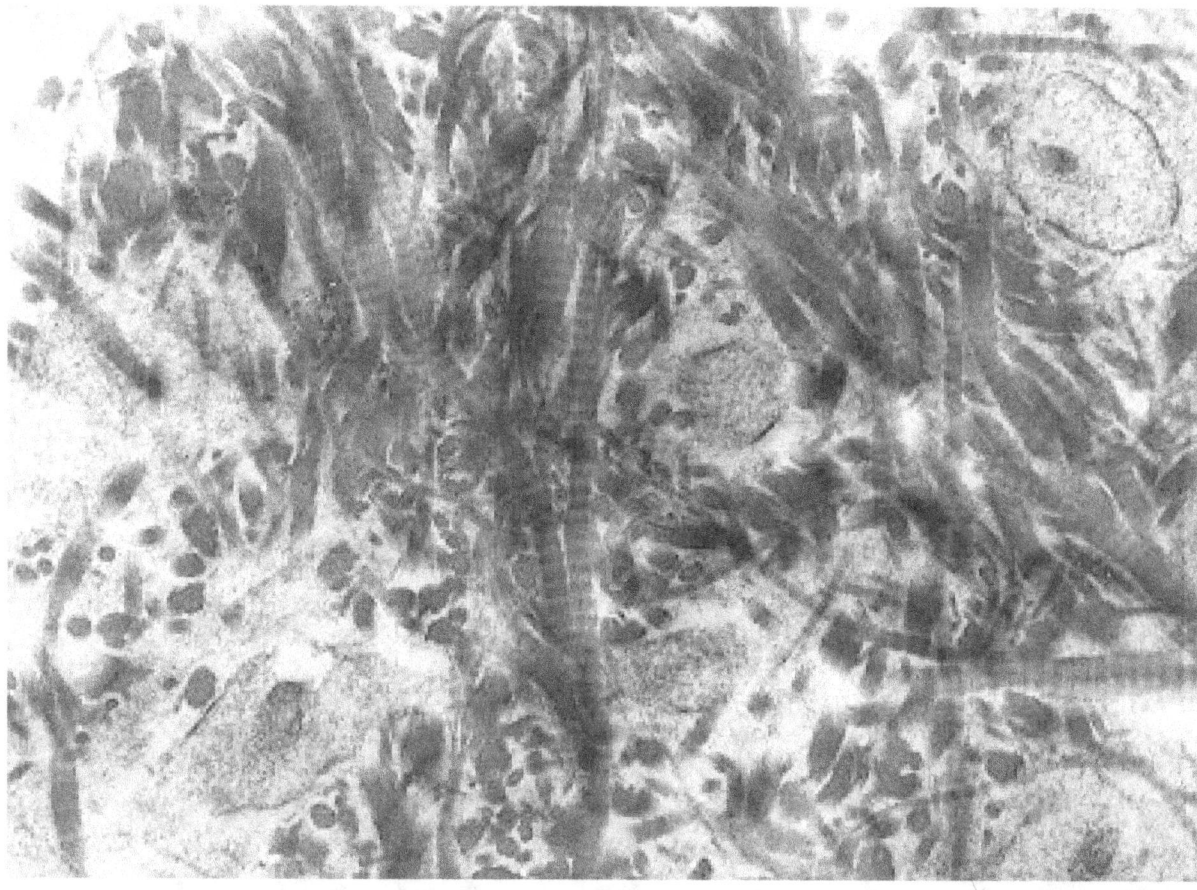

Collagen fibers of woven bone

an electrical device that can store energy in the electric field between a pair of closely spaced conductors (called 'plates'). When voltage is applied to the capacitor, electric charges of equal magnitude, but opposite polarity, build up on each plate. Capacitors are used in electrical circuits as energy-storage devices, as well as in electronic filters to differentiate between high-frequency and low-frequency signals.

Electro-mechanical

Piezoelectricity is the ability of crystals to generate a voltage in response to an applied mechanical stress. The piezoelectric effect is reversible in that piezoelectric crystals, when subjected to an externally applied voltage, can change shape by a small amount. Polymer materials like rubber, wool, hair, wood fiber, and silk often behave as electrets. For example, the polymer polyvinylidene fluoride (PVDF) exhibits a piezoelectric response several times larger than the traditional piezoelectric material quartz (crystalline SiO_2). The deformation (~0.1%) lends itself to useful technical applications such as high-voltage sources, loudspeakers, lasers, as well as chemical, biological, and acousto-optic sensors and/or transducers.

3.3.4 Optical

Materials can transmit (e.g. glass) or reflect (e.g. metals) visible light.

Many materials will transmit some wavelengths while blocking others. For example, window glass is transparent to visible light, but much less so to most of the frequencies of ultraviolet light that cause sunburn. This property is used for frequency-selective optical filters, which can alter the color of incident light.

For some purposes, both the optical and mechanical properties of a material can be of interest. For example, the sensors

Granite rock formation in the Chilean Patagonia. Like most inorganic minerals formed by oxidation in the Earth's atmosphere, granite consists primarily of crystalline silica SiO_2 and alumina Al_2O_3.

on an infrared homing ("heat-seeking") missile must be protected by a cover which is transparent to infrared radiation. The current material of choice for high-speed infrared-guided missile domes is single-crystal sapphire. The optical transmission of sapphire does not actually extend to cover the entire mid-infrared range (3–5 μm), but starts to drop off at wavelengths greater than approximately 4.5 μm at room temperature. While the strength of sapphire is better than that of other available mid-range infrared dome materials at room temperature, it weakens above 600 °C. A long-standing trade-off exists between optical bandpass and mechanical durability; new materials such as transparent ceramics or optical nanocomposites may provide improved performance.

Guided lightwave transmission involves the field of fiber optics and the ability of certain glasses to transmit, simultaneously and with low loss of intensity, a range of frequencies (multi-mode optical waveguides) with little interference between them. Optical waveguides are used as components in integrated optical circuits or as the transmission medium in optical communication systems.

Opto-electronic

Main article: Solar cell

A solar cell or photovoltaic cell is a device that converts light energy into electrical energy. Fundamentally, the device needs to fulfill only two functions: photo-generation of charge carriers (electrons and holes) in a light-absorbing material, and separation of the charge carriers to a conductive contact that will transmit the electricity (simply put, carrying electrons off through a metal contact into an external circuit). This conversion is called the photoelectric effect, and the field of

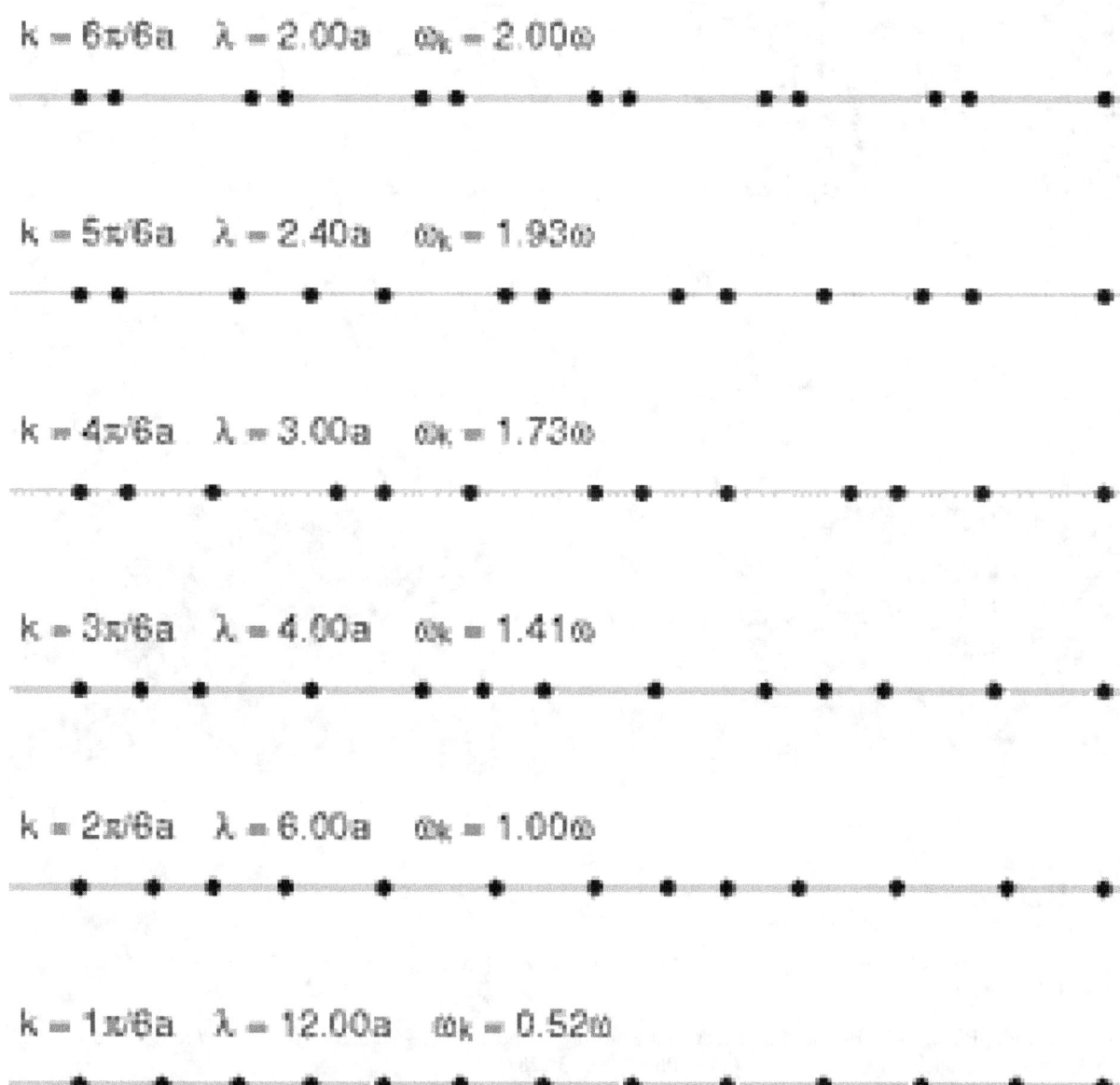

Normal modes of atomic vibration in a crystalline solid.

research related to solar cells is known as photovoltaics.

Solar cells have many applications. They have long been used in situations where electrical power from the grid is un-available, such as in remote area power systems, Earth-orbiting satellites and space probes, handheld calculators, wrist watches, remote radiotelephones and water pumping applications. More recently, they are starting to be used in assem-blies of solar modules (photovoltaic arrays) connected to the electricity grid through an inverter, that is not to act as a sole supply but as an additional electricity source.

All solar cells require a light absorbing material contained within the cell structure to absorb photons and generate electrons via the photovoltaic effect. The materials used in solar cells tend to have the property of preferentially absorbing the wavelengths of solar light that reach the earth surface. However, some solar cells are optimized for light absorption beyond Earth's atmosphere as well.

Video of superconducting levitation of YBCO

3.4   References

[1] Mortimer, Charles E. (1975). *Chemistry: A Conceptual Approach* (3rd ed.). New York:: D. Van Nostrad Company. ISBN 0-442-25545-4.

[2] Buffat, Ph.; Borel, J.-P. (1976). "Size effect on the melting temperature of gold particles". *Physical Review A* **13** (6): 2287. Bibcode:1976PhRvA..13.2287B. doi:10.1103/PhysRevA.13.2287.

[3] Walter H. Kohl (1995). *Handbook of materials and techniques for vacuum devices*. Springer. pp. 164–167. ISBN 1-56396-387-6.

[4] Shpak, Anatoly P; Kotrechko, Sergiy O; Mazilova, Tatjana I; Mikhailovskij, Igor M (2009). "Inherent tensile strength of molybdenum nanocrystals". *Science and Technology of Advanced Materials* **10** (4): 045004. Bibcode:2009STAdM..10d5004S. doi:10.1088/1468-6996/10/4/045004.

3.5   External links

- Wiki on equipment for handling and processing Bulk Solids

Chapter 4

Liquid

For other uses, see Liquid (disambiguation).

A **liquid** is a nearly incompressible fluid that conforms to the shape of its container but retains a (nearly) constant volume

The formation of a spherical droplet of liquid water minimizes the surface area, which is the natural result of surface tension in liquids.

independent of pressure. As such, it is one of the four fundamental states of matter (the others being solid, gas, and plasma), and is the only state with a definite volume but no fixed shape. A liquid is made up of tiny vibrating particles of matter, such as atoms, held together by intermolecular bonds. Water is, by far, the most common liquid on Earth. Like a gas, a liquid is able to flow and take the shape of a container. Most liquids resist compression, although others can be compressed. Unlike a gas, a liquid does not disperse to fill every space of a container, and maintains a fairly constant density. A distinctive property of the liquid state is surface tension, leading to wetting phenomena.

The density of a liquid is usually close to that of a solid, and much higher than in a gas. Therefore, liquid and solid are both termed condensed matter. On the other hand, as liquids and gases share the ability to flow, they are both called fluids. Although liquid water is abundant on Earth, this state of matter is actually the least common in the known universe, because liquids require a relatively narrow temperature/pressure range to exist. Most known matter in the universe is in gaseous form (with traces of detectable solid matter) as interstellar clouds or in plasma form within stars.

4.1   Introduction

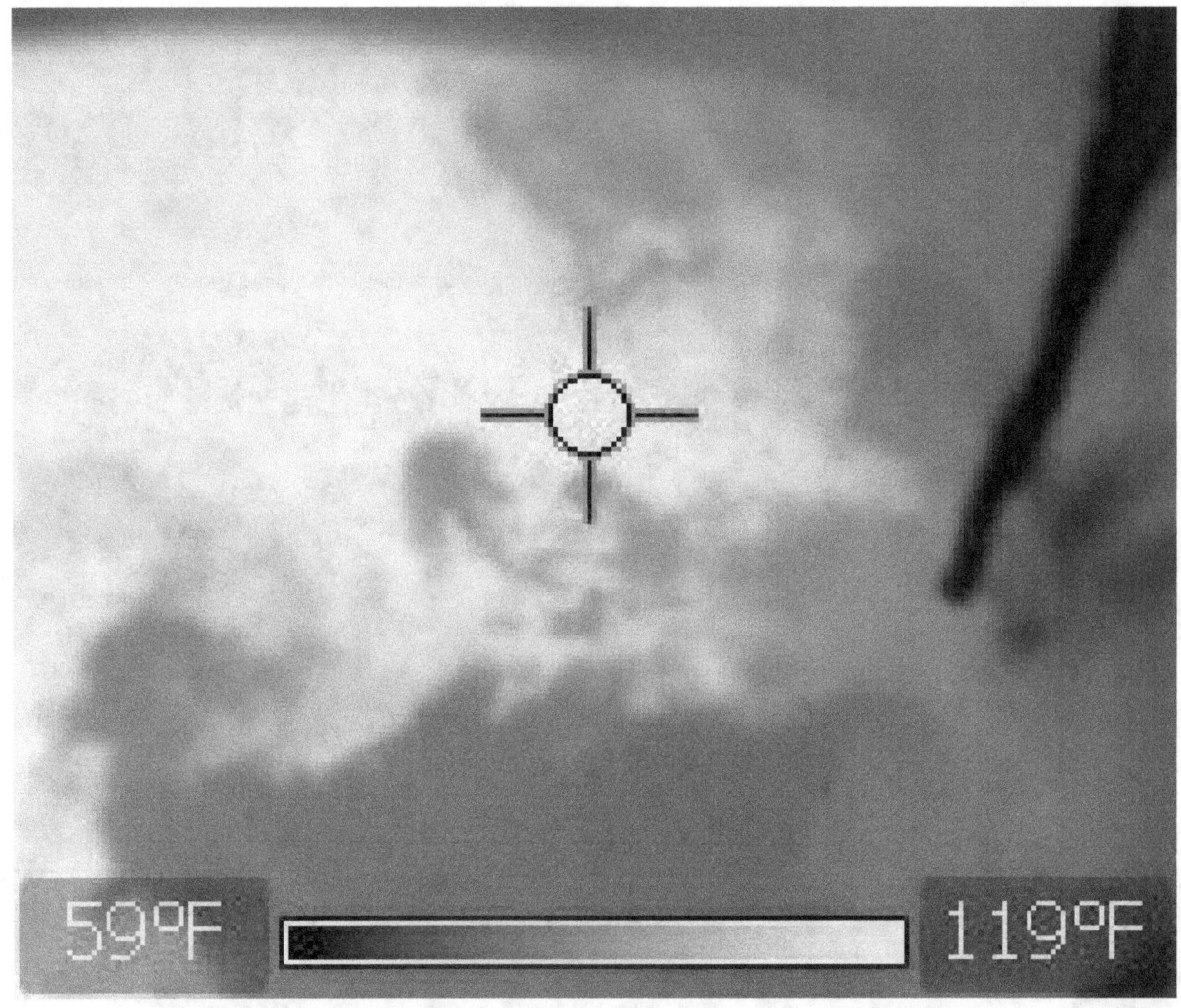

Thermal image of a sink full of hot water with cold water being added, showing how the hot and the cold water flow into each other.

Liquid is one of the four primary states of matter, with the others being solid, gas and plasma. A liquid is a fluid. Unlike a solid, the molecules in a liquid have a much greater freedom to move. The forces that bind the molecules together in a solid are only temporary in a liquid, allowing a liquid to flow while a solid remains rigid.

A liquid, like a gas, displays the properties of a fluid. A liquid can flow, assume the shape of a container, and, if placed in a sealed container, will distribute applied pressure evenly to every surface in the container. If you place the liquid in a bag, you can squeeze it into any shape you want. Unlike a gas, a liquid may not always mix readily with another liquid, will not always fill every space in the container, forming its own surface, and will not compress significantly, except under extremely high pressures. These properties make a liquid suitable for applications such as hydraulics.

Liquid particles are bound firmly but not rigidly. They are able to move around one another freely, resulting in a limited degree of particle mobility. As the temperature increases, the increased vibrations of the molecules causes distances between the molecules to increase. When a liquid reaches its boiling point, the cohesive forces that bind the molecules closely together break, and the liquid changes to its gaseous state (unless superheating occurs). If the temperature is decreased, the distances between the molecules become smaller. When the liquid reaches its freezing point the molecules will usually lock into a very specific order, called crystallizing, and the bonds between them become more rigid, changing the liquid into its solid state (unless supercooling occurs).

4.2 Examples

Only two elements are liquid at standard conditions for temperature and pressure: mercury and bromine. Four more elements have melting points slightly above room temperature: francium, caesium, gallium and rubidium.[1] Metal alloys that are liquid at room temperature include NaK, a sodium-potassium metal alloy, galinstan, a fusible alloy liquid, and some amalgams (alloys involving mercury).

Pure substances that are liquid under normal conditions include water, ethanol and many other organic solvents. Liquid water is of vital importance in chemistry and biology; it is believed to be a necessity for the existence of life.

Inorganic liquids include water, magma, inorganic nonaqueous solvents and many acids.

Important everyday liquids include aqueous solutions like household bleach, other mixtures of different substances such as mineral oil and gasoline, emulsions like vinaigrette or mayonnaise, suspensions like blood, and colloids like paint and milk.

Many gases can be liquefied by cooling, producing liquids such as liquid oxygen, liquid nitrogen, liquid hydrogen and liquid helium. Not all gases can be liquified at atmospheric pressure, for example carbon dioxide can only be liquified at pressures above 5.1 atm.

Some materials cannot be classified within the classical three states of matter; they possess solid-like and liquid-like properties. Examples include liquid crystals, used in LCD displays, and biological membranes.

4.3 Applications

Liquids have a variety of uses, as lubricants, solvents, and coolants. In hydraulic systems, liquid is used to transmit power.

In tribology, liquids are studied for their properties as lubricants. Lubricants such as oil are chosen for viscosity and flow characteristics that are suitable throughout the operating temperature range of the component. Oils are often used in engines, gear boxes, metalworking, and hydraulic systems for their good lubrication properties.[2]

Many liquids are used as solvents, to dissolve other liquids or solids. Solutions are found in a wide variety of applications, including paints, sealants, and adhesives. Naphtha and acetone are used frequently in industry to clean oil, grease, and tar from parts and machinery. Body fluids are water based solutions.

Surfactants are commonly found in soaps and detergents. Solvents like alcohol are often used as antimicrobials. They are found in cosmetics, inks, and liquid dye lasers. They are used in the food industry, in processes such as the extraction of vegetable oil.[3]

Liquids tend to have better thermal conductivity than gases, and the ability to flow makes a liquid suitable for removing excess heat from mechanical components. The heat can be removed by channeling the liquid through a heat exchanger, such as a radiator, or the heat can be removed with the liquid during evaporation.[4] Water or glycol coolants are used to keep engines from overheating.[5] The coolants used in nuclear reactors include water or liquid metals, such as sodium or bismuth.[6] Liquid propellant films are used to cool the thrust chambers of rockets.[7] In machining, water and oils are used to remove the excess heat generated, which can quickly ruin both the work piece and the tooling. During perspiration, sweat removes heat from the human body by evaporating. In the heating, ventilation, and air-conditioning industry (HVAC), liquids such as water are used to transfer heat from one area to another.[8]

Liquid is the primary component of hydraulic systems, which take advantage of Pascal's law to provide fluid power. Devices such as pumps and waterwheels have been used to change liquid motion into mechanical work since ancient times. Oils are forced through hydraulic pumps, which transmit this force to hydraulic cylinders. Hydraulics can be found in many applications, such as automotive brakes and transmissions, heavy equipment, and airplane control systems. Various hydraulic presses are used extensively in repair and manufacturing, for lifting, pressing, clamping and forming.[9]

Liquids are sometimes used in measuring devices. A thermometer often uses the thermal expansion of liquids, such as mercury, combined with their ability to flow to indicate temperature. A manometer uses the weight of the liquid to indicate air pressure.[10]

4.4   Mechanical properties

4.4.1   Volume

Quantities of liquids are commonly measured in units of volume. These include the SI unit cubic metre (m^3) and its divisions, in particular the cubic decimeter, more commonly called the litre (1 dm^3 = 1 L = 0.001 m^3), and the cubic centimetre, also called millilitre (1 cm^3 = 1 mL = 0.001 L = 10^{-6} m^3).

The volume of a quantity of liquid is fixed by its temperature and pressure. Liquids generally expand when heated, and contract when cooled. Water between 0 °C and 4 °C is a notable exception. Liquids have little compressibility. Water, for example, will compress by only 46.4 parts per million for every unit increase in atmospheric pressure (bar).[11] At around 4000 bar (58,000 psi) of pressure, at room temperature, water only experiences an 11% decrease in volume.[12] In the study of fluid dynamics, liquids are often treated as incompressible, especially when studying incompressible flow. This incompressible nature makes a liquid suitable for transmitting hydraulic power, because very little of the energy is lost in the form of compression.[12] However, the very slight compressibility does lead to other phenomena. The banging of pipes, called water hammer, occurs when a valve is suddenly closed, creating a huge pressure-spike at the valve that travels backward through the system. Another phenomenon caused by liquid's incompressibility is cavitation, where liquid in an area of low pressure vaporizes and forms bubbles, which then collapse as they enter high pressure areas. This causes liquid to fill the cavity left by the bubble with tremendous, localized force, eroding any adjacent solid surface.[13]

4.4.2   Pressure and buoyancy

Main article: fluid statics

In a gravitational field, liquids exert pressure on the sides of a container as well as on anything within the liquid itself. This pressure is transmitted in all directions and increases with depth. If a liquid is at rest in a uniform gravitational field, the pressure, p, at any depth, z, is given by

$$p = \rho g z$$

where:

> ρ is the density of the liquid (assumed constant)
>
> g is the gravitational acceleration.

Note that this formula assumes that the pressure *at* the free surface is zero, and that surface tension effects may be neglected.

Objects immersed in liquids are subject to the phenomenon of buoyancy. (Buoyancy is also observed in other fluids, but is especially strong in liquids due to their high density.)

4.4.3   Surfaces

Main article: surface science

Unless the volume of a liquid exactly matches the volume of its container, one or more surfaces are observed. The

Surface waves in water

surface of a liquid behaves like an elastic membrane in which surface tension appears, allowing the formation of drops and bubbles. Surface waves, capillary action, wetting, and ripples are other consequences of surface tension.

Free surface

Main article: Free surface

A **free surface** is the surface of a fluid that is subject to both zero perpendicular normal stress and parallel shear stress, such as the boundary between, e.g., liquid water and the air in the Earth's atmosphere.

Level

The **liquid level** (as in, e.g., water level) is the height associated with the liquid free surface, especially when it's the top-most surface. It may be measured with a level sensor.

4.4.4 Flow

Main articles: fluid mechanics and fluid dynamics

Viscosity measures the resistance of a liquid which is being deformed by either shear stress or extensional stress.

When a liquid is supercooled towards the glass transition, the viscosity increases dramatically. The liquid then becomes a viscoelastic medium that shows both the elasticity of a solid and the fluidity of a liquid, depending on the time scale of observation or on the frequency of perturbation.

4.4.5   Sound propagation

Main article: speed of sound § Speed of sound in liquids

Hence the speed of sound in a fluid is given by $c = \sqrt{K/\rho}$ where K is the bulk modulus of the fluid, and ϱ the density. To give a typical value, in fresh water c=1497 m/s at 25 °C.

4.5   Thermodynamics

4.5.1   Phase transitions

Main articles: boiling, boiling point, melting and melting point
At a temperature below the boiling point, any matter in liquid form will evaporate until the condensation of gas above reach

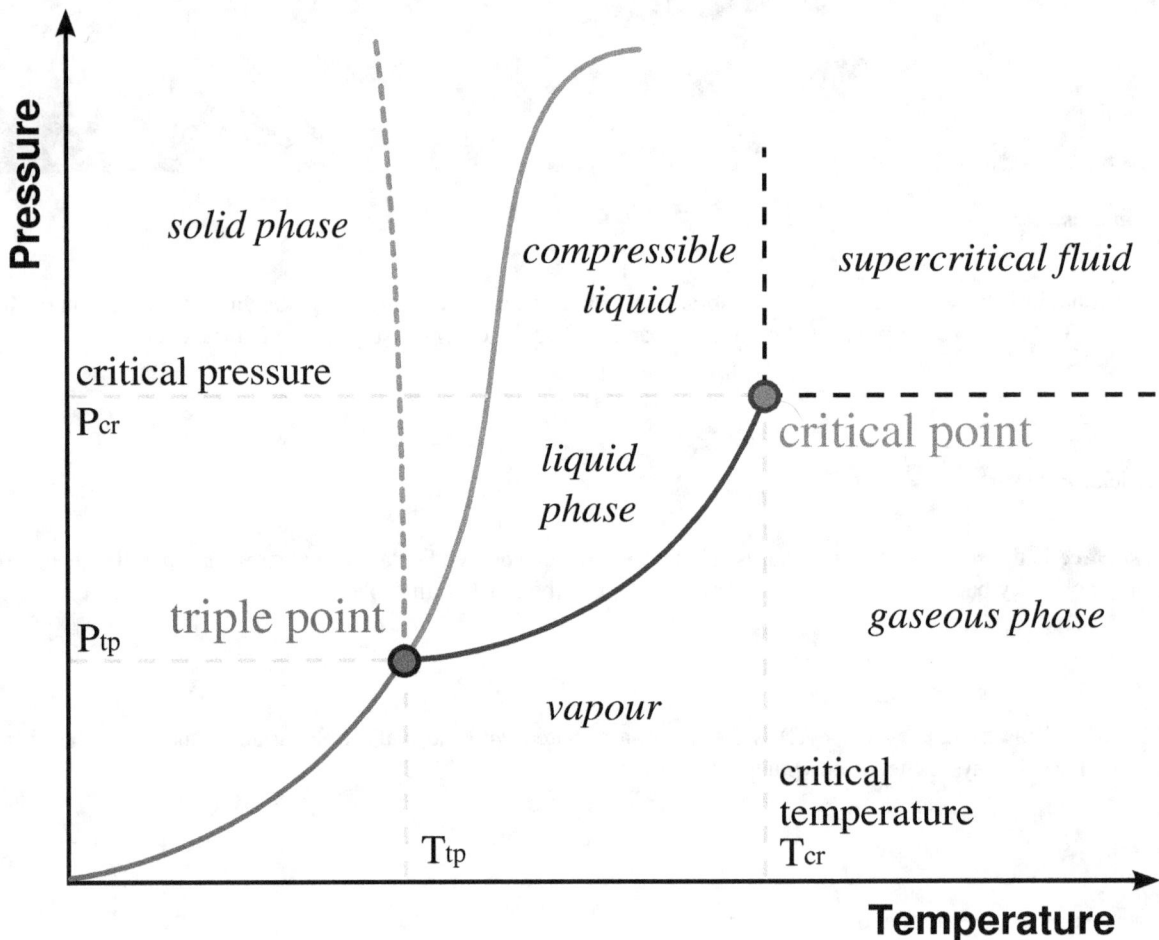

A typical phase diagram. The dotted line gives the anomalous behaviour of water. The green lines show how the freezing point can vary with pressure, and the blue line shows how the boiling point can vary with pressure. The red line shows the boundary where sublimation or deposition can occur.

an equilibrium. At this point the gas will condense at the same rate as the liquid evaporates. Thus, a liquid cannot exist permanently if the evaporated liquid is continually removed. A liquid at its boiling point will evaporate more quickly than the gas can condense at the current pressure. A liquid at or above its boiling point will normally boil, though superheating can prevent this in certain circumstances.

At a temperature below the freezing point, a liquid will tend to crystallize, changing to its solid form. Unlike the transition to gas, there is no equilibrium at this transition under constant pressure, so unless supercooling occurs, the liquid will eventually completely crystallize. Note that this is only true under constant pressure, so e.g. water and ice in a closed, strong container might reach an equilibrium where both phases coexist. For the opposite transition from solid to liquid, see melting.

4.5.2   Liquids in space

The phase diagram explains why liquids do not exist in space or any other vacuum. Since the pressure is zero (except on surfaces or interiors of planets and moons) water and other liquids exposed to space will either immediately boil or freeze depending on the temperature. In regions of space near the earth, water will freeze if the sun is not shining directly on it and vapourize (sublime) as soon as it is in sunlight. If water exists as ice on the moon, it can only exist in shadowed holes where the sun never shines and where the surrounding rock doesn't heat it up too much. At some point near the orbit of Saturn, the light from the sun is too faint to sublime ice to water vapour. This is evident from the longevity of the ice that composes Saturn's rings.

4.5.3   Solutions

Main article: solution

Liquids can display immiscibility. The most familiar mixture of two immiscible liquids in everyday life is the vegetable oil and water in Italian salad dressing. A familiar set of miscible liquids is water and alcohol. Liquid components in a mixture can often be separated from one another via fractional distillation.

4.6   Microscopic properties

4.6.1   Static structure factor

Main article: structure of liquids and glasses
In a liquid, atoms do not form a crystalline lattice, nor do they show any other form of long-range order. This is evidenced by the absence of Bragg peaks in X-ray and neutron diffraction. Under normal conditions, the diffraction pattern has circular symmetry, expressing the isotropy of the liquid. In radial direction, the diffraction intensity smoothly oscillates. This is usually described by the static structure factor $S(q)$, with wavenumber $q=(4\pi/\lambda)\sin\theta$ given by the wavelength λ of the probe (photon or neutron) and the Bragg angle θ. The oscillations of $S(q)$ express the *near order* of the liquid, i.e. the correlations between an atom and a few shells of nearest, second nearest, ... neighbors.

A more intuitive description of these correlations is given by the radial distribution function $g(r)$, which is basically the Fourier transform of $S(q)$. It represents a spatial average of a temporal snapshot of pair correlations in the liquid.

4.6.2   Sound dispersion and structural relaxation

The above expression for the sound velocity $c = \sqrt{K/\rho}$ contains the bulk modulus K. If K is frequency independent then the liquid behaves as a linear medium, so that sound propagates without dissipation and without mode coupling. In reality, any liquid shows some dispersion: with increasing frequency, K crosses over from the low-frequency, liquid-like limit K_0 to the high-frequency, solid-like limit K_∞. In normal liquids, most of this cross over takes place at frequencies between GHz and THz, sometimes called hypersound.

Structure of a classical monatomic liquid. Atoms have many nearest neighbors in contact, yet no long-range order is present.

At sub-GHz frequencies, a normal liquid cannot sustain shear waves: the zero-frequency limit of the shear modulus is $G_0 = 0$. This is sometimes seen as the defining property of a liquid.[14][15] However, just as the bulk modulus K, the shear modulus G is frequency dependent, and at hypersound frequencies it shows a similar cross over from the liquid-like limit G_0 to a solid-like, non-zero limit G_∞ .

According to the Kramers-Kronig relation, the dispersion in the sound velocity (given by the real part of K or G) goes along with a maximum in the sound attenuation (dissipation, given by the imaginary part of K or G). According to linear response theory, the Fourier transform of K or G describes how the system returns to equilibrium after an external perturbation; for this reason, the dispersion step in the GHz..THz region is also called structural relaxation. According the fluctuation-dissipation theorem, relaxation *towards* equilibrium is intimately connected to fluctuations *in* equilibrium. The density fluctuations associated with sound waves can be experimentally observed by Brillouin scattering.

On supercooling a liquid towards the glass transition, the crossover from liquid-like to solid-like response moves from GHz to MHz, kHz, Hz, ...; equivalently, the characteristic time of structural relaxation increases from ns to μs, ms, s, ... This is the microscopic explanation for the above-mentioned viscoelastic behaviour of glass-forming liquids.

4.6.3   Effects of association

The mechanisms of atomic/molecular diffusion (or particle displacement) in solids are closely related to the mechanisms of viscous flow and solidification in liquid materials. Descriptions of viscosity in terms of molecular "free space" within the liquid[16] were modified as needed in order to account for liquids whose molecules are known to be "associated" in the liquid state at ordinary temperatures. When various molecules combine together to form an associated molecule, they enclose within a semi-rigid system a certain amount of space which before was available as free space for mobile molecules. Thus, increase in viscosity upon cooling due to the tendency of most substances to become *associated* on cooling.[17]

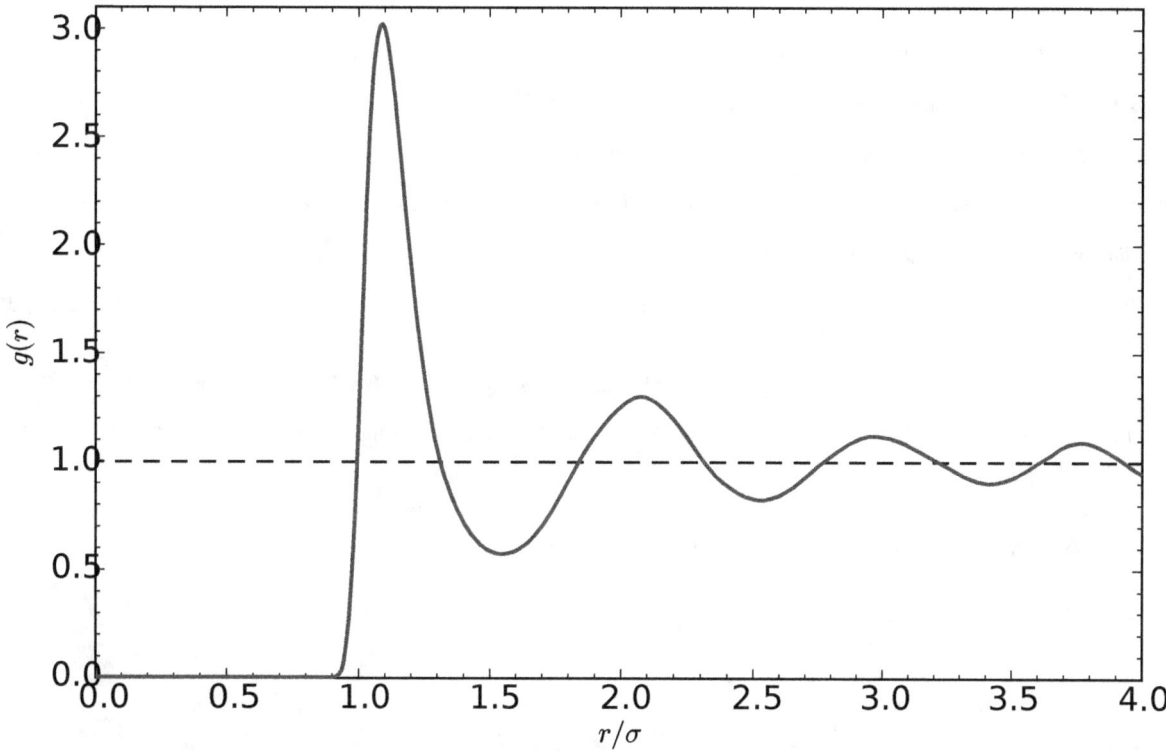

Radial distribution function of the Lennard-Jones model fluid.

Similar arguments could be used to describe the effects of pressure on viscosity, where it may be assumed that the viscosity is chiefly a function of the volume for liquids with a finite compressibility. An increasing viscosity with rise of pressure is therefore expected. In addition, if the volume is expanded by heat but reduced again by pressure, the viscosity remains the same.

The local tendency to orientation of molecules in small groups lends the liquid (as referred to previously) a certain degree of association. This association results in a considerable "internal pressure" within a liquid, which is due almost entirely to those molecules which, on account of their temporary low velocities (following the Maxwell distribution) have coalesced with other molecules. The internal pressure between several such molecules might correspond to that between a group of molecules in the solid form.

4.7 References

[1] Theodore Gray, The Elements: A Visual Exploration of Every Known Atom in the Universe New York: Workman Publishing, 2009 p. 127 ISBN 1-57912-814-9

[2] Theo Mang, Wilfried Dressel "Lubricants and lubrication", Wiley-VCH 2007 ISBN 3-527-31497-0

[3] George Wypych "Handbook of solvents" William Andrew Publishing 2001 pp. 847–881 ISBN 1-895198-24-0

[4] N. B. Vargaftik "Handbook of thermal conductivity of liquids and gases" CRC Press 1994 ISBN 0-8493-9345-0

[5] Jack Erjavec "Automotive technology: a systems approach" Delmar Learning 2000 p. 309 ISBN 1-4018-4831-1

[6] Gerald Wendt "The prospects of nuclear power and technology" D. Van Nostrand Company 1957 p. 266

[7] "Modern engineering for design of liquid-propellant rocket engines" by Dieter K. Huzel, David H. Huang – American Institute of Aeronautics and Astronautics 1992 p. 99 ISBN 1-56347-013-6

[8]   Thomas E Mull "HVAC principles and applications manual" McGraw-Hill 1997 ISBN 0-07-044451-X

[9]   R. Keith Mobley *Fluid power dynamics* Butterworth-Heinemann 2000 p. vii ISBN 0-7506-7174-2

[10] Bela G. Liptak "Instrument engineers' handbook: process control" CRC Press 1999 p. 807 ISBN 0-8493-1081-4

[11] http://hyperphysics.phy-astr.gsu.edu/hbase/tables/compress.html

[12] *Intelligent Energy Field Manufacturing: Interdisciplinary Process Innovations* By Wenwu Zhang -- CRC Press 2011 Page 144

[13] *Fluid Mechanics and Hydraulic Machines* by S. C. Gupta -- Dorling-Kindersley 2006 Page 85

[14] Born, Max (1940). "On the stability of crystal lattices". *Mathematical Proceedings* (Cambridge Philosophical Society) **36** (2): 160–172. doi:10.1017/S0305004100017138.

[15] Born, Max (1939)."Thermodynamics of Crystals and Melting".*Journal of Chemical Physics***7**(8): 591–604.doi:10.1063/1.

[16] D.B. Macleod (1923). "On a relation between the viscosity of a liquid and its coefficient of expansion". *Trans. Farad. Soc.* **19**: 6. doi:10.1039/tf9231900006.

[17] G.W Stewart (1930). "The Cybotactic (Molecular Group) Condition in Liquids; the Association of Molecules". *Phys. Rev.* **35** (7): 726. Bibcode:1930PhRv...35..726S. doi:10.1103/PhysRev.35.726.

Chapter 5

Gas

This article is about the physical properties of gas as a state of matter. For the automotive fuel, see gasoline. For the uses of gases, and other meanings, see Gas (disambiguation).

Gas is one of the four fundamental states of matter (the others being solid, liquid, and plasma). A pure gas may be

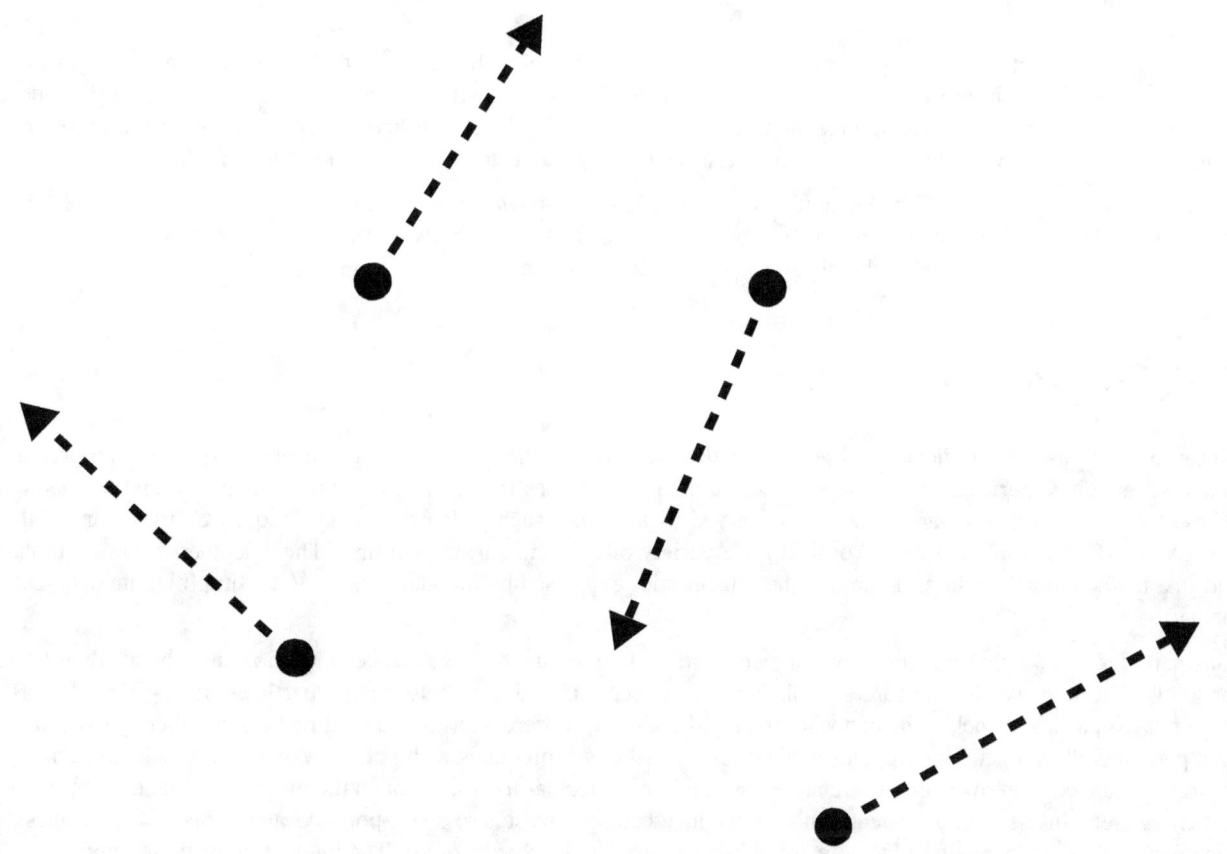

Gas phase particles (atoms, molecules, or ions) move around freely in the absence of an applied electric field.

made up of individual atoms (e.g. a noble gas like neon), elemental molecules made from one type of atom (e.g. oxygen), or compound molecules made from a variety of atoms (e.g. carbon dioxide). A gas mixture would contain a variety of pure gases much like the air. What distinguishes a gas from liquids and solids is the vast separation of the individual gas particles. This separation usually makes a colorless gas invisible to the human observer. The interaction of gas particles in the presence of electric and gravitational fields are considered negligible as indicated by the constant velocity vectors in the image. One type of commonly known gas is steam.

The gaseous state of matter is found between the liquid and plasma states,[1] the latter of which provides the upper temperature boundary for gases. Bounding the lower end of the temperature scale lie degenerative quantum gases[2] which are gaining increasing attention.[3] High-density atomic gases super cooled to incredibly low temperatures are classified by their statistical behavior as either a Bose gas or a Fermi gas. For a comprehensive listing of these exotic states of matter see list of states of matter.

5.1 Elemental gases

The only chemical elements which are stable multi atom homonuclear molecules at standard temperature and pressure (STP), are hydrogen (H_2), nitrogen (N_2) and oxygen (O_2); plus two halogens, fluorine (F_2) and chlorine (Cl_2). These gases, when grouped together with the monatomic noble gases; which are helium (He), neon (Ne), argon (Ar), krypton (Kr), xenon (Xe) and radon (Rn) ; are called "elemental gases". Alternatively they are sometimes known as "molecular gases" to distinguish them from molecules that are also chemical compounds.

5.2 Etymology

The word *gas* is a neologism first used by the early 17th-century Flemish chemist J.B. van Helmont.[4] Van Helmont's word appears to have been simply a phonetic transcription of the Greek word χάος *Chaos* – the *g* in Dutch being pronounced like *ch* in "loch" – in which case Van Helmont was simply following the established alchemical usage first attested in the works of Paracelsus. According to Paracelsus's terminology, *chaos* meant something like "ultra-rarefied water".[5]

An alternative story[6] is that Van Helmont's word is corrupted from *gahst* (or *geist*), signifying a ghost or spirit. This was because certain gases suggested a supernatural origin, such as from their ability to cause death, extinguish flames, and to occur in "mines, bottom of wells, churchyards and other lonely places".

5.3 Physical characteristics

Because most gases are difficult to observe directly, they are described through the use of four physical properties or macroscopic characteristics: pressure, volume, number of particles (chemists group them by moles) and temperature. These four characteristics were repeatedly observed by scientists such as Robert Boyle, Jacques Charles, John Dalton, Joseph Gay-Lussac and Amedeo Avogadro for a variety of gases in various settings. Their detailed studies ultimately led to a mathematical relationship among these properties expressed by the ideal gas law (see simplified models section below).

Gas particles are widely separated from one another, and consequently have weaker intermolecular bonds than liquids or solids. These intermolecular forces result from electrostatic interactions between gas particles. Like-charged areas of different gas particles repel, while oppositely charged regions of different gas particles attract one another; gases that contain permanently charged ions are known as plasmas. Gaseous compounds with polar covalent bonds contain permanent charge imbalances and so experience relatively strong intermolecular forces, although the molecule while the compound's net charge remains neutral. Transient, randomly induced charges exist across non-polar covalent bonds of molecules and electrostatic interactions caused by them are referred to as Van der Waals forces. The interaction of these intermolecular forces varies within a substance which determines many of the physical properties unique to each gas.[7][8] A comparison of *boiling points* for compounds formed by ionic and covalent bonds leads us to this conclusion.[9] The drifting smoke particles in the image provides some insight into low pressure gas behavior.

Compared to the other states of matter, gases have low density and viscosity. Pressure and temperature influence the particles within a certain volume. This variation in particle separation and speed is referred to as *compressibility*. This particle separation and size influences optical properties of gases as can be found in the following list of refractive indices. Finally, gas particles spread apart or diffuse in order to homogeneously distribute themselves throughout any container.

Drifting smoke particles provide clues to the movement of the surrounding gas.

5.4 Macroscopic

When observing a gas, it is typical to specify a frame of reference or length scale. A *larger* length scale corresponds to a macroscopic or global point of view of the gas. This region (referred to as a volume) must be sufficient in size to contain a large sampling of gas particles. The resulting statistical analysis of this sample size produces the **"average"** behavior (i.e. velocity, temperature or pressure) of all the gas particles within the region. In contrast, a *smaller* length scale corresponds to a microscopic or particle point of view.

Macroscopically, the gas characteristics measured are either in terms of the gas particles themselves (velocity, pressure, or temperature) or their surroundings (volume). For example, Robert Boyle studied pneumatic chemistry for a small portion of his career. One of his experiments related the macroscopic properties of pressure and volume of a gas. His experiment used a J-tube manometer which looks like a test tube in the shape of the letter J. Boyle trapped an inert gas in the closed end of the test tube with a column of mercury, thereby making the number of particles and the temperature constant. He observed that when the pressure was increased in the gas, by adding more mercury to the column, the trapped gas' volume decreased (this is known as an inverse relationship). Furthermore, when Boyle multiplied the pressure and volume of each observation, the product was constant. This relationship held for every gas that Boyle observed leading to the law, (PV=k), named to honor his work in this field.

There are many mathematical tools available for analyzing gas properties. As gases are subjected to extreme conditions, these tools become a bit more complex, from the Euler equations for inviscid flow to the Navier–Stokes equations[10] that fully account for viscous effects. These equations are adapted to the conditions of the gas system in question. Boyle's lab equipment allowed the use of algebra to obtain his analytical results. His results were possible because he was studying gases in relatively low pressure situations where they behaved in an "ideal" manner. These ideal relationships apply to

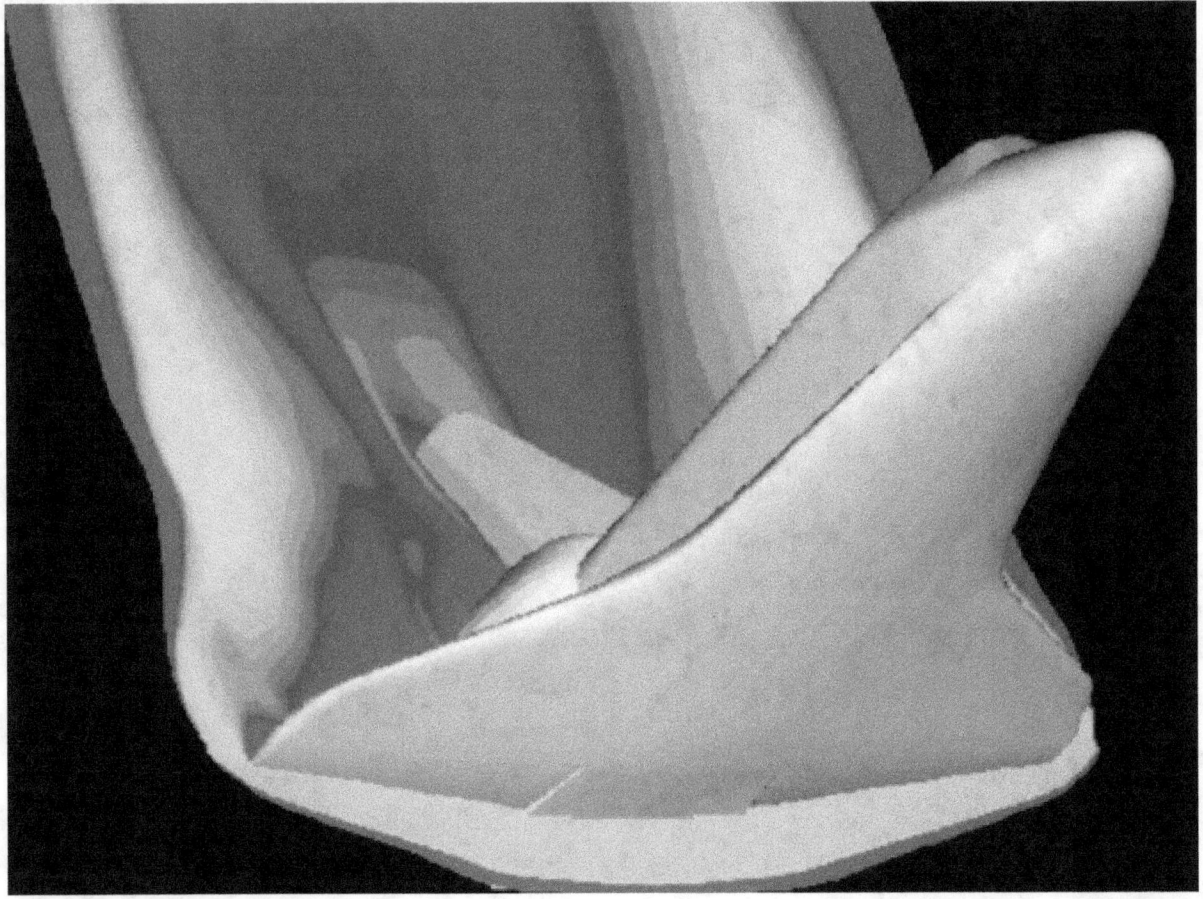

Shuttle imagery of re-entry phase.

safety calculations for a variety of flight conditions on the materials in use. The high technology equipment in use today was designed to help us safely explore the more exotic operating environments where the gases no longer behave in an "ideal" manner. This advanced math, including statistics and multivariable calculus, makes possible the solution to such complex dynamic situations as space vehicle reentry. An example is the analysis of the space shuttle reentry pictured to ensure the material properties under this loading condition are appropriate. In this flight regime, the gas is no longer behaving ideally.

5.4.1   Pressure

Main article: Pressure

The symbol used to represent *pressure* in equations is **"p"** or **"P"** with SI units of pascals.

When describing a container of gas, the term pressure (or absolute pressure) refers to the average force per unit area that the gas exerts on the surface of the container. Within this volume, it is sometimes easier to visualize the gas particles moving in straight lines until they collide with the container (see diagram at top of the article). The force imparted by a gas particle into the container during this collision is the change in momentum of the particle.[11] During a collision only the normal component of velocity changes. A particle traveling parallel to the wall does not change its momentum. Therefore, the average force on a surface must be the average change in linear momentum from all of these gas particle collisions.

Pressure is the sum of all the normal components of force exerted by the particles impacting the walls of the container divided by the surface area of the wall.

5.4.2   Temperature

Air balloon shrinks after submersion in liquid nitrogen

Main article: Thermodynamic temperature

The symbol used to represent *temperature* in equations is T with SI units of kelvins.

The speed of a gas particle is proportional to its absolute temperature. The volume of the balloon in the video shrinks when the trapped gas particles slow down with the addition of extremely cold nitrogen. The temperature of any physical system is related to the motions of the particles (molecules and atoms) which make up the [gas] system.[12] In statistical mechanics, temperature is the measure of the average kinetic energy stored in a particle. The methods of storing this energy are dictated by the degrees of freedom of the particle itself (energy modes). Kinetic energy added (endothermic process) to gas particles by way of collisions produces linear, rotational, and vibrational motion. In contrast, a molecule in a solid can only increase its vibrational modes with the addition of heat as the lattice crystal structure prevents both linear and rotational motions. These heated gas molecules have a greater speed range which constantly varies due to constant collisions with other particles. The speed range can be described by the Maxwell–Boltzmann distribution. Use of this distribution implies ideal gases near thermodynamic equilibrium for the system of particles being considered.

5.4.3   Specific volume

Main article: Specific volume

The symbol used to represent *specific volume* in equations is "**v**" with SI units of cubic meters per kilogram.

See also: Gas volume

The symbol used to represent **volume** in equations is "**V**" with SI units of cubic meters.

When performing a thermodynamic analysis, it is typical to speak of intensive and extensive properties. Properties which depend on the amount of gas (either by mass or volume) are called *extensive* properties, while properties that do not depend on the amount of gas are called *intensive* properties. **Specific volume** is an example of an *intensive* property because it is the ratio of volume occupied by a *unit of mass* of a gas that is identical throughout a system at equilibrium.[13] 1000 atoms a gas occupy the same space as any other 1000 atoms for any given temperature and pressure. This concept is easier to visualize for solids such as iron which are incompressible compared to gases. Since a gas fills any container in which it is placed, **volume** is an *extensive property*.

5.4.4   Density

Main article: Density

The symbol used to represent **density** in equations is ρ (rho) with SI units of kilograms per cubic meter. This term is the reciprocal of specific volume.

Since gas molecules can move freely within a container, their mass is normally characterized by **density**. Density is the amount of mass per unit volume of a substance, or the inverse of specific volume. For gases, the density can vary over a wide range because the particles are free to move closer together when constrained by pressure or volume. This variation of density is referred to as compressibility. Like pressure and temperature, density is a state variable of a gas and the change in density during any process is governed by the laws of thermodynamics. For a static gas, the density is the same throughout the entire container. Density is therefore a scalar quantity. It can be shown by **kinetic theory** that the density is *inversely* proportional to the size of the container in which a fixed mass of gas is confined. In this case of a fixed mass, the density decreases as the volume increases.

5.5   Microscopic

If one could observe a gas under a powerful microscope, one would see a collection of particles (molecules, atoms, ions, electrons, etc.) without any definite shape or volume that are in more or less random motion. These neutral gas particles only change direction when they collide with another particle or with the sides of the container. In an ideal gas, these collisions are perfectly elastic. This particle or microscopic view of a gas is described by the Kinetic-molecular theory. The assumptions behind this theory can be found in the postulates section of Kinetic Theory.

5.5.1   Kinetic theory

Main article: Kinetic theory

Kinetic theory provides insight into the macroscopic properties of gases by considering their molecular composition and motion. Starting with the definitions of momentum and kinetic energy,[14] one can use the conservation of momentum and geometric relationships of a cube to relate macroscopic system properties of temperature and pressure to the microscopic property of kinetic energy per molecule. The theory provides averaged values for these two properties.

The theory also explains how the gas system responds to change. For example, as a gas is heated from absolute zero, when it is (in theory) perfectly still, its internal energy (temperature) is increased. As a gas is heated, the particles speed up and its temperature rises. This results in greater numbers of collisions with the container per unit time due to the higher particle speeds associated with elevated temperatures. The pressure increases in proportion to the number of collisions per unit time.

5.5.2 Brownian motion

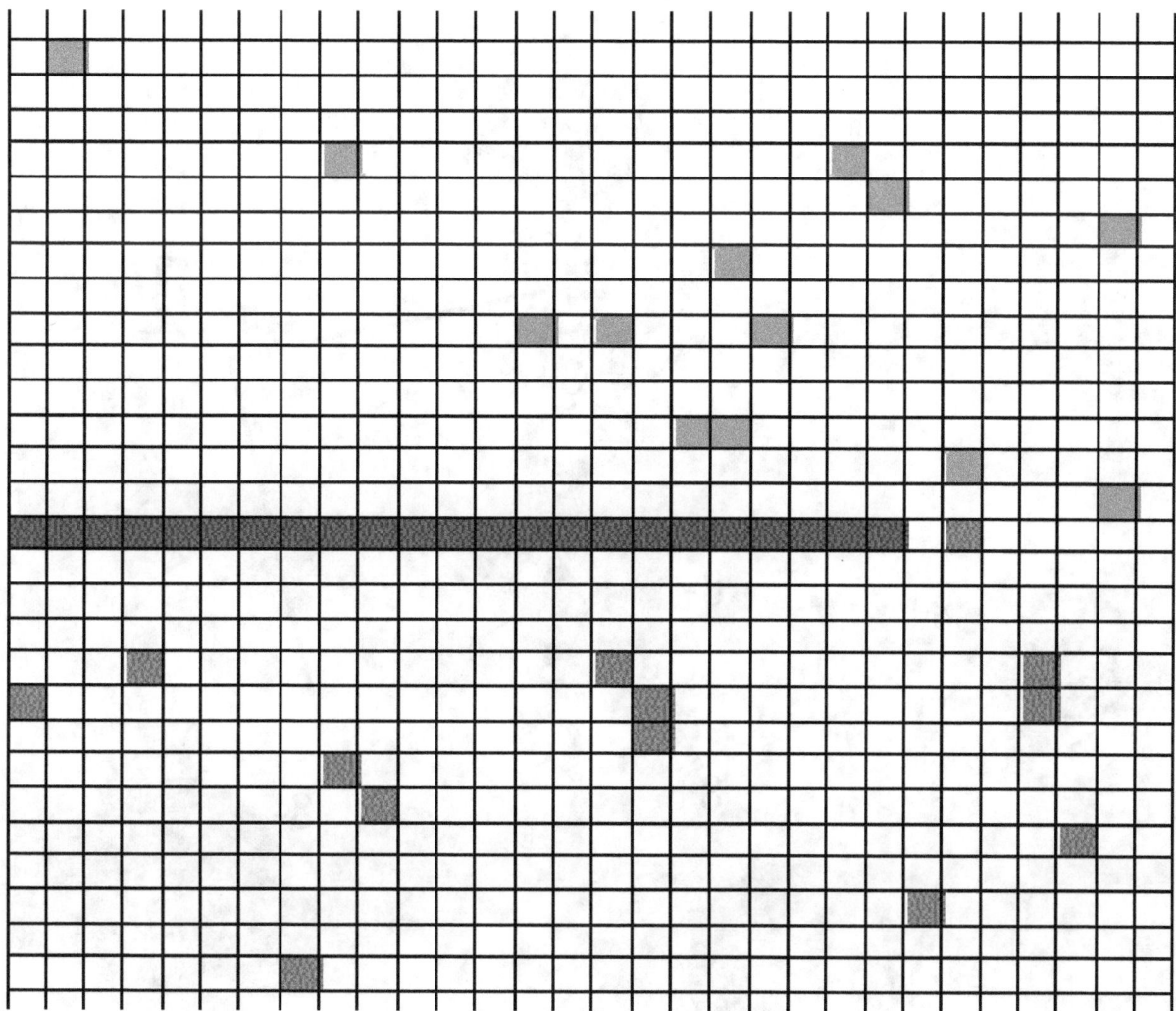

Random motion of gas particles results in diffusion.

Main article: Brownian motion

Brownian motion is the mathematical model used to describe the random movement of particles suspended in a fluid. The gas particle animation, using pink and green particles, illustrates how this behavior results in the spreading out of gases (entropy). These events are also described by particle theory.

Since it is at the limit of (or beyond) current technology to observe individual gas particles (atoms or molecules), only theoretical calculations give suggestions about how they move, but their motion is different from Brownian motion because Brownian motion involves a smooth drag due to the frictional force of many gas molecules, punctuated by violent collisions of an individual (or several) gas molecule(s) with the particle. The particle (generally consisting of millions or billions of atoms) thus moves in a jagged course, yet not so jagged as would be expected if an individual gas molecule were examined.

5.5.3   Intermolecular forces

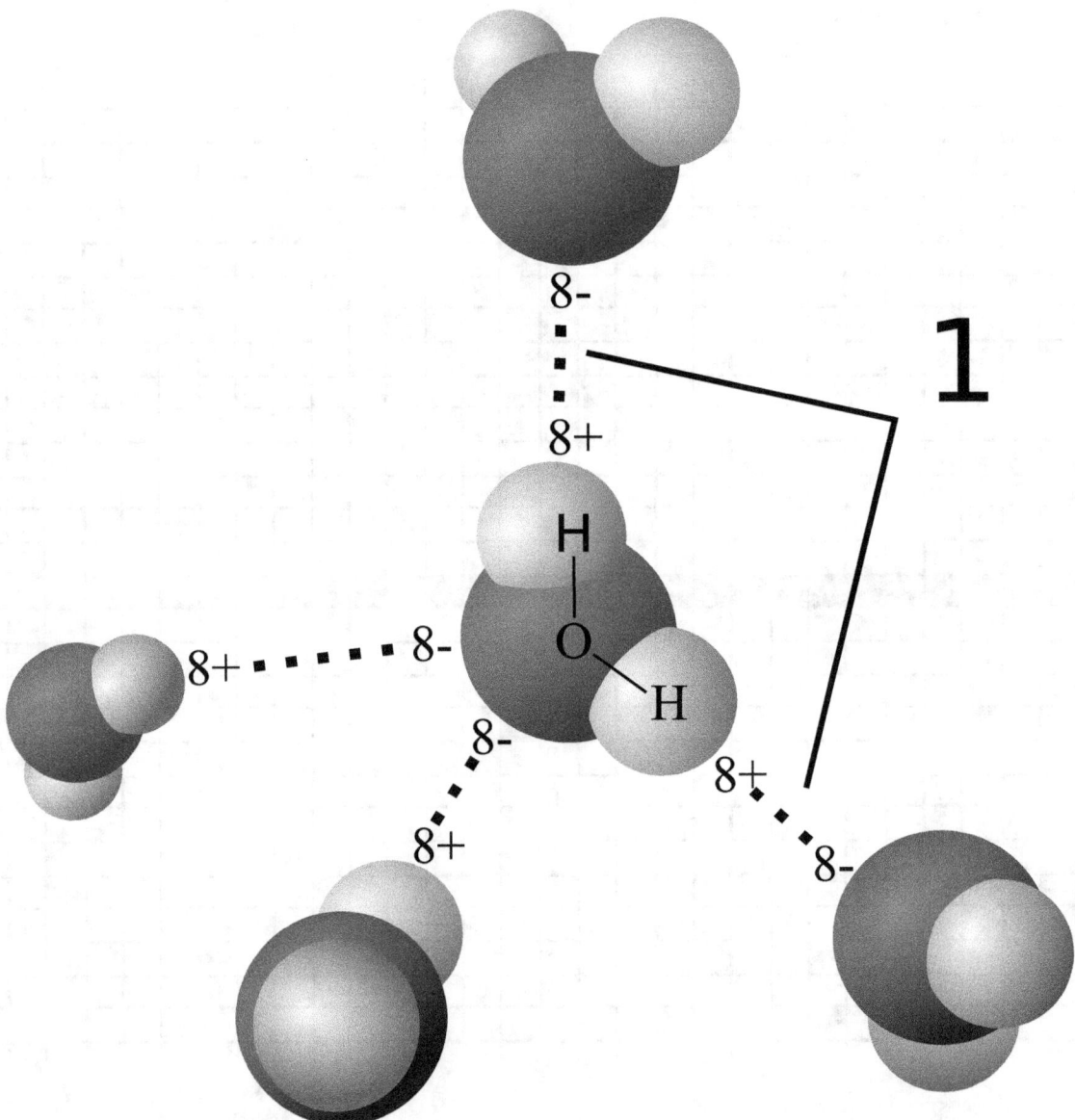

When gases are compressed, intermolecular forces like those shown here start to play a more active role.

Main articles: van der Waals force and Intermolecular force

As discussed earlier, momentary attractions (or repulsions) between particles have an effect on gas dynamics. In physical chemistry, the name given to these intermolecular forces is *van der Waals force*. These forces play a key role in determining physical properties of a gas such as viscosity and flow rate (see physical characteristics section). Ignoring these forces in certain conditions (see Kinetic-molecular theory) allows a real gas to be treated like an ideal gas. This assumption allows the use of ideal gas laws which greatly simplifies calculations.

Proper use of these gas relationships requires the Kinetic-molecular theory (KMT). When gas particles possess a magnetic charge or Intermolecular force they gradually influence one another as the spacing between them is reduced (the hydrogen bond model illustrates one example). In the absence of any charge, at some point when the spacing between gas particles is greatly reduced they can no longer avoid collisions between themselves at normal gas temperatures. Another case for increased collisions among gas particles would include a fixed volume of gas, which upon heating would contain very fast particles. *This means that these ideal equations provide reasonable results* except *for extremely high pressure (compressible) or high temperature (ionized) conditions.* Notice that all of these excepted conditions allow energy transfer to take place within the gas system. The absence of these internal transfers is what is referred to as ideal conditions in which the energy exchange occurs only at the boundaries of the system. Real gases experience some of these collisions and intermolecular forces. When these collisions are statistically negligible (incompressible), results from these ideal equations are still meaningful. If the gas particles are compressed into close proximity they behave more like a liquid (see fluid dynamics).

5.6 Simplified models

Main article: Equation of state

An *equation of state* (for gases) is a mathematical model used to roughly describe or predict the state properties of a gas. At present, there is no single equation of state that accurately predicts the properties of all gases under all conditions. Therefore, a number of much more accurate equations of state have been developed for gases in specific temperature and pressure ranges. The "gas models" that are most widely discussed are "perfect gas", "ideal gas" and "real gas". Each of these models has its own set of assumptions to facilitate the analysis of a given thermodynamic system.[15] Each successive model expands the temperature range of coverage to which it applies.

5.6.1 Ideal and perfect gas models

Main article: Perfect gas

The equation of state for an ideal or perfect gas is the ideal gas law and reads

$$PV = nRT,$$

where P is the pressure, V is the volume, n is amount of gas (in mol units), R is the universal gas constant, 8.314 J/(mol K), and T is the temperature. Written this way, it is sometimes called the "chemist's version", since it emphasizes the number of molecules n. It can also be written as

$$P = \rho R_s T,$$

where R_s is the specific gas constant for a particular gas, in units J/(kg K), and ρ = m/V is density. This notation is the "gas dynamicist's" version, which is more practical in modeling of gas flows involving acceleration without chemical reactions.

The ideal gas law does not make an assumption about the specific heat of a gas. In the most general case, the specific heat is a function of both temperature and pressure. If the pressure-dependence is neglected (and possibly the temperature-dependence as well) in a particular application, sometimes the gas is said to be a perfect gas, although the exact assumptions may vary depending on the author and/or field of science.

For an ideal gas, the ideal gas law applies without restrictions on the specific heat. An ideal gas is a simplified "real gas" with the assumption that the compressibility factor Z is set to 1 meaning that this pneumatic ratio remains constant. A compressibility factor of one also requires the four state variables to follow the ideal gas law.

This approximation is more suitable for applications in engineering although simpler models can be used to produce a "ball-park" range as to where the real solution should lie. An example where the "ideal gas approximation" would be suitable would be inside a combustion chamber of a jet engine.[16] It may also be useful to keep the elementary reactions and chemical dissociations for calculating emissions.

5.6.2   Real gas

21 April 1990 eruption of Mount Redoubt, Alaska, illustrating real gases not in thermodynamic equilibrium.

Main article: Real gas

Each one of the assumptions listed below adds to the complexity of the problem's solution. As the density of a gas increases with rising pressure, the intermolecular forces play a more substantial role in gas behavior which results in the ideal gas law no longer providing "reasonable" results. At the upper end of the engine temperature ranges (e.g. combustor sections – 1300 K), the complex fuel particles absorb internal energy by means of rotations and vibrations that cause their specific heats to vary from those of diatomic molecules and noble gases. At more than double that temperature, electronic excitation and dissociation of the gas particles begins to occur causing the pressure to adjust to a greater number of particles (transition from gas to plasma).[17] Finally, all of the thermodynamic processes were presumed to describe uniform gases whose velocities varied according to a fixed distribution. Using a non-equilibrium situation implies the flow field must be characterized in some manner to enable a solution. One of the first attempts to expand the boundaries of the ideal gas law was to include coverage for different thermodynamic processes by adjusting the equation to read $pV^n = constant$ and then varying the n through different values such as the specific heat ratio, γ.

Real gas effects include those adjustments made to account for a greater range of gas behavior:

- Compressibility effects (Z allowed to vary from 1.0)

- Variable heat capacity (specific heats vary with temperature)

- Van der Waals forces (related to compressibility, can substitute other equations of state)

- Non-equilibrium thermodynamic effects

- Issues with molecular dissociation and elementary reactions with variable composition.

For most applications, such a detailed analysis is excessive. Examples where "Real Gas effects" would have a significant impact would be on the Space Shuttle re-entry where extremely high temperatures and pressures are present or the gases produced during geological events as in the image of the 1990 eruption of Mount Redoubt.

5.7 Historical synthesis

See also: Gas laws

5.7.1 Boyle's law

Main article: Boyle's law

Boyle's Law was perhaps the first expression of an equation of state. In 1662 Robert Boyle performed a series of experiments employing a J-shaped glass tube, which was sealed on one end. Mercury was added to the tube, trapping a fixed quantity of air in the short, sealed end of the tube. Then the volume of gas was carefully measured as additional mercury was added to the tube. The pressure of the gas could be determined by the difference between the mercury level in the short end of the tube and that in the long, open end. The image of Boyle's Equipment shows some of the exotic tools used by Boyle during his study of gases.

Through these experiments, Boyle noted that the pressure exerted by a gas held at a constant temperature varies inversely with the volume of the gas.[18] For example, if the volume is halved, the pressure is doubled; and if the volume is doubled, the pressure is halved. Given the inverse relationship between pressure and volume, the product of pressure (P) and volume (V) is a constant (k) for a given mass of confined gas as long as the temperature is constant. Stated as a formula, thus is:

$$PV = k$$

Because the before and after volumes and pressures of the fixed amount of gas, where the before and after temperatures are the same both equal the constant k, they can be related by the equation:

$$P_1 V_1 = P_2 V_2.$$

5.7.2 Charles's Law

Main article: Charles's law

In 1787, the French physicist and balloon pioneer, Jacques Charles, found that oxygen, nitrogen, hydrogen, carbon dioxide, and air expand to the same extent over the same 80 kelvin interval. He noted that, for an ideal gas at constant pressure, the volume is directly proportional to its temperature:

$$\frac{V_1}{T_1} = \frac{V_2}{T_2}$$

5.7.3 Gay-Lussac's Law

Main article: Gay-Lussac's Law

In 1802, Joseph Louis Gay-Lussac published results of similar, though more extensive experiments.[19] Gay-Lussac credited Charle's earlier work by naming the law in his honor. Gay-Lussac himself is credited with the law describing pressure, which he found in 1809. It states that the pressure exerted on a container's sides by an ideal gas is proportional to its temperature.

$$\frac{P_1}{T_1} = \frac{P_2}{T_2}$$

5.7.4 Avogadro's law

Main article: Avogadro's law

In 1811, Amedeo Avogadro verified that equal volumes of pure gases contain the same number of particles. His theory was not generally accepted until 1858 when another Italian chemist Stanislao Cannizzaro was able to explain non-ideal exceptions. For his work with gases a century prior, the number that bears his name Avogadro's constant represents the number of atoms found in 12 grams of elemental carbon-12 (6.022×10^{23} mol^{-1}). This specific number of gas particles, at standard temperature and pressure (ideal gas law) occupies 22.40 liters, which is referred to as the molar volume.

Avogadro's law states that the volume occupied by an ideal gas is proportional to the number of moles (or molecules) present in the container. This gives rise to the molar volume of a gas, which at STP is 22.4 dm^3 (or litres). The relation is given by

$$\frac{V_1}{n_1} = \frac{V_2}{n_2}$$

where n is equal to the number of moles of gas (the number of molecules divided by Avogadro's Number).

5.7.5 Dalton's law

Main article: Dalton's law

In 1801, John Dalton published the **Law of Partial Pressures** from his work with ideal gas law relationship: The pressure of a mixture of non reactive gases is equal to the sum of the pressures of all of the constituent gases alone. Mathematically, this can be represented for *n* species as:

Pressuretotal = Pressure$_1$ + Pressure$_2$ + ... + Pressuren

The image of Dalton's journal depicts symbology he used as shorthand to record the path he followed. Among his key journal observations upon mixing unreactive "elastic fluids" (gases) were the following:[20]

- Unlike liquids, heavier gases did not drift to the bottom upon mixing.

- Gas particle identity played no role in determining final pressure (they behaved as if their size was negligible).

5.8 Special topics

5.8.1 Compressibility

Main article: Compressibility factor

Thermodynamicists use this factor (Z) to alter the ideal gas equation to account for compressibility effects of real gases. This factor represents the ratio of actual to ideal specific volumes. It is sometimes referred to as a "fudge-factor" or correction to expand the useful range of the ideal gas law for design purposes. *Usually* this Z value is very close to unity. The compressibility factor image illustrates how Z varies over a range of very cold temperatures.

5.8.2 Reynolds number

Main article: Reynolds number

In fluid mechanics, the Reynolds number is the ratio of inertial forces ($vs\varrho$) to viscous forces (μ/L). It is one of the most important dimensionless numbers in fluid dynamics and is used, usually along with other dimensionless numbers, to provide a criterion for determining dynamic similitude. As such, the Reynolds number provides the link between modeling results (design) and the full-scale actual conditions. It can also be used to characterize the flow.

5.8.3 Viscosity

Main article: Viscosity

Viscosity, a physical property, is a measure of how well adjacent molecules stick to one another. A solid can withstand a shearing force due to the strength of these sticky intermolecular forces. A fluid will continuously deform when subjected to a similar load. While a gas has a lower value of viscosity than a liquid, it is still an observable property. If gases had no viscosity, then they would not stick to the surface of a wing and form a boundary layer. A study of the delta wing in the Schlieren image reveals that the gas particles stick to one another (see Boundary layer section).

5.8.4 Turbulence

Main article: Turbulence

In fluid dynamics, **turbulence** or turbulent flow is a flow regime characterized by chaotic, stochastic property changes. This includes low momentum diffusion, high momentum convection, and rapid variation of pressure and velocity in space and time. The Satellite view of weather around Robinson Crusoe Islands illustrates just one example.

5.8.5 Boundary layer

Main article: Boundary layer

Particles will, in effect, "stick" to the surface of an object moving through it. This layer of particles is called the **boundary layer**. At the surface of the object, it is essentially static due to the friction of the surface. The object, with its boundary layer is effectively the new shape of the object that the rest of the molecules "see" as the object approaches. This boundary layer *can* separate from the surface, essentially creating a new surface and completely changing the flow path. The classical example of this is a stalling airfoil. The delta wing image clearly shows the boundary layer thickening as the gas flows from right to left along the leading edge.

5.8.6 Maximum entropy principle

Main article: Principle of maximum entropy

As the total number of degrees of freedom approaches infinity, the system will be found in the macrostate that corresponds to the highest multiplicity. In order to illustrate this principle, observe the skin temperature of a frozen metal bar. Using a thermal image of the skin temperature, note the temperature distribution on the surface. This initial observation of temperature represents a "microstate." At some future time, a second observation of the skin temperature produces a second microstate. By continuing this observation process, it is possible to produce a series of microstates that illustrate the thermal history of the bar's surface. Characterization of this historical series of microstates is possible by choosing the macrostate that successfully classifies them all into a single grouping.

5.8.7   Thermodynamic equilibrium

Main article: Thermodynamic equilibrium

When energy transfer ceases from a system, this condition is referred to as thermodynamic equilibrium. Usually this condition implies the system and surroundings are at the same temperature so that heat no longer transfers between them. It also implies that external forces are balanced (volume does not change), and all chemical reactions within the system are complete. The timeline varies for these events depending on the system in question. A container of ice allowed to melt at room temperature takes hours, while in semiconductors the heat transfer that occurs in the device transition from an on to off state could be on the order of a few nanoseconds.

5.9   See also

- Quasi-solid

- Greenhouse gas

- Natural gas

- Volcanic gas

- Breathing

- Wind

5.10   Notes

[1] This early 20th century discussion infers what is regarded as the plasma state. See page 137 of American Chemical Society, Faraday Society, Chemical Society (Great Britain) *The Journal of physical chemistry, Volume 11* Cornell (1907).

[2] The work by T. Zelevinski provides another link to latest research about Strontium in this new field of study. See Tanya Zelevinsky (2009). "84Sr—just right for forming a Bose-Einstein condensate". *Physics* **2**: 94. Bibcode:2009PhyOJ...2...94Z. doi:10.1103/physics.2.94.

[3] for links material on the Bose–Einstein condensate see Quantum Gas Microscope Offers Glimpse Of Quirky Ultracold Atoms. ScienceDaily. 4 November 2009.

[4] J. B. van Helmont, *Ortus medicinae.* ... (Amsterdam, (Netherlands): Louis Elzevir, 1652 (first edition: 1648)). The word "gas" first appears on page 58, where he mentions: "... Gas (meum scil. inventum) ..." (... gas (namely, my discovery) ...). On page 59, he states: "... in nominis egestate, halitum illum, Gas vocavi, non longe a Chao ..." (... in need of a name, I called this vapor "gas", not far from "chaos" ...)

[5] Harper, Douglas. "gas". *Online Etymology Dictionary*.

[6] Draper, John William (1861). *A textbook on chemistry*. New York: Harper and Sons. p. 178.

[7] The authors make the connection between molecular forces of metals and their corresponding physical properties. By extension, this concept would apply to gases as well, though not universally. Cornell (1907) pp. 164–5.

[8] One noticeable exception to this physical property connection is conductivity which varies depending on the state of matter (ionic compounds in water) as described by Michael Faraday in the 1833 when he noted that ice does not conduct a current. See page 45 of John Tyndall's *Faraday as a Discoverer* (1868).

[9] John S. Hutchinson (2008). *Concept Development Studies in Chemistry*. p. 67.

[10] Anderson, p.501

[11] J. Clerk Maxwell (1904). *Theory of Heat*. Mineola: Dover Publications. pp. 319–20. ISBN 0-486-41735-2.

[12] See pages 137–8 of Society, Cornell (1907).

[13] Kenneth Wark (1977). *Thermodynamics* (3 ed.). McGraw-Hill. p. 12. ISBN 0-07-068280-1.

[14] For assumptions of Kinetic Theory see McPherson, pp.60–61

[15] Anderson, pp. 289–291

[16] John, p.205

[17] John, pp. 247–56

[18] McPherson, pp.52–55

[19] McPherson, pp.55–60

[20] John P. Millington (1906). *John Dalton*. pp. 72, 77–78.

5.11 References

- Anderson, John D. (1984). *Fundamentals of Aerodynamics*. McGraw-Hill Higher Education. ISBN 0-07-001656-9.

- John, James (1984). *Gas Dynamics*. Allyn and Bacon. ISBN 0-205-08014-6.

- McPherson, William and Henderson, William (1917). *An Elementary study of chemistry*.

5.12 Further reading

- Philip Hill and Carl Peterson. *Mechanics and Thermodynamics of Propulsion: Second Edition* Addison-Wesley, 1992. ISBN 0-201-14659-2

- National Aeronautics and Space Administration (NASA). Animated Gas Lab. Accessed February 2008.

- Georgia State University. HyperPhysics. Accessed February 2008.

- Antony Lewis WordWeb. Accessed February 2008.

- Northwestern Michigan College The Gaseous State. Accessed February 2008.

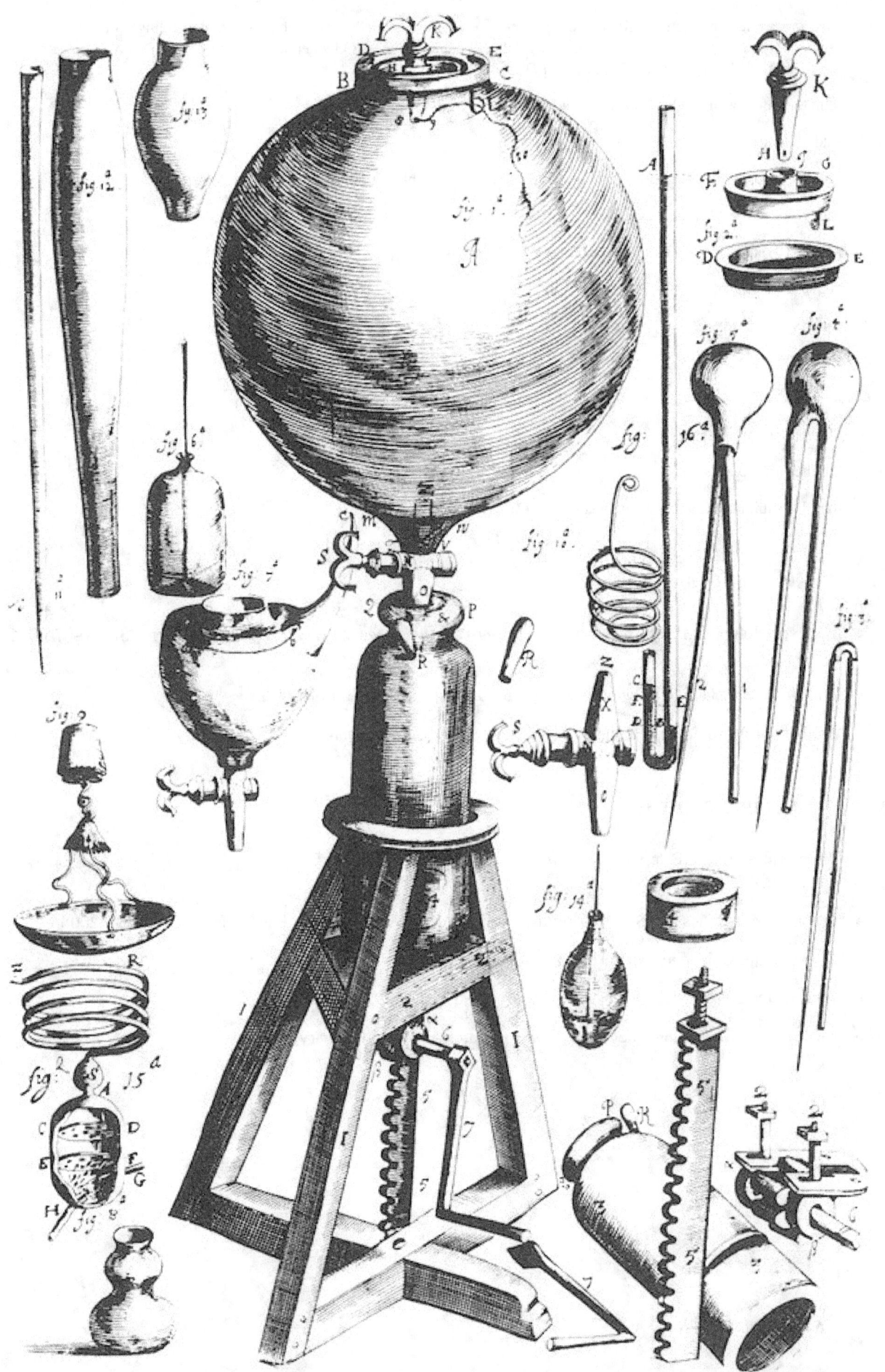

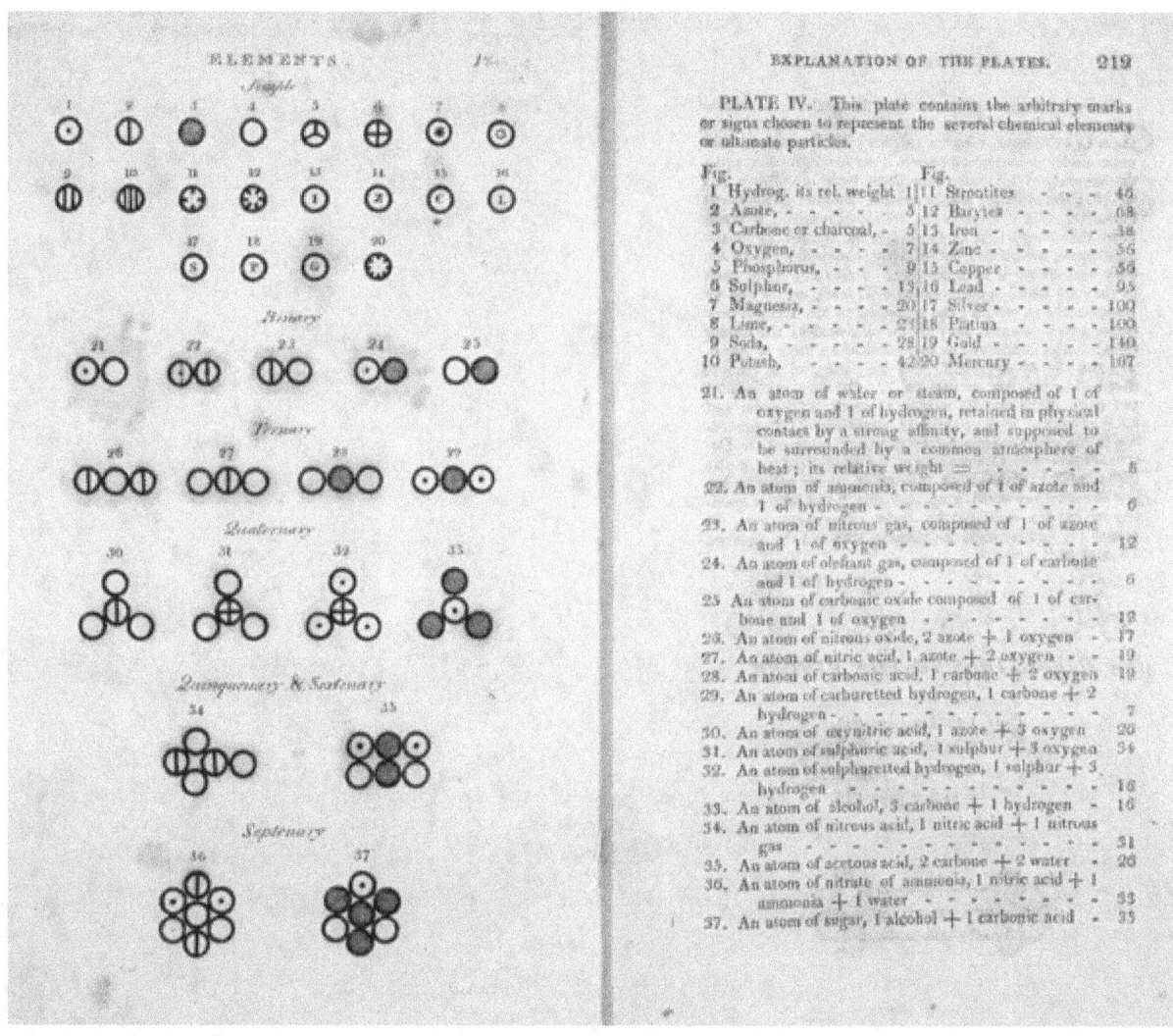

Dalton's notation.

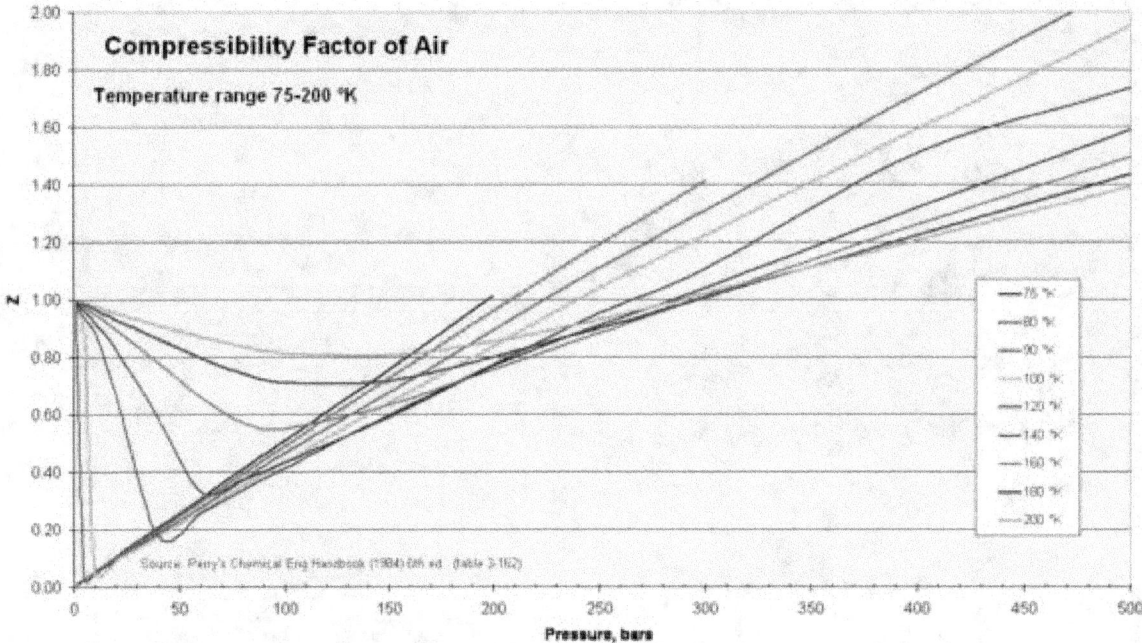

Compressibility factors for air.

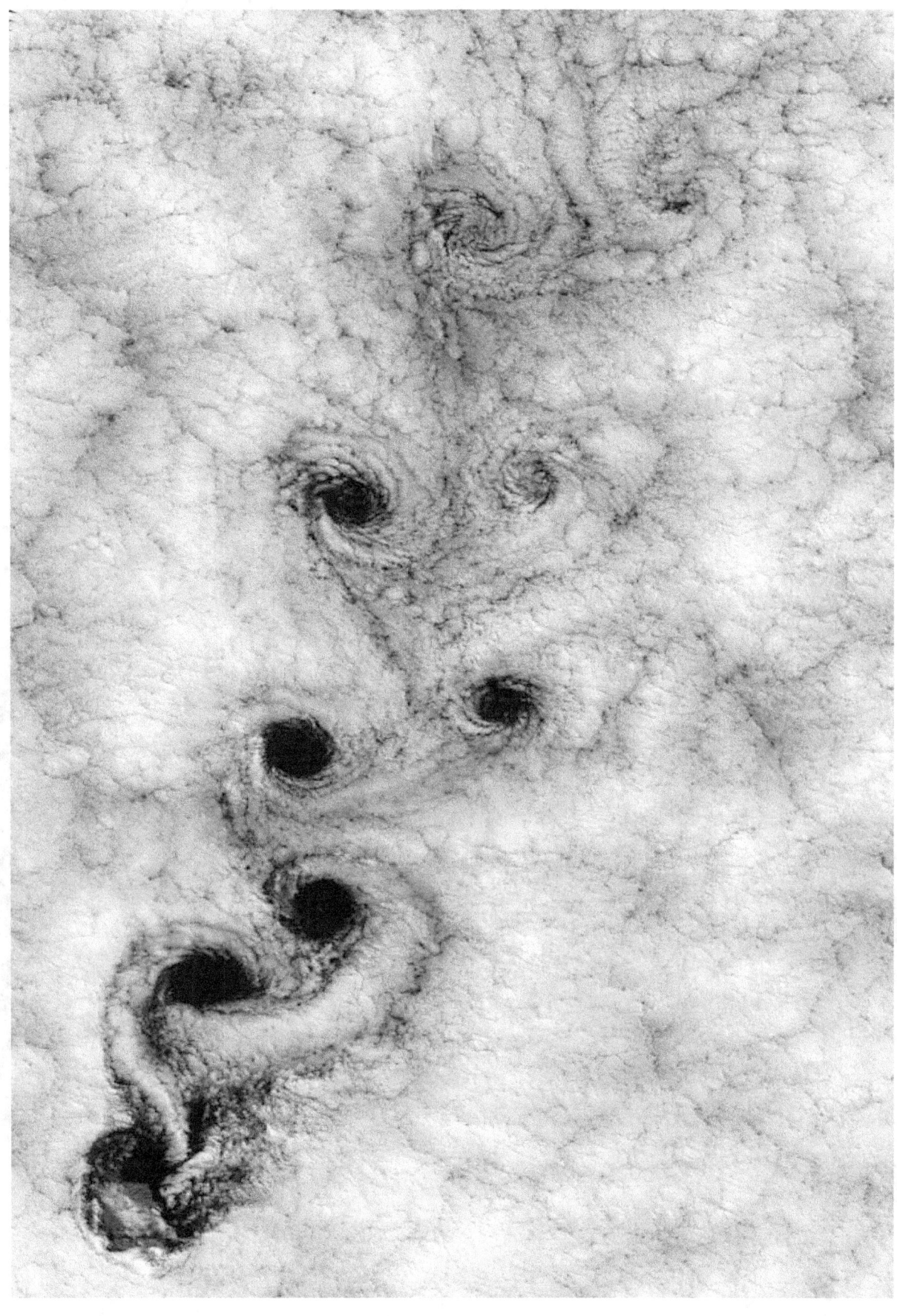

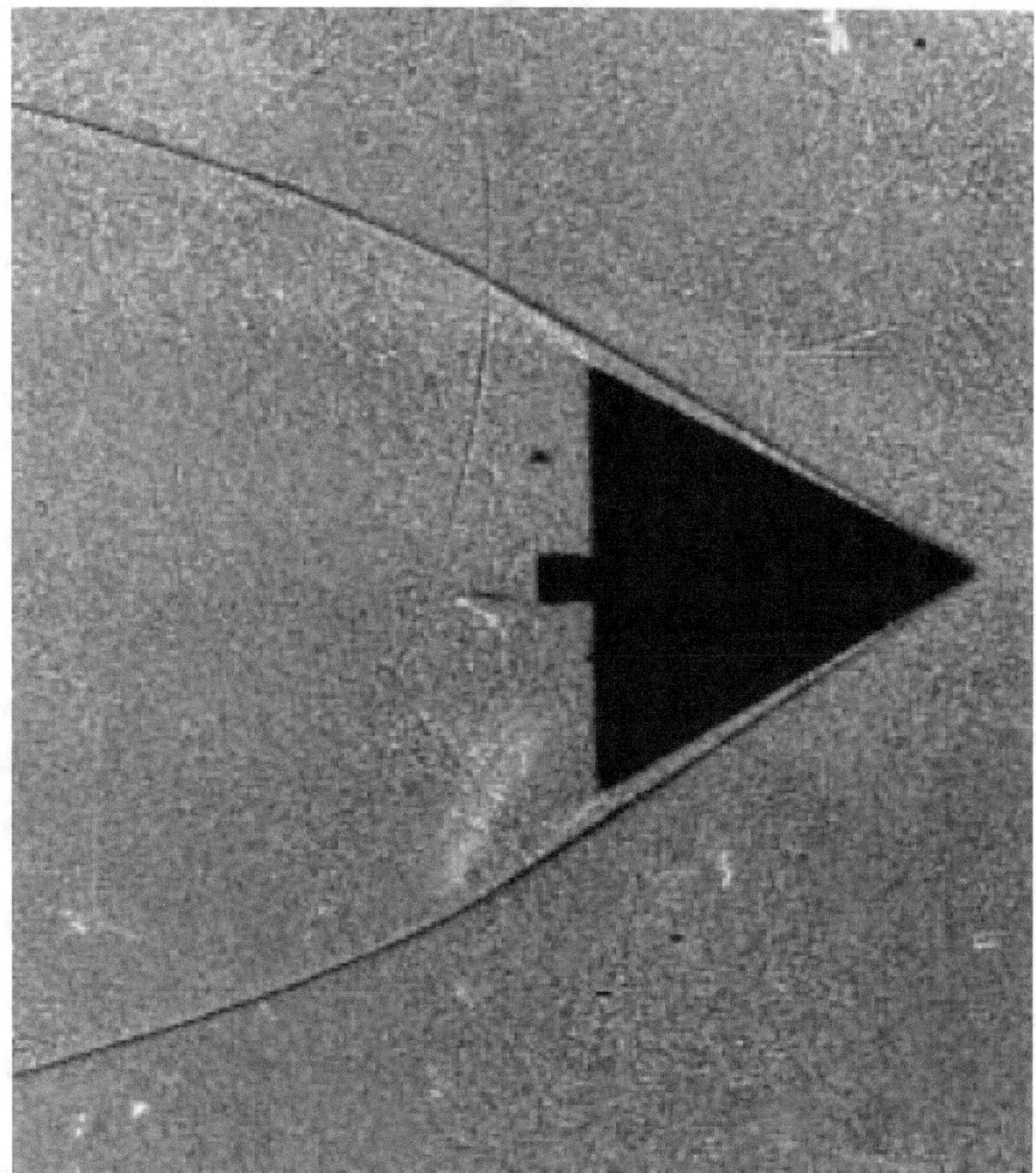

Delta wing in wind tunnel. The shadows form as the indices of refraction change within the gas as it compresses on the leading edge of this wing.

Chapter 6

Plasma (physics)

For other uses, see Plasma.

Plasma (from Greek πλάσμα, "anything formed"[1]) is one of the four fundamental states of matter, the others being solid, liquid, and gas. A plasma has properties unlike those of the other states.

A plasma can be created by heating a gas or subjecting it to a strong electromagnetic field applied with a laser or microwave generator. This decreases or increases the number of electrons, creating positive or negative charged particles called ions,[2] and is accompanied by the dissociation of molecular bonds, if present.[3]

The presence of a significant number of charge carriers makes plasma electrically conductive so that it responds strongly to electromagnetic fields. Like gas, plasma does not have a definite shape or a definite volume unless enclosed in a container. Unlike gas, under the influence of a magnetic field, it may form structures such as filaments, beams and double layers.

Plasma is the most abundant form of ordinary matter in the Universe (the only matter known to exist for sure, the more abundant dark matter is hypothetical and may or may not be explained by ordinary matter), most of which is in the rarefied intergalactic regions, particularly the intracluster medium, and in stars, including the Sun.[4][5] A common form of plasmas on Earth is seen in neon signs.

Much of the understanding of plasmas has come from the pursuit of controlled nuclear fusion and fusion power, for which plasma physics provides the scientific basis.

6.1 Properties and parameters

6.1.1 Definition

Plasma is loosely described as an electrically neutral medium of unbound positive and negative particles (i.e. the overall charge of a plasma is roughly zero). It is important to note that although they are unbound, these particles are not 'free' in the sense of not experiencing forces. When the charges move, they generate electric currents with magnetic fields, and as a result, they are affected by each other's fields. This governs their collective behavior with many degrees of freedom.[3][7] A definition can have three criteria:[8][9]

1. **The plasma approximation**: Charged particles must be close enough together that each particle influences many nearby charged particles, rather than just interacting with the closest particle (these collective effects are a distinguishing feature of a plasma). The plasma approximation is valid when the number of charge carriers within the sphere of influence (called the *Debye sphere* whose radius is the Debye screening length) of a particular particle is higher than unity to provide collective behavior of the charged particles. The average number of particles in the Debye sphere is given by the plasma parameter, "Λ" (the Greek uppercase letter Lambda).

2. **Bulk interactions**: The Debye screening length (defined above) is short compared to the physical size of the

Artist's rendition of the Earth's plasma fountain, showing oxygen, helium, and hydrogen ions that gush into space from regions near the Earth's poles. The faint yellow area shown above the north pole represents gas lost from Earth into space; the green area is the aurora borealis, where plasma energy pours back into the atmosphere.[6]

plasma. This criterion means that interactions in the bulk of the plasma are more important than those at its edges, where boundary effects may take place. When this criterion is satisfied, the plasma is quasineutral.

3. **Plasma frequency**: The electron plasma frequency (measuring plasma oscillations of the electrons) is large compared to the electron-neutral collision frequency (measuring frequency of collisions between electrons and neutral particles). When this condition is valid, electrostatic interactions dominate over the processes of ordinary gas kinetics.

6.1.2   Ranges of parameters

Plasma parameters can take on values varying by many orders of magnitude, but the properties of plasmas with apparently disparate parameters may be very similar (see plasma scaling). The following chart considers only conventional atomic plasmas and not exotic phenomena like quark gluon plasmas:

6.1.3   Degree of ionization

For plasma to exist, ionization is necessary. The term "plasma density" by itself usually refers to the "electron density", that is, the number of free electrons per unit volume. The degree of ionization of a plasma is the proportion of atoms that have lost or gained electrons, and is controlled mostly by the temperature. Even a partially ionized gas in which as little as 1% of the particles are ionized can have the characteristics of a plasma (i.e., response to magnetic fields and high electrical conductivity). The degree of ionization, α, is defined as $\alpha = \frac{n_i}{n_i+n_n}$, where n_i is the number density of ions and n_n is the number density of neutral atoms. The *electron density* is related to this by the average charge state $\langle Z \rangle$ of the ions through $n_e = \langle Z \rangle n_i$, where n_e is the number density of electrons.

6.1.4   Temperatures

See also: Nonthermal plasma

Plasma temperature is commonly measured in kelvins or electronvolts and is, informally, a measure of the thermal kinetic energy per particle. Very high temperatures are usually needed to sustain ionization, which is a defining feature of a plasma. The degree of plasma ionization is determined by the electron temperature relative to the ionization energy (and more weakly by the density), in a relationship called the Saha equation. At low temperatures, ions and electrons tend to recombine into bound states—atoms[12]—and the plasma will eventually become a gas.

In most cases the electrons are close enough to thermal equilibrium that their temperature is relatively well-defined, even when there is a significant deviation from a Maxwellian energy distribution function, for example, due to UV radiation, energetic particles, or strong electric fields. Because of the large difference in mass, the electrons come to thermodynamic equilibrium amongst themselves much faster than they come into equilibrium with the ions or neutral atoms. For this reason, the ion temperature may be very different from (usually lower than) the electron temperature. This is especially common in weakly ionized technological plasmas, where the ions are often near the ambient temperature.

Thermal vs. nonthermal plasmas

Based on the relative temperatures of the electrons, ions and neutrals, plasmas are classified as "thermal" or "non-thermal". Thermal plasmas have electrons and the heavy particles at the same temperature, i.e. they are in thermal equilibrium with each other. Nonthermal plasmas on the other hand have the ions and neutrals at a much lower temperature (sometimes room temperature), whereas electrons are much "hotter" ($T_e \gg T_n$).

A plasma is sometimes referred to as being "hot" if it is nearly fully ionized, or "cold" if only a small fraction (for example 1%) of the gas molecules are ionized, but other definitions of the terms "hot plasma" and "cold plasma" are common. Even in a "cold" plasma, the electron temperature is still typically several thousand degrees Celsius. Plasmas utilized in

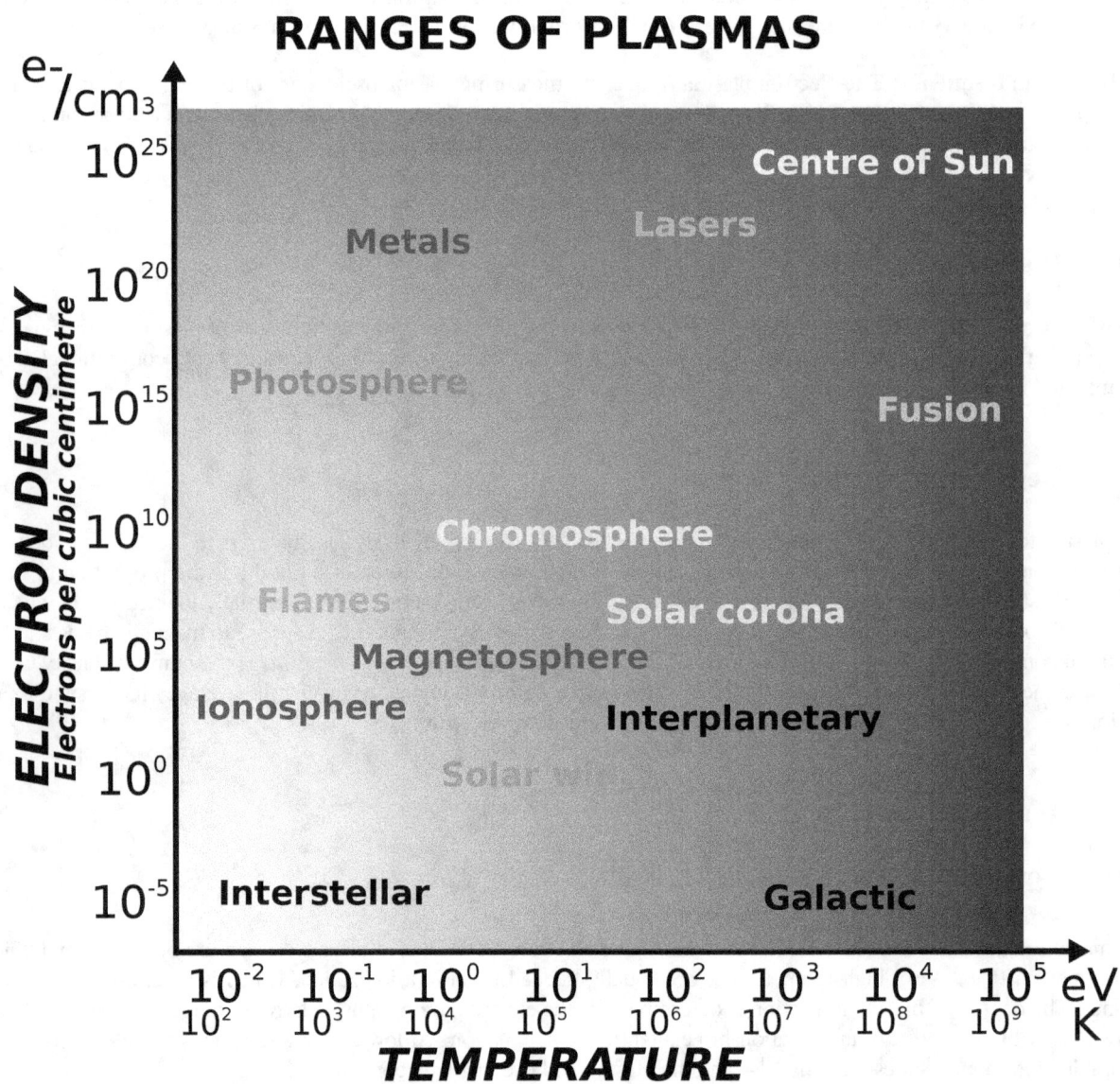

Range of plasmas. *Density increases upwards, temperature increases towards the right. The free electrons in a metal may be considered an electron plasma.*[10]

"plasma technology" ("technological plasmas") are usually cold plasmas in the sense that only a small fraction of the gas molecules are ionized.

6.1.5 Plasma potential

Since plasmas are very good electrical conductors, electric potentials play an important role. The potential as it exists on average in the space between charged particles, independent of the question of how it can be measured, is called the "plasma potential", or the "space potential". If an electrode is inserted into a plasma, its potential will generally lie considerably below the plasma potential due to what is termed a Debye sheath. The good electrical conductivity of plasmas makes their electric fields very small. This results in the important concept of "quasineutrality", which says the density of negative charges is approximately equal to the density of positive charges over large volumes of the plasma ($n_e = \langle Z \rangle n_i$

), but on the scale of the Debye length there can be charge imbalance. In the special case that *double layers* are formed, the charge separation can extend some tens of Debye lengths.

The magnitude of the potentials and electric fields must be determined by means other than simply finding the net charge density. A common example is to assume that the electrons satisfy the Boltzmann relation:

$$n_e \propto e^{e\Phi/k_B T_e}.$$

Differentiating this relation provides a means to calculate the electric field from the density:

$$\vec{E} = (k_B T_e/e)(\nabla n_e/n_e).$$

It is possible to produce a plasma that is not quasineutral. An electron beam, for example, has only negative charges. The density of a non-neutral plasma must generally be very low, or it must be very small, otherwise it will be dissipated by the repulsive electrostatic force.

In astrophysical plasmas, Debye screening prevents electric fields from directly affecting the plasma over large distances, i.e., greater than the Debye length. However, the existence of charged particles causes the plasma to generate, and be affected by, magnetic fields. This can and does cause extremely complex behavior, such as the generation of plasma double layers, an object that separates charge over a few tens of Debye lengths. The dynamics of plasmas interacting with external and self-generated magnetic fields are studied in the academic discipline of magnetohydrodynamics.

6.1.6   Magnetization

Plasma with a magnetic field strong enough to influence the motion of the charged particles is said to be magnetized. A common quantitative criterion is that a particle on average completes at least one gyration around the magnetic field before making a collision, i.e., $\omega_{ce}/v_{coll} > 1$, where ω_{ce} is the "electron gyrofrequency" and v_{coll} is the "electron collision rate". It is often the case that the electrons are magnetized while the ions are not. Magnetized plasmas are *anisotropic*, meaning that their properties in the direction parallel to the magnetic field are different from those perpendicular to it. While electric fields in plasmas are usually small due to the high conductivity, the electric field associated with a plasma moving in a magnetic field is given by $\mathbf{E} = -v \times \mathbf{B}$ (where $\mathbf{E}$ is the electric field, $\mathbf{v}$ is the velocity, and $\mathbf{B}$ is the magnetic field), and is not affected by Debye shielding.[14]

6.1.7   Comparison of plasma and gas phases

Plasma is often called the *fourth state of matter* after solid, liquids and gases.[15][16] It is distinct from these and other lower-energy states of matter. Although it is closely related to the gas phase in that it also has no definite form or volume, it differs in a number of ways, including the following:

6.2   Common plasmas

Further information: Astrophysical plasma, Interstellar medium and Intergalactic space

Plasmas are by far the most common phase of ordinary matter in the universe, both by mass and by volume.[18] Essentially, all of the visible light from space comes from stars, which are plasmas with a temperature such that they radiate strongly at visible wavelengths. Most of the ordinary (or baryonic) matter in the universe, however, is found in the intergalactic medium, which is also a plasma, but much hotter, so that it radiates primarily as X-rays.

In 1937, Hannes Alfvén argued that if plasma pervaded the universe, it could then carry electric currents capable of generating a galactic magnetic field.[19] After winning the Nobel Prize, he emphasized that:

In order to understand the phenomena in a certain plasma region, it is necessary to map not only the magnetic but also the electric field and the electric currents. Space is filled with a network of currents which transfer energy and momentum over large or very large distances. The currents often pinch to filamentary or surface currents. The latter are likely to give space, as also interstellar and intergalactic space, a cellular structure.[20]

By contrast the current scientific consensus is that about 96% of the total energy density in the universe is not plasma or any other form of ordinary matter, but a combination of cold dark matter and dark energy. Our Sun, and all stars, are made of plasma, much of interstellar space is filled with a plasma, albeit a very sparse one, and intergalactic space too. Even black holes, which are not directly visible, are thought to be fuelled by accreting ionising matter (i.e. plasma),[21] and they are associated with astrophysical jets of luminous ejected plasma,[22] such as M87's jet that extends 5,000 light-years.[23]

In our solar system, interplanetary space is filled with the plasma of the Solar Wind that extends from the Sun out to the heliopause. However, the density of ordinary matter is much higher than average and much higher than that of either dark matter or dark energy. The planet Jupiter accounts for most of the *non*-plasma, only about 0.1% of the mass and $10^{-15}\%$ of the volume within the orbit of Pluto.

Dust and small grains within a plasma will also pick up a net negative charge, so that they in turn may act like a very heavy negative ion component of the plasma (see dusty plasmas).

6.3 Complex plasma phenomena

Although the underlying equations governing plasmas are relatively simple, plasma behavior is extraordinarily varied and subtle: the emergence of unexpected behavior from a simple model is a typical feature of a complex system. Such systems lie in some sense on the boundary between ordered and disordered behavior and cannot typically be described either by simple, smooth, mathematical functions, or by pure randomness. The spontaneous formation of interesting spatial features on a wide range of length scales is one manifestation of plasma complexity. The features are interesting, for example, because they are very sharp, spatially intermittent (the distance between features is much larger than the features themselves), or have a fractal form. Many of these features were first studied in the laboratory, and have subsequently been recognized throughout the universe. Examples of complexity and complex structures in plasmas include:

6.3.1 Filamentation

Striations or string-like structures,[27] also known as birkeland currents, are seen in many plasmas, like the plasma ball, the aurora,[28] lightning,[29] electric arcs, solar flares,[30] and supernova remnants.[31] They are sometimes associated with larger current densities, and the interaction with the magnetic field can form a magnetic rope structure.[32] High power microwave breakdown at atmospheric pressure also leads to the formation of filamentary structures.[33] (See also Plasma pinch)

Filamentation also refers to the self-focusing of a high power laser pulse. At high powers, the nonlinear part of the index of refraction becomes important and causes a higher index of refraction in the center of the laser beam, where the laser is brighter than at the edges, causing a feedback that focuses the laser even more. The tighter focused laser has a higher peak brightness (irradiance) that forms a plasma. The plasma has an index of refraction lower than one, and causes a defocusing of the laser beam. The interplay of the focusing index of refraction, and the defocusing plasma makes the formation of a long filament of plasma that can be micrometers to kilometers in length.[34] One interesting aspect of the filamentation generated plasma is the relatively low ion density due to defocusing effects of the ionized electrons.[35] (See also Filament propagation)

6.3.2 Shocks or double layers

Plasma properties change rapidly (within a few Debye lengths) across a two-dimensional sheet in the presence of a (moving) shock or (stationary) double layer. Double layers involve localized charge separation, which causes a large potential difference across the layer, but does not generate an electric field outside the layer. Double layers separate adjacent plasma regions with different physical characteristics, and are often found in current carrying plasmas. They accelerate both ions and electrons.

6.3.3 Electric fields and circuits

Quasineutrality of a plasma requires that plasma currents close on themselves in electric circuits. Such circuits follow Kirchhoff's circuit laws and possess a resistance and inductance. These circuits must generally be treated as a strongly coupled system, with the behavior in each plasma region dependent on the entire circuit. It is this strong coupling between system elements, together with nonlinearity, which may lead to complex behavior. Electrical circuits in plasmas store inductive (magnetic) energy, and should the circuit be disrupted, for example, by a plasma instability, the inductive energy will be released as plasma heating and acceleration. This is a common explanation for the heating that takes place in the solar corona. Electric currents, and in particular, magnetic-field-aligned electric currents (which are sometimes generically referred to as "Birkeland currents"), are also observed in the Earth's aurora, and in plasma filaments.

6.3.4 Cellular structure

Narrow sheets with sharp gradients may separate regions with different properties such as magnetization, density and temperature, resulting in cell-like regions. Examples include the magnetosphere, heliosphere, and heliospheric current sheet. Hannes Alfvén wrote: "From the cosmological point of view, the most important new space research discovery is probably the cellular structure of space. As has been seen in every region of space accessible to in situ measurements, there are a number of 'cell walls', sheets of electric currents, which divide space into compartments with different magnetization, temperature, density, etc."[36]

6.3.5 Critical ionization velocity

The critical ionization velocity is the relative velocity between an ionized plasma and a neutral gas, above which a runaway ionization process takes place. The critical ionization process is a quite general mechanism for the conversion of the kinetic energy of a rapidly streaming gas into ionization and plasma thermal energy. Critical phenomena in general are typical of complex systems, and may lead to sharp spatial or temporal features.

6.3.6 Ultracold plasma

Ultracold plasmas are created in a magneto-optical trap (MOT) by trapping and cooling neutral atoms, to temperatures of 1 mK or lower, and then using another laser to ionize the atoms by giving each of the outermost electrons just enough energy to escape the electrical attraction of its parent ion.

One advantage of ultracold plasmas are their well characterized and tunable initial conditions, including their size and electron temperature. By adjusting the wavelength of the ionizing laser, the kinetic energy of the liberated electrons can be tuned as low as 0.1 K, a limit set by the frequency bandwidth of the laser pulse. The ions inherit the millikelvin temperatures of the neutral atoms, but are quickly heated through a process known as disorder induced heating (DIH). This type of non-equilibrium ultracold plasma evolves rapidly, and displays many other interesting phenomena.[37]

One of the metastable states of a strongly nonideal plasma is Rydberg matter, which forms upon condensation of excited atoms.

6.3.7 Non-neutral plasma

The strength and range of the electric force and the good conductivity of plasmas usually ensure that the densities of positive and negative charges in any sizeable region are equal ("quasineutrality"). A plasma with a significant excess of charge density, or, in the extreme case, is composed of a single species, is called a non-neutral plasma. In such a plasma, electric fields play a dominant role. Examples are charged particle beams, an electron cloud in a Penning trap and positron plasmas.[38]

6.3.8   Dusty plasma/grain plasma

A dusty plasma contains tiny charged particles of dust (typically found in space). The dust particles acquire high charges and interact with each other. A plasma that contains larger particles is called grain plasma. Under laboratory conditions, dusty plasmas are also called *complex plasmas*.[39]

6.3.9   Impermeable plasma

Impermeable plasma is a type of thermal plasma which acts like an impermeable solid with respect to gas or cold plasma and can be physically pushed. Interaction of cold gas and thermal plasma was briefly studied by a group led by Hannes Alfvén in 1960s and 1970s for its possible applications in insulation of fusion plasma from the reactor walls.[40] However, later it was found that the external magnetic fields in this configuration could induce kink instabilities in the plasma and subsequently lead to an unexpectedly high heat loss to the walls.[41] In 2013, a group of materials scientists reported that they have successfully generated stable impermeable plasma with no magnetic confinement using only an ultrahigh-pressure blanket of cold gas. While spectroscopic data on the characteristics of plasma were claimed to be difficult to obtain due to the high pressure, the passive effect of plasma on synthesis of different nanostructures clearly suggested the effective confinement. They also showed that upon maintaining the impermeability for a few tens of seconds, screening of ions at the plasma-gas interface could give rise to a strong secondary mode of heating (known as viscous heating) leading to different kinetics of reactions and formation of complex nanomaterials.[42]

6.4   Mathematical descriptions

Main article: Plasma modeling

To completely describe the state of a plasma, we would need to write down all the particle locations and velocities and describe the electromagnetic field in the plasma region. However, it is generally not practical or necessary to keep track of all the particles in a plasma. Therefore, plasma physicists commonly use less detailed descriptions, of which there are two main types:

6.4.1   Fluid model

Fluid models describe plasmas in terms of smoothed quantities, like density and averaged velocity around each position (see Plasma parameters). One simple fluid model, magnetohydrodynamics, treats the plasma as a single fluid governed by a combination of Maxwell's equations and the Navier–Stokes equations. A more general description is the two-fluid plasma picture, where the ions and electrons are described separately. Fluid models are often accurate when collisionality is sufficiently high to keep the plasma velocity distribution close to a Maxwell–Boltzmann distribution. Because fluid models usually describe the plasma in terms of a single flow at a certain temperature at each spatial location, they can neither capture velocity space structures like beams or double layers, nor resolve wave-particle effects.

6.4.2   Kinetic model

Kinetic models describe the particle velocity distribution function at each point in the plasma and therefore do not need to assume a Maxwell–Boltzmann distribution. A kinetic description is often necessary for collisionless plasmas. There are two common approaches to kinetic description of a plasma. One is based on representing the smoothed distribution function on a grid in velocity and position. The other, known as the particle-in-cell (PIC) technique, includes kinetic information by following the trajectories of a large number of individual particles. Kinetic models are generally more computationally intensive than fluid models. The Vlasov equation may be used to describe the dynamics of a system of charged particles interacting with an electromagnetic field. In magnetized plasmas, a gyrokinetic approach can substantially reduce the computational expense of a fully kinetic simulation.

6.5 Artificial plasmas

Most artificial plasmas are generated by the application of electric and/or magnetic fields. Plasma generated in a laboratory setting and for industrial use can be generally categorized by:

- The type of power source used to generate the plasma—DC, RF and microwave

- The pressure they operate at—vacuum pressure (< 10 mTorr or 1 Pa), moderate pressure (~ 1 Torr or 100 Pa), atmospheric pressure (760 Torr or 100 kPa)

- The degree of ionization within the plasma—fully, partially, or weakly ionized

- The temperature relationships within the plasma—thermal plasma ($T_e = T_i = T_{gas}$), non-thermal or "cold" plasma ($T_e \gg T_i = T_{gas}$)

- The electrode configuration used to generate the plasma

- The magnetization of the particles within the plasma—magnetized (both ion and electrons are trapped in Larmor orbits by the magnetic field), partially magnetized (the electrons but not the ions are trapped by the magnetic field), non-magnetized (the magnetic field is too weak to trap the particles in orbits but may generate Lorentz forces)

- The application.

6.5.1 Generation of artificial plasma

Just like the many uses of plasma, there are several means for its generation, however, one principle is common to all of them: there must be energy input to produce and sustain it.[44] For this case, plasma is generated when an electric current is applied across a dielectric gas or fluid (an electrically non-conducting material) as can be seen in the image to the right, which shows a discharge tube as a simple example (DC used for simplicity).

The potential difference and subsequent electric field pull the bound electrons (negative) toward the anode (positive electrode) while the cathode (negative electrode) pulls the nucleus.[45] As the voltage increases, the current stresses the material (by electric polarization) beyond its dielectric limit (termed strength) into a stage of electrical breakdown, marked by an electric spark, where the material transforms from being an insulator into a conductor (as it becomes increasingly ionized). The underlying process is the Townsend avalanche, where collisions between electrons and neutral gas atoms create more ions and electrons (as can be seen in the figure on the right). The first impact of an electron on an atom results in one ion and two electrons. Therefore, the number of charged particles increases rapidly (in the millions) only "after about 20 successive sets of collisions",[46] mainly due to a small mean free path (average distance travelled between collisions).

Electric arc

With ample current density and ionization, this forms a luminous electric arc (a continuous electric discharge similar to lightning) between the electrodes.[Note 1] Electrical resistance along the continuous electric arc creates heat, which dissociates more gas molecules and ionizes the resulting atoms (where degree of ionization is determined by temperature), and as per the sequence: solid-liquid-gas-plasma, the gas is gradually turned into a thermal plasma.[Note 2] A thermal plasma is in thermal equilibrium, which is to say that the temperature is relatively homogeneous throughout the heavy particles (i.e. atoms, molecules and ions) and electrons. This is so because when thermal plasmas are generated, electrical energy is given to electrons, which, due to their great mobility and large numbers, are able to disperse it rapidly and by elastic collision (without energy loss) to the heavy particles.[47][Note 3]

6.5.2 Examples of industrial/commercial plasma

Because of their sizable temperature and density ranges, plasmas find applications in many fields of research, technology and industry. For example, in: industrial and extractive metallurgy,[47] surface treatments such as plasma spraying (coating), etching in microelectronics,[48] metal cutting[49] and welding; as well as in everyday vehicle exhaust cleanup and fluorescent/luminescent lamps,[44] while even playing a part in supersonic combustion engines for aerospace engineering.[50]

Low-pressure discharges

- *Glow discharge plasmas*: non-thermal plasmas generated by the application of DC or low frequency RF (<100 kHz) electric field to the gap between two metal electrodes. Probably the most common plasma; this is the type of plasma generated within fluorescent light tubes.[51]

- *Capacitively coupled plasma (CCP)*: similar to glow discharge plasmas, but generated with high frequency RF electric fields, typically 13.56 MHz. These differ from glow discharges in that the sheaths are much less intense. These are widely used in the microfabrication and integrated circuit manufacturing industries for plasma etching and plasma enhanced chemical vapor deposition.[52]

- *Cascaded Arc Plasma Source*: a device to produce low temperature (~1eV) high density plasmas (HDP).

- *Inductively coupled plasma (ICP)*: similar to a CCP and with similar applications but the electrode consists of a coil wrapped around the chamber where plasma is formed.[53]

- *Wave heated plasma*: similar to CCP and ICP in that it is typically RF (or microwave). Examples include helicon discharge and electron cyclotron resonance (ECR).[54]

Atmospheric pressure

- *Arc discharge:* this is a high power thermal discharge of very high temperature (~10,000 K). It can be generated using various power supplies. It is commonly used in metallurgical processes. For example, it is used to smelt minerals containing Al_2O_3 to produce aluminium.

- *Corona discharge:* this is a non-thermal discharge generated by the application of high voltage to sharp electrode tips. It is commonly used in ozone generators and particle precipitators.

- *Dielectric barrier discharge (DBD):* this is a non-thermal discharge generated by the application of high voltages across small gaps wherein a non-conducting coating prevents the transition of the plasma discharge into an arc. It is often mislabeled 'Corona' discharge in industry and has similar application to corona discharges. It is also widely used in the web treatment of fabrics.[55] The application of the discharge to synthetic fabrics and plastics functionalizes the surface and allows for paints, glues and similar materials to adhere.[56]

- *Capacitive discharge:* this is a nonthermal plasma generated by the application of RF power (e.g., 13.56 MHz) to one powered electrode, with a grounded electrode held at a small separation distance on the order of 1 cm. Such discharges are commonly stabilized using a noble gas such as helium or argon.[57]

- "Piezoelectric direct discharge plasma:" is a nonthermal plasma generated at the high-side of a piezoelectric transformer (PT). This generation variant is particularly suited for high efficient and compact devices where a separate high voltage power supply is not desired.

6.6   History

Plasma was first identified in a Crookes tube, and so described by Sir William Crookes in 1879 (he called it "radiant matter").[58] The nature of the Crookes tube "cathode ray" matter was subsequently identified by British physicist Sir J.J. Thomson in 1897.[59] The term "plasma" was coined by Irving Langmuir in 1928,[60] perhaps because the glowing discharge molds itself to the shape of the Crookes tube (Gr. πλάσμα – a thing moulded or formed).[61] Langmuir described his observations as:

Except near the electrodes, where there are *sheaths* containing very few electrons, the ionized gas contains ions and electrons in about equal numbers so that the resultant space charge is very small. We shall use the name *plasma* to describe this region containing balanced charges of ions and electrons.[60]

6.7 Fields of active research

This is just a partial list of topics. See list of plasma (physics) articles. A more complete and organized list can be found on web sites on plasma science and technology.[62]

6.8 See also

- Plasma torch

- Ambipolar diffusion

- Hannes Alfvén Prize

- Plasma channel

- Plasma parameters

- Plasma nitriding

- Magnetohydrodynamics (MHD)

- Electric field screening

- List of plasma physicists

- List of plasma (physics) articles

- Important publications in plasma physics

- IEEE Nuclear and Plasma Sciences Society

- Quark-gluon plasma

- Nikola Tesla

- Space physics

- Total electron content

6.9 Notes

[1] The material undergoes various 'regimes' or stages (e.g. saturation, breakdown, glow, transition and thermal arc) as the voltage is increased under the voltage-current relationship. The voltage rises to its maximum value in the saturation stage, and thereafter it undergoes fluctuations of the various stages; while the current progressively increases throughout.[46]

[2] Across literature, there appears to be no strict definition on where the boundary is between a gas and plasma. Nevertheless, it is enough to say that at 2,000°C the gas molecules become atomized, and ionized at 3,000 °C and "in this state, [the] gas has a liquid like viscosity at atmospheric pressure and the free electric charges confer relatively high electrical conductivities that can approach those of metals."[47]

[3] Note that non-thermal, or non-equilibrium plasmas are not as ionized and have lower energy densities, and thus the temperature is not dispersed evenly among the particles, where some heavy ones remain 'cold'.

6.10 References

[1] πλάσμα, Henry George Liddell, Robert Scott, *A Greek–English Lexicon*, on Perseus

[2] Luo, Q-Z; D'Angelo, N; Merlino, R. L. (1998). "Shock formation in a negative ion plasma" (PDF) **5** (8). Department of Physics and Astronomy. Retrieved 2011-11-20.

[3] Sturrock, Peter A. (1994). *Plasma Physics: An Introduction to the Theory of Astrophysical, Geophysical & Laboratory Plasmas.* Cambridge University Press. ISBN 978-0-521-44810-9.

[4] "Ionization and Plasmas". The University of Tennessee, Knoxville Department of Physics and Astronomy.

[5] "How Lightning Works". HowStuffWorks.

[6] Plasma fountain Source, press release: Solar Wind Squeezes Some of Earth's Atmosphere into Space

[7] Hazeltine, R.D.; Waelbroeck, F.L. (2004). *The Framework of Plasma Physics.* Westview Press. ISBN 978-0-7382-0047-7.

[8] Dendy, R. O. (1990). *Plasma Dynamics.* Oxford University Press. ISBN 978-0-19-852041-2.

[9] Hastings, Daniel & Garrett, Henry (2000). *Spacecraft-Environment Interactions.* Cambridge University Press. ISBN 978-0-521-47128-2.

[10] Peratt, A. L. (1996). "Advances in Numerical Modeling of Astrophysical and Space Plasmas". *Astrophysics and Space Science* **242** (1–2): 93–163. Bibcode:1996Ap&SS.242...93P. doi:10.1007/BF00645112.

[11] See The Nonneutral Plasma Group at the University of California, San Diego

[12] Nicholson, Dwight R. (1983). *Introduction to Plasma Theory.* John Wiley & Sons. ISBN 978-0-471-09045-8.

[13] See Flashes in the Sky: Earth's Gamma-Ray Bursts Triggered by Lightning

[14] Richard Fitzpatrick, *Introduction to Plasma Physics*, Magnetized plasmas

[15] Yaffa Eliezer, Shalom Eliezer, *The Fourth State of Matter: An Introduction to the Physics of Plasma*, Publisher: Adam Hilger, 1989, ISBN 978-0-85274-164-1, 226 pages, page 5

[16] Bittencourt, J.A. (2004). *Fundamentals of Plasma Physics.* Springer. p. 1. ISBN 9780387209753.

[17] Hong, Alice (2000). "Dielectric Strength of Air". *The Physics Factbook.*

[18] It is often stated that more than 99% of the material in the visible universe is plasma. See, for example, Gurnett, D. A. & Bhattacharjee, A. (2005). *Introduction to Plasma Physics: With Space and Laboratory Applications.* Cambridge, UK: Cambridge University Press. p. 2. ISBN 978-0-521-36483-6. and Scherer, K; Fichtner, H & Heber, B (2005). *Space Weather: The Physics Behind a Slogan.* Berlin: Springer. p. 138. ISBN 978-3-540-22907-0..

[19] Alfvén, Hannes (1937). "Cosmic Radiation as an Intra-galactic Phenomenon". *Ark. f. mat., astr. o. fys.* **25B**: 29.

[20] Hannes, A (1990). "Cosmology in the Plasma Universe: An Introductory Exposition". *IEEE Transactions on Plasma Science* **18**: 5–10. Bibcode:1990ITPS...18....5P. doi:10.1109/27.45495. ISSN 0093-3813.

[21] Mészáros, Péter (2010) *The High Energy Universe: Ultra-High Energy Events in Astrophysics and Cosmology*, Publisher Cambridge University Press, ISBN 978-0-521-51700-3, p. 99.

[22] Raine, Derek J. and Thomas, Edwin George (2010) *Black Holes: An Introduction*, Publisher: Imperial College Press, ISBN 978-1-84816-382-9, p. 160

[23] Nemiroff, Robert and Bonnell, Jerry (11 December 2004) Astronomy Picture of the Day, nasa.gov

[24] IPPEX Glossary of Fusion Terms. Ippex.pppl.gov. Retrieved on 2011-11-19.

[25] "Plasma and Flames – The Burning Question", from the Coalition for Plasma Science, retrieved 8 November 2012

[26] von Engel, A. and Cozens, J.R. (1976) "Flame Plasma" in *Advances in electronics and electron physics*, L. L. Marton (ed.), Academic Press, ISBN 978-0-12-014520-1, p. 99

[27] Dickel, J. R. (1990). "The Filaments in Supernova Remnants: Sheets, Strings, Ribbons, or?". *Bulletin of the American Astronomical Society* **22**: 832. Bibcode:1990BAAS...22..832D.

[28] Grydeland, T. (2003). "Interferometric observations of filamentary structures associated with plasma instability in the auroral ionosphere". *Geophysical Research Letters* **30** (6). doi:10.1029/2002GL016362.

[29] Moss, G. D.; Pasko, V. P.; Liu, N.; Veronis, G. (2006). "Monte Carlo model for analysis of thermal runaway electrons in streamer tips in transient luminous events and streamer zones of lightning leaders". *Journal of Geophysical Research* **111**. doi:10.1029/2005JA011350.

[30] Doherty, Lowell R.; Menzel, Donald H. (1965). "Filamentary Structure in Solar Prominences". *The Astrophysical Journal* **141**: 251. Bibcode:1965ApJ...141..251D. doi:10.1086/148107.

[31] Hubble views the Crab Nebula M1: The Crab Nebula Filaments at the Wayback Machine (archived 5 October 2009). The University of Arizona

[32] Zhang, Y. A.; Song, M. T.; Ji, H. S. (2002). "A rope-shaped solar filament and a IIIb flare". *Chinese Astronomy and Astrophysics* **26** (4): 442. doi:10.1016/S0275-1062(02)00095-4.

[33] Boeuf, J. P.; Chaudhury, B.; Zhu, G. Q. (2010). "Theory and Modeling of Self-Organization and Propagation of Filamentary Plasma Arrays in Microwave Breakdown at Atmospheric Pressure". *Physical Review Letters* **104**. doi:10.1103/PhysRevLett..

[34] Chin, S. L. (2006). "Some Fundamental Concepts of Femtosecond Laser Filamentation" (PDF). *Journal of the Korean Physical Society* **49**: 281.

[35] Talebpour, A.; Abdel-Fattah, M.; Chin, S. L. (2000). "Focusing limits of intense ultrafast laser pulses in a high pressure gas: Road to new spectroscopic source". *Optics Communications* **183** (5–6): 479. doi:10.1016/S0030-4018(00)00903-2.

[36] Alfvén, Hannes (1981). "section VI.13.1. Cellular Structure of Space". *Cosmic Plasma*. Dordrecht. ISBN 978-90-277-1151-9.

[37] National Research Council (U.S.). Plasma 2010 Committee (2007). *Plasma science: advancing knowledge in the national interest*. National Academies Press. pp. 190–193. ISBN 978-0-309-10943-7.

[38] Greaves, R. G.; Tinkle, M. D.; Surko, C. M. (1994). "Creation and uses of positron plasmas". *Physics of Plasmas* **1** (5): 1439. doi:10.1063/1.870693.

[39] Morfill, G. E.; Ivlev, Alexei V. (2009). "Complex plasmas: An interdisciplinary research field". *Review of Modern Physics* **81** (4): 1353–1404. Bibcode:2009RvMP...81.1353M. doi:10.1103/RevModPhys.81.1353.

[40] Alfvén, H.; Smårs, E. (1960). "Gas-Insulation of a Hot Plasma". *Nature* **188** (4753): 801–802. Bibcode:1960Natur.188..801A. doi:10.1038/188801a0.

[41] Braams, C.M. (1966). "Stability of Plasma Confined by a Cold-Gas Blanket". *Physical Review Letters* **17** (9): 470–471. Bibcode:1966PhRvL..17..470B. doi:10.1103/PhysRevLett.17.470.

[42] Yaghoubi, A.; Mélinon, P. (2013). "Tunable synthesis and in situ growth of silicon-carbon mesostructures using impermeable plasma". *Scientific Reports* **3**. Bibcode:2013NatSR...3E1083Y. doi:10.1038/srep01083. PMC 3547321. PMID 23330064.

[43] See Evolution of the Solar System, *1976*

[44] Hippler, R.; Kersten, H.; Schmidt, M.; Schoenbach, K.M., eds. (2008). "Plasma Sources". *Low Temperature Plasmas: Fundamentals, Technologies, and Techniques* (2nd ed.). Wiley-VCH. ISBN 978-3-527-40673-9.

[45] Chen, Francis F. (1984). *Plasma Physics and Controlled Fusion*. Plenum Press. ISBN 978-0-306-41332-2.

[46] Leal-Quirós, Edbertho (2004). "Plasma Processing of Municipal Solid Waste". *Brazilian Journal of Physics* **34** (4B): 1587. Bibcode:2004BrJPh..34.1587L. doi:10.1590/S0103-97332004000800015.

[47] Gomez, E.; Rani, D. A.; Cheeseman, C. R.; Deegan, D.; Wise, M.; Boccaccini, A. R. (2009). "Thermal plasma technology for the treatment of wastes: A critical review". *Journal of Hazardous Materials* **161** (2–3): 614–626. doi:10.1016/j.jhazmat.. PMID 18499345.

[48] National Research Council (1991). *Plasma Processing of Materials : Scientific Opportunities and Technological Challenges*. National Academies Press. ISBN 978-0-309-04597-1.

[49] Nemchinsky, V. A.; Severance, W. S. (2006). "What we know and what we do not know about plasma arc cutting". *Journal of Physics D: Applied Physics* **39** (22): R423. doi:10.1088/0022-3727/39/22/R01.

[50] Peretich, M.A.; O'Brien, W.F.; Schetz, J.A. (2007). "Plasma torch power control for scramjet application" (PDF). Virginia Space Grant Consortium. Retrieved 12 April 2010.

[51] Stern, David P. "The Fluorescent Lamp: A plasma you can use". Retrieved 2010-05-19.

[52] Sobolewski, M.A.; Langan & Felker, J.G. & B.S. (1997). "Electrical optimization of plasma-enhanced chemical vapor deposition chamber cleaning plasmas" (PDF) **16** (1). Journal of Vacuum Science and Technology B. pp. 173–182. Archived from the original (PDF) on January 18, 2009.

[53] Okumura, T. (2010). "Inductively Coupled Plasma Sources and Applications". *Physics Research International* **2010**: 1. doi:10.1155/2010/164249.

[54] *Plasma Chemistry*. Cambridge University Press. 2008. p. 229. ISBN 9781139471732.

[55] Leroux, F.; Perwuelz, A.; Campagne, C.; Behary, N. (2006). "Atmospheric air-plasma treatments of polyester textile structures". *Journal of Adhesion Science and Technology* **20** (9): 939. doi:10.1163/156856106777657788.

[56] Leroux, F. D. R.; Campagne, C.; Perwuelz, A.; Gengembre, L. O. (2008). "Polypropylene film chemical and physical modifications by dielectric barrier discharge plasma treatment at atmospheric pressure". *Journal of Colloid and Interface Science* **328** (2): 412–420. doi:10.1016/j.jcis.2008.09.062. PMID 18930244.

[57] Park, J.; Henins, I.; Herrmann, H. W.; Selwyn, G. S.; Hicks, R. F. (2001). "Discharge phenomena of an atmospheric pressure radio-frequency capacitive plasma source". *Journal of Applied Physics* **89**: 20. doi:10.1063/1.1323753.

[58] Crookes presented a lecture to the British Association for the Advancement of Science, in Sheffield, on Friday, 22 August 1879

[59] Announced in his evening lecture to the Royal Institution on Friday, 30 April 1897, and published in "J. J. Thomson (1856–1940)". *Philosophical Magazine* **44**: 293. 1897. doi:10.1080/14786449708621070.

[60] Langmuir, I. (1928). "Oscillations in Ionized Gases". *Proceedings of the National Academy of Sciences* **14** (8): 627. doi:

[61] Brown, Sanborn C. (1978). "Chapter 1: A Short History of Gaseous Electronics". In HIRSH, Merle N. e OSKAM, H. J. *Gaseous Electronics* **1**. Academic Press. ISBN 978-0-12-349701-7.

[62] Web site for Plasma science and technology

[63] "High-tech dentistry – "St Elmo's frier" – Using a plasma torch to clean your teeth". The Economist print edition. Jun 17, 2009. Retrieved 2009-09-07.

6.11   External links

- Free plasma physics books and notes

- Plasmas: the Fourth State of Matter

- Plasma Science and Technology

- Plasma on the Internet – a list of plasma related links.

- Introduction to Plasma Physics: Graduate course given by Richard Fitzpatrick|M.I.T. Introduction by I.H.Hutchinson

- Plasma Material Interaction

- How to make a glowing ball of plasma in your microwave with a grape|More (Video)

- How to make plasma in your microwave with only one match (video)

- OpenPIC3D – 3D Hybrid Particle-In-Cell simulation of plasma dynamics

- Plasma Formulary Interactive

Lightning is an example of plasma present at Earth's surface. Typically, lightning discharges 30,000 amperes at up to 100 million volts, and emits light, radio waves, X-rays and even gamma rays. Plasma temperatures in lightning can approach 28,000 K (28,000 °C;

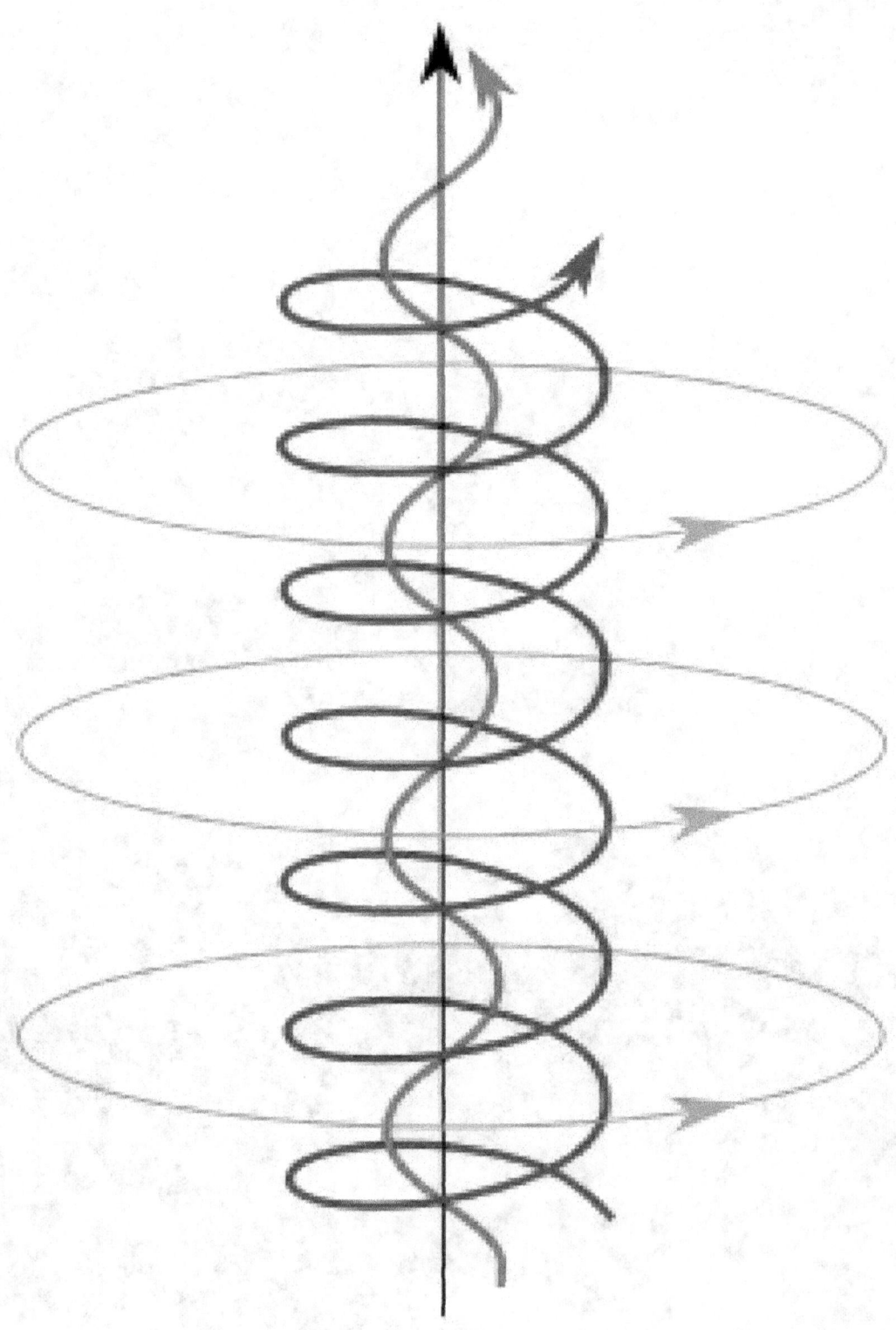

The complex self-constricting magnetic field lines and current paths in a field-aligned Birkeland current that can develop in a plasma.

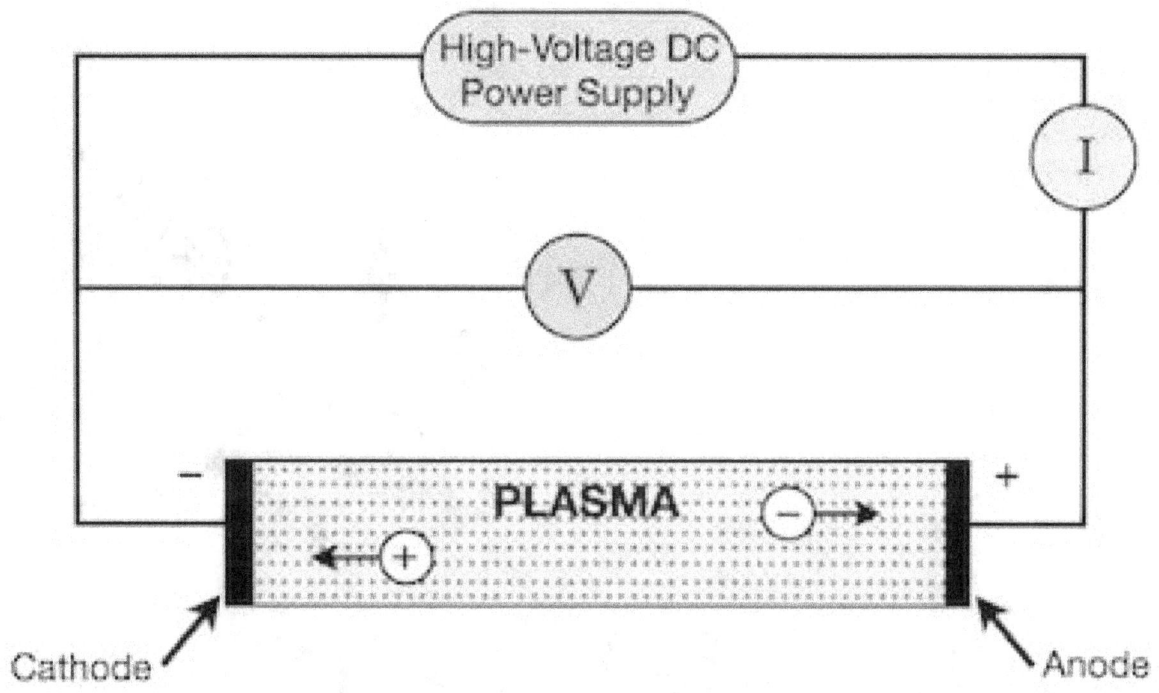

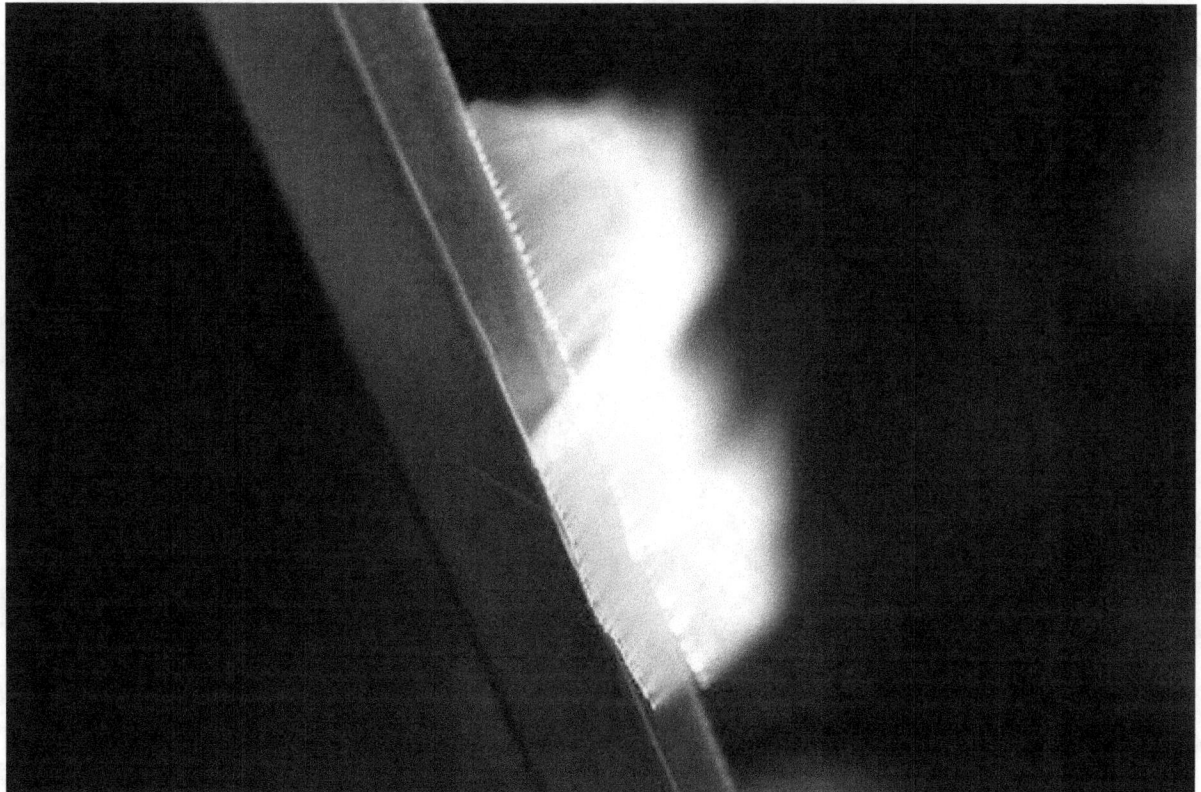

Artificial plasma produced in air by a Jacob's Ladder

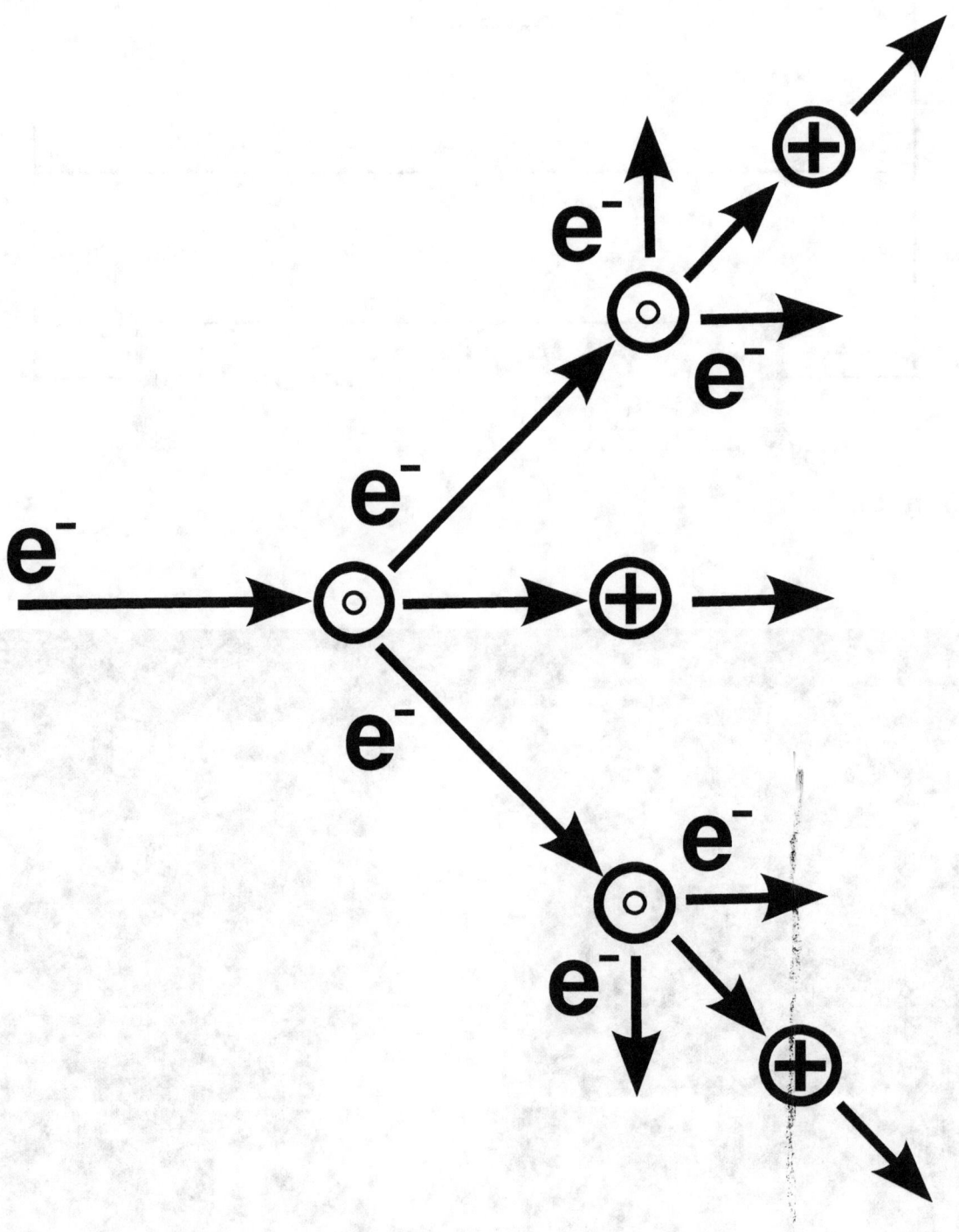

Cascade process of ionization. Electrons are 'e−', neutral atoms 'o', and cations '+'.

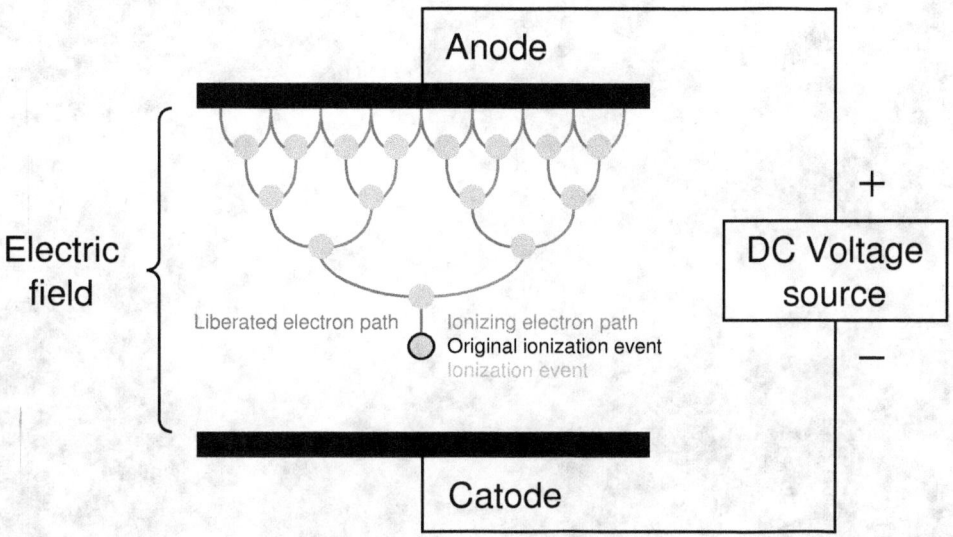

Avalanche effect between two electrodes. The original ionisation event liberates one electron, and each subsequent collision liberates a further electron, so two electrons emerge from each collision: the ionising electron and the liberated electron.

Hall effect thruster. The electric field in a plasma double layer is so effective at accelerating ions that electric fields are used in ion drives.

Chapter 7

Bose–Einstein condensate

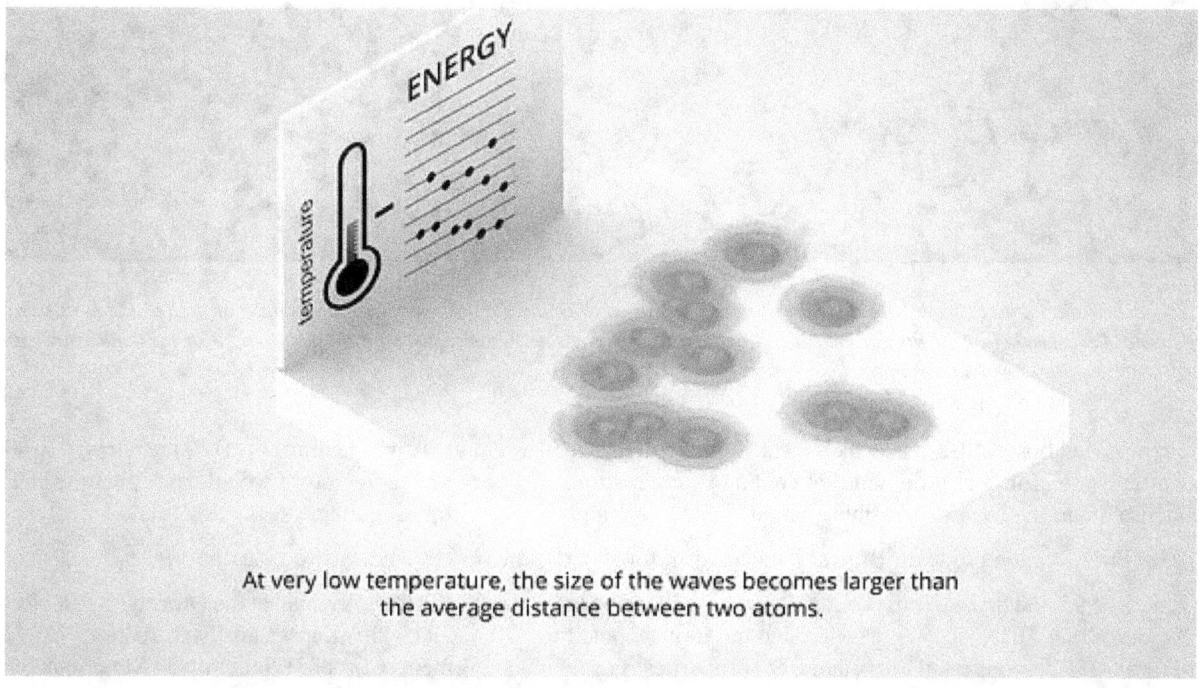

At very low temperature, the size of the waves becomes larger than the average distance between two atoms.

Schematic Bose-Einstein Condensation versus temperature and the energy diagram

A **Bose–Einstein condensate** (**BEC**) is a state of matter of a dilute gas of bosons cooled to temperatures very close to absolute zero (that is, very near 0 K or −273.15 °C). Under such conditions, a large fraction of bosons occupy the lowest quantum state, at which point macroscopic quantum phenomena become apparent.

This state was first predicted, generally, in 1924–25 by Satyendra Nath Bose and Albert Einstein.

7.1   History

Bose first sent a paper to Einstein on the quantum statistics of light quanta (now called photons). Einstein was impressed, translated the paper himself from English to German and submitted it for Bose to the *Zeitschrift für Physik*, which published it. (The Einstein manuscript, once believed to be lost, was found in a library at Leiden University in 2005.[1]). Einstein then extended Bose's ideas to matter in two other papers.[2] The result of their efforts is the concept of a Bose gas, governed by Bose–Einstein statistics, which describes the statistical distribution of identical particles with integer

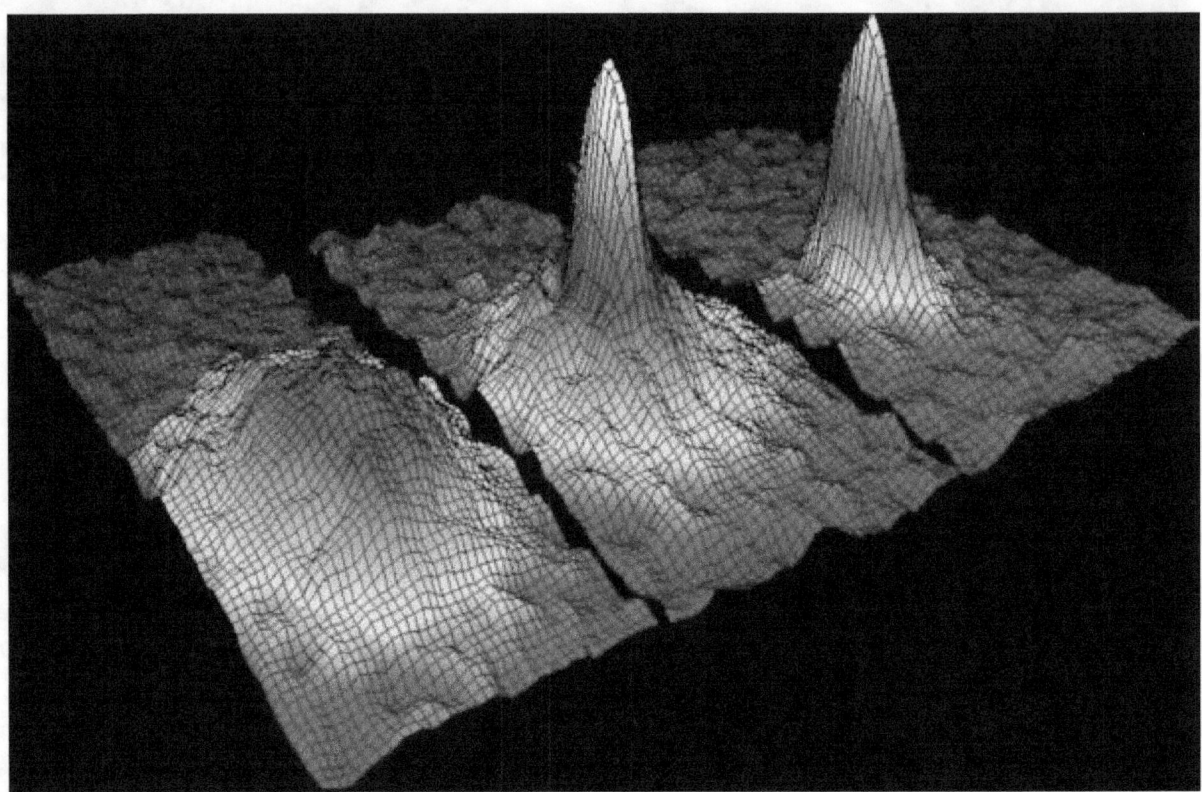

Velocity-distribution data (3 views) for a gas of rubidium atoms, confirming the discovery of a new phase of matter, the Bose–Einstein condensate. Left: just before the appearance of a Bose–Einstein condensate. Center: just after the appearance of the condensate. Right: after further evaporation, leaving a sample of nearly pure condensate.

spin, now called bosons. Bosons, which include the photon as well as atoms such as helium-4 (^{4}He), are allowed to share a quantum state. Einstein proposed that cooling bosonic atoms to a very low temperature would cause them to fall (or "condense") into the lowest accessible quantum state, resulting in a new form of matter.

In 1938, Fritz London proposed BEC as a mechanism for superfluidity in ^{4}He and superconductivity.[3][4]

On June 5, 1995, the first gaseous condensate was produced by Eric Cornell and Carl Wieman at the University of Colorado at Boulder NIST–JILA lab, in a gas of rubidium atoms cooled to 170 nanokelvin (nK).[5] Shortly thereafter, Wolfgang Ketterle at MIT demonstrated important BEC properties. For their achievements Cornell, Wieman, and Ketterle received the 2001 Nobel Prize in Physics.[6]

Many isotopes were soon condensed, then molecules, quasi-particles, and photons in 2010.[7]

7.2   Critical temperature

This transition to BEC occurs below a critical temperature, which for a uniform three-dimensional gas consisting of non-interacting particles with no apparent internal degrees of freedom is given by:

$$T_c = \left(\frac{n}{\zeta(3/2)} \right)^{2/3} \frac{2\pi\hbar^2}{mk_B} \approx 3.3125 \, \frac{\hbar^2 n^{2/3}}{mk_B}$$

where:

Interactions shift the value and the corrections can be calculated by mean-field theory.

7.3 Models

7.3.1 Einstein's non-interacting gas

Consider a collection of N noninteracting particles, which can each be in one of two quantum states, $|0\rangle$ and $|1\rangle$. If the two states are equal in energy, each different configuration is equally likely.

If we can tell which particle is which, there are 2^N different configurations, since each particle can be in $|0\rangle$ or $|1\rangle$ independently. In almost all of the configurations, about half the particles are in $|0\rangle$ and the other half in $|1\rangle$. The balance is a statistical effect: the number of configurations is largest when the particles are divided equally.

If the particles are indistinguishable, however, there are only $N+1$ different configurations. If there are K particles in state $|1\rangle$, there are $N - K$ particles in state $|0\rangle$. Whether any particular particle is in state $|0\rangle$ or in state $|1\rangle$ cannot be determined, so each value of K determines a unique quantum state for the whole system.

Suppose now that the energy of state $|1\rangle$ is slightly greater than the energy of state $|0\rangle$ by an amount E. At temperature T, a particle will have a lesser probability to be in state $|1\rangle$ by $e^{-E/kT}$. In the distinguishable case, the particle distribution will be biased slightly towards state $|0\rangle$. But in the indistinguishable case, since there is no statistical pressure toward equal numbers, the most-likely outcome is that most of the particles will collapse into state $|0\rangle$.

In the distinguishable case, for large N, the fraction in state $|0\rangle$ can be computed. It is the same as flipping a coin with probability proportional to $p = \exp(-E/T)$ to land tails.

In the indistinguishable case, each value of K is a single state, which has its own separate Boltzmann probability. So the probability distribution is exponential:

$$P(K) = Ce^{-KE/T} = Cp^K.$$

For large N, the normalization constant C is $(1 - p)$. The expected total number of particles not in the lowest energy state, in the limit that $N{\to}\infty$, is equal to $\sum_{n>0} Cnp^n = p/(1-p)$. It does not grow when N is large; it just approaches a constant. This will be a negligible fraction of the total number of particles. So a collection of enough Bose particles in thermal equilibrium will mostly be in the ground state, with only a few in any excited state, no matter how small the energy difference.

Consider now a gas of particles, which can be in different momentum states labeled $|k\rangle$. If the number of particles is less than the number of thermally accessible states, for high temperatures and low densities, the particles will all be in different states. In this limit, the gas is classical. As the density increases or the temperature decreases, the number of accessible states per particle becomes smaller, and at some point, more particles will be forced into a single state than the maximum allowed for that state by statistical weighting. From this point on, any extra particle added will go into the ground state.

To calculate the transition temperature at any density, integrate, over all momentum states, the expression for maximum number of excited particles, $p/(1 - p)$:

$$N = V \int \frac{d^3k}{(2\pi)^3} \frac{p(k)}{1 - p(k)} = V \int \frac{d^3k}{(2\pi)^3} \frac{1}{e^{\frac{k^2}{2mT}} - 1}$$

$$p(k) = e^{\frac{-k^2}{2mT}}.$$

When the integral is evaluated with factors of kB and $\hbar$ restored by dimensional analysis, it gives the critical temperature formula of the preceding section. Therefore, this integral defines the critical temperature and particle number corresponding to the conditions of negligible chemical potential. In Bose–Einstein statistics distribution, μ is actually still nonzero for BECs; however, μ is less than the ground state energy. Except when specifically talking about the ground state, μ can be approximated for most energy or momentum states as $\mu \approx 0$.

7.3.2   Bogoliubov theory for weakly interacting gas

Bogoliubov considered perturbations on the limit of dilute gas,[9] finding a finite pressure at zero temperature and positive chemical potential. This leads to corrections for the ground state. The Bogoliubov state has pressure(T=0): $P = g/2n^2$.

The original interacting system can be converted to a system of non-interacting particles with a dispersion law.

7.3.3   Gross–Pitaevskii equation

Main article: Gross–Pitaevskii equation

In some simplest cases, the state of condensed particles can be described with a nonlinear Schrödinger equation, also known as Gross-Pitaevskii or Ginzburg-Landau equation. The validity of this approach is actually limited to the case of ultracold temperatures, which fits well for the most alkali atoms experiments.

This approach originates from the assumption that the state of the BEC can be described by the unique wavefunction of the condensate $\psi(\vec{r})$. For a system of this nature, $|\psi(\vec{r})|^2$ is interpreted as the particle density, so the total number of atoms is $N = \int d\vec{r} |\psi(\vec{r})|^2$

Provided essentially all atoms are in the condensate (that is, have condensed to the ground state), and treating the bosons using mean field theory, the energy (E) associated with the state $\psi(\vec{r})$ is:

$$E = \int d\vec{r} \left[\frac{\hbar^2}{2m} |\nabla \psi(\vec{r})|^2 + V(\vec{r})|\psi(\vec{r})|^2 + \frac{1}{2} U_0 |\psi(\vec{r})|^4 \right]$$

Minimizing this energy with respect to infinitesimal variations in $\psi(\vec{r})$, and holding the number of atoms constant, yields the Gross–Pitaevski equation (GPE) (also a non-linear Schrödinger equation):

$$i\hbar \frac{\partial \psi(\vec{r})}{\partial t} = \left(-\frac{\hbar^2 \nabla^2}{2m} + V(\vec{r}) + U_0 |\psi(\vec{r})|^2 \right) \psi(\vec{r})$$

where:

In the case of zero external potential, the dispersion law of interacting Bose-Einstein-condensed particles is given by so-called Bogoliubov spectrum (for $T = 0$):

$$\omega_p = \sqrt{\frac{p^2}{2m} \left(\frac{p^2}{2m} + 2U_0 n_0 \right)}$$

The Gross-Pitaevskii equation (GPE) provides a relatively good description of the behavior of atomic BEC's. However, GPE does not take into account the temperature dependence of dynamical variables, and is therefore valid only for $T = 0$. It is not applicable, for example, for the condensates of excitons, magnons and photons, where the critical temperature is up to room one.

Weaknesses of Gross–Pitaevskii model

The Gross–Pitaevskii model of BEC is a physical approximation valid for certain classes of BECs. By construction, the GPE uses the following simplifications: it assumes that interactions between condensate particles are of the contact two-body type and also neglects anomalous contributions to self-energy.[10] These assumptions are suitable mostly for the dilute three-dimensional condensates. If one relaxes any of these assumptions, the equation for the condensate wavefunction acquires the terms containing higher-order powers of the wavefunction. Moreover, for some physical systems the amount of such terms turns out to be infinite, therefore, the equation becomes essentially non-polynomial. The examples where this could happen are the Bose–Fermi composite condensates,[11][12][13][14] effectively lower-dimensional condensates,[15] and dense condensates and superfluid clusters and droplets.[16]

7.3.4 Other

However, it is clear that in a general case the behaviour of Bose–Einstein condensate can be described by coupled evolution equations for condensate density, superfluid velocity and distribution function of elementary excitations. This problem was in 1977 by Peletminskii et al. in microscopical approach. The Peletminskii equations are valid for any finite temperatures below the critical point. Years after, in 1985, Kirkpatrick and Dorfman obtained similar equations using another microscopical approach. The Peletminskii equations also reproduce Khalatnikov hydrodynamical equations for superfluid as a limiting case.

7.3.5 Superfluidity of BEC and Landau criterion

The phenomena of superfluidity of a Bose gas and superconductivity of a strongly-correlated Fermi gas (a gas of Cooper pairs) are tightly connected to Bose-Einstein condensation. Under corresponding conditions, below the temperature of phase transition, these phenomena were observed in helium-4 and different classes of superconductors. In this sense, the superconductivity is often called the superfluidity of Fermi gas. In the simplest form, the origin of superfluidity can be seen from the weakly interacting bosons model.

7.4 Experimental observation

7.4.1 Superfluid He-4

In 1938, Pyotr Kapitsa, John Allen and Don Misener discovered that helium-4 became a new kind of fluid, now known as a superfluid, at temperatures less than 2.17 K (the lambda point). Superfluid helium has many unusual properties, including zero viscosity (the ability to flow without dissipating energy) and the existence of quantized vortices. It was quickly believed that the superfluidity was due to partial Bose–Einstein condensation of the liquid. In fact, many properties of superfluid helium also appear in gaseous condensates created by Cornell, Wieman and Ketterle (see below). Superfluid helium-4 is a liquid rather than a gas, which means that the interactions between the atoms are relatively strong; the original theory of Bose–Einstein condensation must be heavily modified in order to describe it. Bose–Einstein condensation remains, however, fundamental to the superfluid properties of helium-4. Note that helium-3, a fermion, also enters a superfluid phase at low temperature, which can be explained by the formation of bosonic Cooper pairs of two atoms (see also fermionic condensate).

7.4.2 Gaseous

The first "pure" Bose–Einstein condensate was created by Eric Cornell, Carl Wieman, and co-workers at JILA on 5 June 1995. They cooled a dilute vapor of approximately two thousand rubidium-87 atoms to below 170 nK using a combination of laser cooling (a technique that won its inventors Steven Chu, Claude Cohen-Tannoudji, and William D. Phillips the 1997 Nobel Prize in Physics) and magnetic evaporative cooling. About four months later, an independent effort led by Wolfgang Ketterle at MIT condensed sodium-23. Ketterle's condensate had a hundred times more atoms, allowing important results such as the observation of quantum mechanical interference between two different condensates. Cornell, Wieman and Ketterle won the 2001 Nobel Prize in Physics for their achievements.[17]

A group led by Randall Hulet at Rice University announced a condensate of lithium atoms only one month following the JILA work.[18] Lithium has attractive interactions, causing the condensate to be unstable and collapse for all but a few atoms. Hulet's team subsequently showed the condensate could be stabilized by confinement quantum pressure for up to about 1000 atoms. Various isotopes have since been condensed.

Velocity-distribution data graph

In the image accompanying this article, the velocity-distribution data indicates the formation of a Bose–Einstein condensate out of a gas of rubidium atoms. The false colors indicate the number of atoms at each velocity, with red being the fewest and white being the most. The areas appearing white and light blue are at the lowest velocities. The peak is not infinitely narrow because of the Heisenberg uncertainty principle: spatially confined atoms have a minimum width velocity distribution. This width is given by the curvature of the magnetic potential in the given direction. More tightly confined directions have bigger widths in the ballistic velocity distribution. This anisotropy of the peak on the right is a purely quantum-mechanical effect and does not exist in the thermal distribution on the left. This graph served as the cover design for the 1999 textbook *Thermal Physics* by Ralph Baierlein.[19]

7.4.3   Quasiparticles

Main article: Bose-Einstein condensation of quasiparticles

Bose–Einstein condensation also applies to quasiparticles in solids. Magnons, Excitons, and Polaritons have integer spin and form condensates.

Magnons, electron spin waves, can be controlled by a magnetic field. Densities from the limit of a dilute gas to a strongly interacting Bose liquid are possible. Magnetic ordering is the analog of superfluidity. In 1999 condensation was demonstrated in antiferromagnetic $TlCuCl_3$,[20] at temperatures as large as 14 K. The high transition temperature (relative to atomic gases) is due to the magnons small mass (near an electron) and greater achievable density. In 2006, condensation in a ferromagnetic Yttrium-iron-garnet thin film was seen even at room temperature,[21][22] with optical pumping.

Excitons, electron-hole pairs, were predicted to condense at low temperature and high density by Boer et al. in 1961. Bilayer system experiments first demonstrated condensation in 2003, by Hall voltage disappearance. Fast optical exciton creation was used to form condensates in sub-Kelvin Cu_2O in 2005 on.

Polariton condensation was detected in a 5 K quantum well microcavity.

7.5   Peculiar properties

7.5.1   Vortices

As in many other systems, vortices can exist in BECs. These can be created, for example, by 'stirring' the condensate with lasers, or rotating the confining trap. The vortex created will be a quantum vortex. These phenomena are allowed for by the non-linear $|\psi(\vec{r})|^2$ term in the GPE. As the vortices must have quantized angular momentum the wavefunction may have the form $\psi(\vec{r}) = \phi(\rho, z)e^{i\ell\theta}$ where ρ, z and θ are as in the cylindrical coordinate system, and ℓ is the angular number. This is particularly likely for an axially symmetric (for instance, harmonic) confining potential, which is commonly used. The notion is easily generalized. To determine $\phi(\rho, z)$, the energy of $\psi(\vec{r})$ must be minimized, according to the constraint $\psi(\vec{r}) = \phi(\rho, z)e^{i\ell\theta}$. This is usually done computationally, however in a uniform medium the analytic form

$$\phi = \frac{nx}{\sqrt{2 + x^2}}$$

demonstrates the correct behavior, and is a good approximation.

A singly charged vortex ($\ell = 1$) is in the ground state, with its energy ϵ_v given by

$$\epsilon_v = \pi n \frac{\hbar^2}{m} \ln\left(1.464\frac{b}{\xi}\right)$$

where b is the farthest distance from the vortex considered.(To obtain an energy which is well defined it is necessary to include this boundary b.)

For multiply charged vortices ($\ell > 1$) the energy is approximated by

$$\epsilon_v \approx \ell^2 \pi n \frac{\hbar^2}{m} \ln\left(\frac{b}{\xi}\right)$$

which is greater than that of ℓ singly charged vortices, indicating that these multiply charged vortices are unstable to decay. Research has, however, indicated they are metastable states, so may have relatively long lifetimes.

Closely related to the creation of vortices in BECs is the generation of so-called dark solitons in one-dimensional BECs. These topological objects feature a phase gradient across their nodal plane, which stabilizes their shape even in propagation and interaction. Although solitons carry no charge and are thus prone to decay, relatively long-lived dark solitons have been produced and studied extensively.[23]

7.5.2 Attractive interactions

Experiments led by Randall Hulet at Rice University from 1995 through 2000 showed that lithium condensates with attractive interactions could stably exist up to a critical atom number. Quench cooling the gas, they observed the condensate to grow, then subsequently collapse as the attraction overwhelmed the zero-point energy of the confining potential, in a burst reminiscent of a supernova, with an explosion preceded by an implosion.

Further work on attractive condensates was performed in 2000 by the JILA team, of Cornell, Wieman and coworkers. Their instrumentation now had better control so they used naturally *attracting* atoms of rubidium-85 (having negative atom–atom scattering length). Through Feshbach resonance involving a sweep of the magnetic field causing spin flip collisions, they lowered the characteristic, discrete energies at which rubidium bonds, making their Rb-85 atoms repulsive and creating a stable condensate. The reversible flip from attraction to repulsion stems from quantum interference among wave-like condensate atoms.

When the JILA team raised the magnetic field strength further, the condensate suddenly reverted to attraction, imploded and shrank beyond detection, then exploded, expelling about two-thirds of its 10,000 atoms. About half of the atoms in the condensate seemed to have disappeared from the experiment altogether, not seen in the cold remnant or expanding gas cloud.[17] Carl Wieman explained that under current atomic theory this characteristic of Bose–Einstein condensate could not be explained because the energy state of an atom near absolute zero should not be enough to cause an implosion; however, subsequent mean field theories have been proposed to explain it. Most likely they formed molecules of two rubidium atoms,[24] energy gained by this bond imparts velocity sufficient to leave the trap without being detected.

7.6 Current research

Compared to more commonly encountered states of matter, Bose–Einstein condensates are extremely fragile. The slightest interaction with the external environment can be enough to warm them past the condensation threshold, eliminating their interesting properties and forming a normal gas.

Nevertheless, they have proven useful in exploring a wide range of questions in fundamental physics, and the years since the initial discoveries by the JILA and MIT groups have seen an increase in experimental and theoretical activity. Examples include experiments that have demonstrated interference between condensates due to wave–particle duality,[25] the study of superfluidity and quantized vortices, the creation of bright matter wave solitons from Bose condensates confined to one dimension, and the slowing of light pulses to very low speeds using electromagnetically induced transparency.[26] Vortices in Bose–Einstein condensates are also currently the subject of analogue gravity research, studying the possibility of modeling black holes and their related phenomena in such environments in the laboratory. Experimenters have also realized "optical lattices", where the interference pattern from overlapping lasers provides a periodic potential. These have been used to explore the transition between a superfluid and a Mott insulator,[27] and may be useful in studying Bose–Einstein condensation in fewer than three dimensions, for example the Tonks–Girardeau gas.

Bose–Einstein condensates composed of a wide range of isotopes have been produced.[28]

Cooling fermions to extremely low temperatures has created degenerate gases, subject to the Pauli exclusion principle. To exhibit Bose–Einstein condensation, the fermions must "pair up" to form bosonic compound particles (e.g. molecules or Cooper pairs). The first molecular condensates were created in November 2003 by the groups of Rudolf Grimm at the University of Innsbruck, Deborah S. Jin at the University of Colorado at Boulder and Wolfgang Ketterle at MIT. Jin quickly went on to create the first fermionic condensate composed of Cooper pairs.[29]

In 1999, Danish physicist Lene Hau led a team from Harvard University which slowed a beam of light to about 17 meters per second., using a superfluid.[30] Hau and her associates have since made a group of condensate atoms recoil from a light pulse such that they recorded the light's phase and amplitude, recovered by a second nearby condensate, in what they term "slow-light-mediated atomic matter-wave amplification" using Bose–Einstein condensates: details are discussed in *Nature*.[31]

Researchers in the new field of atomtronics use the properties of Bose–Einstein condensates when manipulating groups of identical cold atoms using lasers.[32] Further, BECs have been proposed by Emmanuel David Tannenbaum for anti-stealth technology.[33]

7.6.1   Isotopes

The effect has mainly been observed on alkaline atoms which have nuclear properties particularly suitable for working with traps. As of 2012, using ultra-low temperatures of 10^{-7} K or below, Bose–Einstein condensates had been obtained for a multitude of isotopes, mainly of alkaline, alkaline earth, and lanthanoid atoms (^{7}Li, ^{23}Na, ^{39}K, ^{41}K, ^{85}Rb, ^{87}Rb, ^{133}Cs, ^{52}Cr, ^{40}Ca, ^{84}Sr, ^{86}Sr, ^{88}Sr, ^{174}Yb, ^{164}Dy, and ^{168}Er). Research was finally successful in hydrogen with aid of special methods. In contrast, the superfluid state of ^{4}He below 2.17 K is not a good example, because the interaction between the atoms is too strong. Only 8% of atoms are in the ground state near absolute zero, rather than the 100% of a true condensate.

The bosonic behavior of some of these alkaline gases appears odd at first sight, because their nuclei have half-integer total spin. It arises from a subtle interplay of electronic and nuclear spins: at ultra-low temperatures and corresponding excitation energies, the half-integer total spin of the electronic shell and half-integer total spin of the nucleus are coupled by a very weak hyperfine interaction. The total spin of the atom, arising from this coupling, is an integer lower value. The chemistry of systems at room temperature is determined by the electronic properties, which is essentially fermionic, since room temperature thermal excitations have typical energies much higher than the hyperfine values.

7.7   See also

- Atom laser

- Atomic coherence

- Bose–Einstein correlations

- Bose–Einstein condensation: a network theory approach

- Bose-Einstein condensation of excitons

- Cold Atom Laboratory

- Electromagnetically induced transparency

- Fermionic condensate

- Gas in a box

- Gross–Pitaevskii equation

- Macroscopic quantum phenomena

- Macroscopic quantum self-trapping

- Slow light

- Superconductivity

- Superfluid film

- Superfluid helium-4

- Supersolid

- Tachyon condensation

- Timeline of low-temperature technology

- Super-heavy atom

- Wiener sausage

7.8 References

[1] "Leiden University Einstein archive". Lorentz.leidenuniv.nl. 27 October 1920. Retrieved 23 March 2011.

[2] Clark, Ronald W. (1971). *Einstein: The Life and Times*. Avon Books. pp. 408–409. ISBN 0-380-01159-X.

[3] London, F. (1938). "The λ-Phenomenon of Liquid Helium and the Bose–Einstein Degeneracy". *Nature* **141** (3571): 643–644. Bibcode:1938Natur.141..643L. doi:10.1038/141643a0.

[4] London, F. *Superfluids* Vol.I and II, (reprinted New York: Dover 1964)

[5] http://www.nist.gov/public_affairs/releases/bec_background.cfm

[6] Levi, Barbara Goss (2001). "Cornell, Ketterle, and Wieman Share Nobel Prize for Bose–Einstein Condensates". *Search & Discovery*. Physics Today online. Archived from the original on 24 October 2007. Retrieved 26 January 2008.

[7] Klaers, Jan; Schmitt, Julian; Vewinger, Frank; Weitz, Martin (2010). "Bose–Einstein condensation of photons in an optical microcavity". *Nature* **468** (7323): 545–548. arXiv:1007.4088. Bibcode:2010Natur.468..545K. doi:10.1038/nature09567. PMID 21107426.

[8] (sequence A078434 in OEIS)

[9] N. N. Bogoliubov (1947). "On the theory of superfluidity.". *J. Phys. (USSR), 11:23*.

[10] Beliaev, S. T. Zh. Eksp. Teor. Fiz. 34, 418–432 (1958); ibid. 433–446 [Soviet Phys. JETP 3, 299 (1957)].

[11] Schick, M. (1971). "Two-Dimensional System of Hard-Core Bosons".*Physical Review A***3**(3): 1067.Bibcode:1971PhRvA. doi:10.1103/PhysRevA.3.1067.

[12] Kolomeisky, E.; Straley, J. (1992). "Renormalization-group analysis of the ground-state properties of dilute Bose systems in d spatial dimensions". *Physical Review B* **46** (18): 11749. Bibcode:1992PhRvB..4611749K. doi:10.1103/PhysRevB.46.11749.

[13] Kolomeisky, E. B.; Newman, T. J.; Straley, J. P.; Qi, X. (2000). "Low-Dimensional Bose Liquids: Beyond the Gross-Pitaevskii Approximation". *Physical Review Letters* **85** (6): 1146–1149. arXiv:cond-mat/0002282. Bibcode:2000PhRvL..85.1146K. doi:10.1103/PhysRevLett.85.1146. PMID 10991498.

[14] Chui, S.; Ryzhov, V. (2004). "Collapse transition in mixtures of bosons and fermions".*Physical Review A***69**(4).Bibcode. doi:10.1103/PhysRevA.69.043607.

[15] Salasnich, L.; Parola, A.; Reatto, L. (2002). "Effective wave equations for the dynamics of cigar-shaped and disk-shaped Bose condensates". *Phys. Rev. A* **65** (4): 043614. arXiv:cond-mat/0201395. Bibcode:2002PhRvA..65d3614S. doi:10.1103/PhysRevA.65.0436

[16] Avdeenkov, A. V.; Zloshchastiev, K. G. (2011). "Quantum Bose liquids with logarithmic nonlinearity: Self-sustainability and emergence of spatial extent". *J. Phys. B: At. Mol. Opt. Phys.* **44** (19): 195303. arXiv:1108.0847. Bibcode:2011JPhB...44s5303A. doi:10.1088/0953-4075/44/19/195303.

[17] "Eric A. Cornell and Carl E. Wieman — Nobel Lecture" (PDF). nobelprize.org.

[18] Bradley, C. C.; Sackett, C. A.; Tollett, J. J.; Hulet, R. G. (1995). "Evidence of Bose-Einstein Condensation in an Atomic Gas with Attractive Interactions" (PDF). *Physical review letters* **75** (9): 1687–1690. doi:10.1103/PhysRevLett.75.1687. PMID 10060366.

[19] Baierlein, Ralph (1999). *Thermal Physics*. Cambridge University Press. ISBN 0-521-65838-1.

[20] Nikuni, T.; Oshikawa, M.; Oosawa, A.; Tanaka, H. (1999). "Bose–Einstein Condensation of Dilute Magnons in $TlCuCl_3$". *Physical Review Letters* **84** (25): 5868–71. arXiv:cond-mat/9908118. Bibcode:2000PhRvL..84.5868N. doi:10.1103/. PMID 10991075.

[21] Demokritov, S.O.; Demidov, VE; Dzyapko, O; Melkov, GA; Serga, AA; Hillebrands, B; Slavin, AN (2006). "Bose–Einstein condensation of quasi-equilibrium magnons at room temperature under pumping". *Nature* **443** (5117. PMID 17006509.

[22] *Magnon Bose Einstein Condensation* made simple. Website of the "Westfählische Wilhelms Universität Münster" Prof.Demokritov. Retrieved 25 June 2012.

[23] Becker, Christoph; Stellmer, Simon; Soltan-Panahi, Parvis; Dörscher, Sören; Baumert, Mathis; Richter, Eva-Maria; Kronjäger, Jochen; Bongs, Kai; Sengstock, Klaus (2008). "Oscillations and interactions of dark and dark–bright solitons in Bose–Einstein condensates". *Nature Physics* **4** (6): 496–501. arXiv:0804.0544. Bibcode:2008NatPh...4..496B. doi:10.1038/nphys962.

[24] van Putten, M.H.P.M. (2010). "Pair condensates produced in bosenovae". *Physics Letters A* **374** (33): 3346. Bibcode:2010. doi:10.1016/j.physleta.2010.06.020.

[25] Gorlitz, Axel. "Interference of Condensates (BEC@MIT)". Cua.mit.edu. Retrieved 13 October 2009.

[26] Dutton, Zachary; Ginsberg, Naomi S.; Slowe, Christopher and Hau, Lene Vestergaard (2004). "The art of taming light: ultra-slow and stopped light" (PDF). *Europhysics News* **35** (2): 33. Bibcode:2004ENews..35...33D. doi:10.1051/epn:2004201.

[27] "From Superfluid to Insulator: Bose–Einstein Condensate Undergoes a Quantum Phase Transition". Qpt.physics.harvard.edu. Retrieved 13 October 2009.

[28] "Ten of the best for BEC". Physicsweb.org. 1 June 2005.

[29] "Fermionic condensate makes its debut". Physicsweb.org. 28 January 2004.

[30] Cromie, William J. (18 February 1999). "Physicists Slow Speed of Light". The Harvard University Gazette. Retrieved 26 January 2008.

[31] Ginsberg, N. S.; Garner, S. R.; Hau, L. V. (2007). "Coherent control of optical information with matter wave dynamics". *Nature* **445** (7128): 623–626. doi:10.1038/nature05493. PMID 17287804.

[32] Weiss, P. (12 February 2000). "Atomtronics may be the new electronics". *Science News Online* **157** (7): 104. doi:10.2307/4012185. Retrieved 12 February 2011.

[33] Tannenbaum, Emmanuel David (1970). "Gravimetric Radar: Gravity-based detection of a point-mass moving in a static background". arXiv:1208.2377 [physics.ins-det].

7.9 Further reading

- Bose, S. N. (1924). "Plancks Gesetz und Lichtquantenhypothese". *Zeitschrift für Physik* **26**: 178. Bibcode: doi:10.1007/BF01327326.

- Einstein, A. (1925). "Quantentheorie des einatomigen idealen Gases". *Sitzungsberichte der Preussischen Akademie der Wissenschaften* **1**: 3.,

- Landau, L. D. (1941). "The theory of Superfluity of Helium 111". *J. Phys. USSR* **5**: 71–90.

- L. Landau(1941). "Theory of the Superfluidity of Helium II". *Physical Review* **60**(4): 356–358.Bibcode:1941P. doi:10.1103/PhysRev.60.356.

- M.H. Anderson, J.R. Ensher, M.R. Matthews, C.E. Wieman, and E.A. Cornell (1995). "Observation of Bose–Einstein Condensation in a Dilute Atomic Vapor". *Science* **269** (5221): 198–201. Bibcode:1995Sci...269..198A. doi:10.1126/science.269.5221.198. JSTOR 2888436. PMID 17789847.

- C. Barcelo, S. Liberati and M. Visser (2001). "Analogue gravity from Bose–Einstein condensates". *Classical and Quantum Gravity* **18** (6): 1137–1156. arXiv:gr-qc/0011026. Bibcode:2001CQGra..18.1137B. doi:10.1088/0264-9381/18/6/312.

- P.G. Kevrekidis, R. Carretero-Gonzlaez, D.J. Frantzeskakis and I.G. Kevrekidis (2006). "Vortices in Bose–Einstein Condensates: Some Recent Developments". *Modern Physics Letters B* **5** (33).

- K.B. Davis, M.-O. Mewes, M.R. Andrews, N.J. van Druten, D.S. Durfee, D.M. Kurn, and W. Ketterle (1995). "Bose–Einstein condensation in a gas of sodium atoms". *Physical Review Letters* **75**(22): 3969–397D.doi: 10.1103/PhysRevLett.75.3969. PMID 10059782..

- D. S. Jin, J. R. Ensher, M. R. Matthews, C. E. Wieman, and E. A. Cornell (1996). "Collective Excitations of a Bose–Einstein Condensate in a Dilute Gas". *Physical Review Letters* **77** (3): 420–423. Bibcode:1996PhRvL..77..420J. doi:10.1103/PhysRevLett.77.420. PMID 10062808.

- M. R. Andrews, C. G. Townsend, H.-J. Miesner, D. S. Durfee, D. M. Kurn, and W. Ketterle (1997). "Observation of interference between two Bose condensates". *Science* **275** (5300): 637–641. doi:10.1126/science.275.5300.637. PMID 9005843..

- Eric A. Cornell and Carl E. Wieman (1998). "The Bose–Einstein Condensate". *Scientific American* **278** (3): 40–45. doi:10.1038/scientificamerican0398-40.

- M. R. Matthews, B. P. Anderson, P. C. Haljan, D. S. Hall, C. E. Wieman, and E. A. Cornell (1999). "Vortices in a Bose–Einstein Condensate". *Physical Review Letters* **83** (13): 2498–2501. arXiv:cond-mat/9908209. Bibcode:1999PhRvL..83.2498M. doi:10.1103/PhysRevLett.83.2498.

- E.A. Donley, N.R. Claussen, S.L. Cornish, J.L. Roberts, E.A. Cornell, and C.E. Wieman (2001). "Dynamics of collapsing and exploding Bose–Einstein condensates". *Nature* **412** (6844): 295–299. arXiv:cond-mat/0105019. Bibcode:2001Natur.412..295D. doi:10.1038/35085500. PMID 11460153.

- A. G. Truscott, K. E. Strecker, W. I. McAlexander, G. B. Partridge, and R. G. Hulet (2001). "Observation of Fermi Pressure in a Gas of Trapped Atoms". *Science* **291** (5513): 2570–2572. Bibcode:2001Sci...291.2570T. doi:10.1126/science.1059318. PMID 11283362.

- M. Greiner, O. Mandel, T. Esslinger, T. W. Hänsch, I. Bloch (2002). "Quantum phase transition from a super-fluid to a Mott insulator in a gas of ultracold atoms". *Nature* **415** (6867): 39–44. Bibcode:2002Natur.415...39G. doi:10.1038/415039a. PMID 11780110..

- S. Jochim, M. Bartenstein, A. Altmeyer, G. Hendl, S. Riedl, C. Chin, J. Hecker Denschlag, and R. Grimm (2003). "Bose–Einstein Condensation of Molecules". *Science* **302** (5653): 2101–2103. Bibcode:2003Sci...302.2101J. doi:10.1126/science.1093280. PMID 14615548.

- Markus Greiner, Cindy A. Regal and Deborah S. Jin (2003). "Emergence of a molecular Bose–Einstein condensate from a Fermi gas". *Nature* **426** (6966): 537–540. Bibcode:2003Natur.426..537G. doi:10.1038/nature02199. PMID 14647340.

- M. W. Zwierlein, C. A. Stan, C. H. Schunck, S. M. F. Raupach, S. Gupta, Z. Hadzibabic, and W. Ketterle (2003). "Observation of Bose–Einstein Condensation of Molecules". *Physical Review Letters* **91** (25): 250401. arXiv:cond-mat/0311617. Bibcode:2003PhRvL..91y0401Z. doi:10.1103/PhysRevLett.91.250401. PMID 14754098.

- C. A. Regal, M. Greiner, and D. S. Jin (2004). "Observation of Resonance Condensation of Fermionic Atom Pairs". *Physical Review Letters* **92** (4): 040403. arXiv:cond-mat/0401554. Bibcode:2004PhRvL..92d0403R. doi:10.1103/PhysRevLett.92.040403. PMID 14995356.

- C. J. Pethick and H. Smith, *Bose–Einstein Condensation in Dilute Gases*, Cambridge University Press, Cambridge, 2001.

- Lev P. Pitaevskii and S. Stringari, *Bose–Einstein Condensation*, Clarendon Press, Oxford, 2003.

- Mackie M, Suominen KA, Javanainen J., "Mean-field theory of Feshbach-resonant interactions in 85Rb condensates." Phys Rev Lett. 2002 Oct 28;89(18):180403.

7.10   External links

- Bose–Einstein Condensation 2009 Conference Bose–Einstein Condensation 2009 – Frontiers in Quantum Gases

- BEC Homepage General introduction to Bose–Einstein condensation

- Nobel Prize in Physics 2001 – for the achievement of Bose–Einstein condensation in dilute gases of alkali atoms, and for early fundamental studies of the properties of the condensates

- Physics Today: Cornell, Ketterle, and Wieman Share Nobel Prize for Bose–Einstein Condensates

- Bose–Einstein Condensates at JILA

- Atomcool at Rice University

- Alkali Quantum Gases at MIT

- Atom Optics at UQ

- Einstein's manuscript on the Bose–Einstein condensate discovered at Leiden University

- Bose–Einstein condensate on arxiv.org

- Bosons – The Birds That Flock and Sing Together

- Easy BEC machine – information on constructing a Bose–Einstein condensate machine.

- Verging on absolute zero – Cosmos Online

- Lecture by W Ketterle at MIT in 2001

- Bose–Einstein Condensation at NIST – NIST resource on BEC

Chapter 8

Fermionic condensate

A **fermionic condensate** is a superfluid phase formed by fermionic particles at low temperatures. It is closely related to the Bose–Einstein condensate, a superfluid phase formed by bosonic atoms under similar conditions. Unlike the Bose–Einstein condensates, fermionic condensates are formed using fermions instead of bosons. The earliest recognized fermionic condensate described the state of electrons in a superconductor; the physics of other examples including recent work with fermionic atoms is analogous. The first atomic fermionic condensate was created by a team led by Deborah S. Jin in 2003. A **chiral condensate** is an example of a fermionic condensate that appears in theories of massless fermions with chiral symmetry breaking.

8.1 Background

8.1.1 Superfluidity

Fermionic condensates are attained at temperatures lower than Bose–Einstein condensates. Fermionic condensates are a type of superfluid. As the name suggests, a superfluid possesses fluid properties similar to those possessed by ordinary liquids and gases, such as the lack of a definite shape and the ability to flow in response to applied forces. However, superfluids possess some properties that do not appear in ordinary matter. For instance, they can flow at low velocities without dissipating any energy—i.e. zero viscosity. At higher velocities, energy is dissipated by the formation of quantized vortices, which act as "holes" in the medium where superfluidity breaks down.

Superfluidity was originally discovered in liquid helium-4, in 1938, by Pyotr Kapitsa, John Allen and Don Misener. Superfluidity in helium-4, which occurs at temperatures below 2.17 kelvins (K), has long been understood to result from Bose condensation, the same mechanism that produces the Bose–Einstein condensates. The primary difference between superfluid helium and a Bose–Einstein condensate is that the former is condensed from a liquid while the latter is condensed from a gas.

8.1.2 Fermionic superfluids

It is far more difficult to produce a fermionic superfluid than a bosonic one, because the Pauli exclusion principle prohibits fermions from occupying the same quantum state. However, there is a well-known mechanism by which a superfluid may be formed from fermions. This is the BCS transition, discovered in 1957 by John Bardeen, Leon Cooper and Robert Schrieffer for describing superconductivity. These authors showed that, below a certain temperature, electrons (which are fermions) can pair up to form bound pairs now known as Cooper pairs. As long as collisions with the ionic lattice of the solid do not supply enough energy to break the Cooper pairs, the electron fluid will be able to flow without dissipation. As a result, it becomes a superfluid, and the material through which it flows a superconductor.

The BCS theory was phenomenally successful in describing superconductors. Soon after the publication of the BCS paper, several theorists proposed that a similar phenomenon could occur in fluids made up of fermions other than electrons, such as helium-3 atoms. These speculations were confirmed in 1971, when experiments performed by Douglas D. Osheroff showed that helium-3 becomes a superfluid below 0.0025 K. It was soon verified that the superfluidity of helium-3 arises from a BCS-like mechanism. (The theory of superfluid helium-3 is a little more complicated than the BCS theory of superconductivity. These complications arise because helium atoms repel each other much more strongly than electrons, but the basic idea is the same.)

8.1.3 Creation of the first fermionic condensates

When Eric Cornell and Carl Wieman produced a Bose–Einstein condensate from rubidium atoms in 1995, there naturally arose the prospect of creating a similar sort of condensate made from fermionic atoms, which would form a superfluid by the BCS mechanism. However, early calculations indicated that the temperature required for producing Cooper pairing in atoms would be too cold to achieve. In 2001, Murray Holland at JILA suggested a way of bypassing this difficulty. He speculated that fermionic atoms could be coaxed into pairing up by subjecting them to a strong magnetic field.

In 2003, working on Holland's suggestion, Deborah Jin at JILA, Rudolf Grimm at the University of Innsbruck, and Wolfgang Ketterle at MIT managed to coax fermionic atoms into forming molecular bosons, which then underwent Bose–Einstein condensation. However, this was not a true fermionic condensate. On December 16, 2003, Jin managed to produce a condensate out of fermionic atoms for the first time. The experiment involved 500,000 potassium−40 atoms cooled to a temperature of 5×10^{-8} K, subjected to a time-varying magnetic field. The findings were published in the online edition of *Physical Review Letters* on January 24, 2004.

8.2 Examples

8.2.1 BCS theory

The BCS theory of superconductivity has a fermion condensate. A pair of electrons in a metal, with opposite spins can form a scalar bound state called a Cooper pair. Then, the bound states themselves form a condensate. Since the Cooper pair has electric charge, this fermion condensate breaks the electromagnetic gauge symmetry of a superconductor, giving rise to the wonderful electromagnetic properties of such states.

8.2.2 QCD

In quantum chromodynamics (QCD) the chiral condensate is also called the **quark condensate**. This property of the QCD vacuum is partly responsible for giving masses to hadrons (along with other condensates like the gluon condensate).

In an approximate version of QCD, which has vanishing quark masses for N quark flavours, there is an exact chiral SU(N) × SU(N) symmetry of the theory. The QCD vacuum breaks this symmetry to SU(N) by forming a quark condensate. The quark condensate is therefore an order parameter of transitions between several phases of quark matter in this limit.

This is very similar to the BCS theory of superconductivity. The Cooper pairs are analogous to the pseudoscalar mesons. However, the vacuum carries no charge. Hence all the gauge symmetries are unbroken. Corrections for the masses of the quarks can be incorporated using chiral perturbation theory.

8.2.3 Helium-3 superfluid

A helium-3 atom is a fermion and at very low temperatures, they form two-atom Cooper pairs which are bosonic and condense into a superfluid. These Cooper pairs are substantially larger than the interatomic separation.

8.3 References

- Guenault, Tony (2003). *Basic superfluids*. Taylor & Francis. ISBN 0-7484-0892-4.

- University of Colorado (January 28, 2004). *NIST/University of Colorado Scientists Create New Form of Matter: A Fermionic Condensate*. Press Release.

- Rodgers, Peter & Dumé, Bell (January 28, 2004). *Fermionic condensate makes its debut*. PhysicWeb.

Chapter 9

Quark–gluon plasma

A **quark–gluon plasma** (**QGP**) or **quark soup**[1] is a state of matter in quantum chromodynamics (QCD) which is hypothesized to exist at extremely high temperature, density, or both temperature and density. This state is thought to consist of asymptotically free quarks and gluons, which are several of the basic building blocks of matter. It is believed that up to a few milliseconds after the Big Bang, known as the Quark epoch, the Universe was in a quark–gluon plasma state. On June, 2015 an international team of physicists have produced quark-gluon plasma at the Large Hadron Collider by colliding protons with lead nuclei at high energy inside the supercollider's Compact Muon Solenoid detector. They also discovered that this new state of matter behaves like a fluid.[2]

The strength of the color force means that unlike the gas-like plasma, quark–gluon plasma behaves as a near-ideal Fermi liquid, although research on flow characteristics is ongoing.[3] In the quark matter phase diagram, QGP is placed in the high-temperature, high-density regime; whereas, ordinary matter is a cold and rarefied mixture of nuclei and vacuum, and the hypothetical quark stars would consist of relatively cold, but dense quark matter.

Experiments at CERN's Super Proton Synchrotron (SPS) first tried to create the QGP in the 1980s and 1990s: the results led CERN to announce indirect evidence for a "new state of matter"[4] in 2000. Current experiments (2011) at the Brookhaven National Laboratory's Relativistic Heavy Ion Collider (RHIC) on Long Island (NY, USA) and at CERN's recent Large Hadron Collider near Geneva (Switzerland) are continuing this effort,[5][6] by colliding relativistically accelerated gold (at RHIC) or lead (at LHC) with each other or with protons. Although the results have yet to be independently verified as of February 2010, scientists at Brookhaven RHIC have tentatively claimed to have created a quark–gluon plasma with an approximate temperature of 4 trillion (4×10^{12}) degrees Kelvin.[6]

As already mentioned, three new experiments running on CERN's Large Hadron Collider (LHC), on the spectrometers ALICE,[7] ATLAS and CMS, will continue studying properties of QGP. Starting in November 2010, CERN temporarily ceased colliding protons, and began colliding lead Ions for the ALICE experiment. They were looking to create a QGP and were expected to stop December 6, colliding protons again in January.[8] A new record breaking temperature was set by ALICE: A Large Ion Collider Experiment at CERN on August, 2012 in the ranges of 5.5 trillion (5.5×10^{12}) degrees Kelvin as claimed in their Nature PR.[9]

9.1   General introduction

Quark–gluon plasma is a state of matter in which the elementary particles that make up the hadrons of baryonic matter are freed of their strong attraction for one another under extremely high energy densities. These particles are the quarks and gluons that compose baryonic matter.[10] In normal matter quarks are *confined*; in the QGP quarks are *deconfined*. In classical QCD quarks are the Fermionic components of mesons and baryons while the gluons are considered the Bosonic components of such particles. The gluons are the force carriers, or bosons, of the QCD color force, while the quarks by themselves are their Fermionic matter counterparts.

Although the experimental high temperatures and densities predicted as producing a quark–gluon plasma have been realized in the laboratory, the resulting matter does *not* behave as a quasi-ideal state of free quarks and gluons, but, rather, as an almost perfect dense fluid.[11] Actually, the fact that the quark–gluon plasma will not yet be "free" at temperatures realized at present accelerators was predicted in 1984 as a consequence of the remnant effects of confinement.[12][13]

9.1.1 Relation to normal plasma

A plasma is matter in which charges are screened due to the presence of other mobile charges; for example: Coulomb's Law is suppressed by the screening to yield a distance-dependent charge ($Q \to Q \times \exp(-r/\alpha)$, i.e, the charge Q is reduced exponentially with the distance divided by a screening length α). In a QGP, the color charge of the quarks and gluons is screened. The QGP has other analogies with a normal plasma. There are also dissimilarities because the color charge is non-abelian, whereas the electric charge is abelian. Outside a finite volume of QGP the color-electric field is not screened, so that a volume of QGP must still be color-neutral. It will therefore, like a nucleus, have integer electric charge.

9.1.2 Theory

One consequence of this difference is that the color charge is too large for perturbative computations which are the mainstay of QED. As a result, the main theoretical tools to explore the theory of the QGP is lattice gauge theory.[14] The transition temperature (approximately 175 MeV) was first predicted by lattice gauge theory. Since then lattice gauge theory has been used to predict many other properties of this kind of matter. The AdS/CFT correspondence conjecture may provide insights in QGP, morever the ultimate goal of the fluid/gravity correspondence is to understand QGP. The QCP is believed to be a phase of QCD which is completely locally thermalized and thus suitable for an effective fluid dynamic description.

9.1.3 Production

The QGP can be created by heating matter up to a temperature of 2×10^{12} K, which amounts to 175 MeV per particle. This can be accomplished by colliding two large nuclei at high energy (note that 175 MeV is not the energy of the colliding beam). Lead and gold nuclei have been used for such collisions at CERN SPS and BNL RHIC, respectively. The nuclei are accelerated to ultrarelativistic speeds (contracting their length) and directed towards each other, creating a "fireball", in the rare event of a collision. Hydrodynamic simulation predicts this fireball will expand under its own pressure, and cool while expanding. By carefully studying the spherical and elliptic flow, experimentalists put the theory to test.

9.1.4 How the QGP fits into the general scheme of physics

QCD is one part of the modern theory of particle physics called the Standard Model. Other parts of this theory deal with electroweak interactions and neutrinos. The theory of electrodynamics has been tested and found correct to a few parts in a billion. The theory of weak interactions has been tested and found correct to a few parts in a thousand. Perturbative forms of QCD have been tested to a few percent. In contrast, non-perturbative forms of QCD have barely been tested. The study of the QGP is part of this effort to consolidate the grand theory of particle physics.

The study of the QGP is also a testing ground for finite temperature field theory, a branch of theoretical physics which seeks to understand particle physics under conditions of high temperature. Such studies are important to understand the early evolution of our universe: the first hundred microseconds or so. It is crucial to the physics goals of a new generation of observations of the universe (WMAP and its successors). It is also of relevance to Grand Unification Theories which seek to unify the three fundamental forces of nature (excluding gravity).

9.2 Expected properties

9.2.1   Thermodynamics

The cross-over temperature from the normal hadronic to the QGP phase is about 175 MeV. This "crossover" may actually *not* be only a qualitative feature, but instead one may have to do with a true (second order) phase transition, e.g. of the universality class of the three-dimensional Ising model, as some theorists say, e.g. Frithjof Karsch and coworkers from the university of Bielefeld. The phenomena involved correspond to an energy density of a little less than 1 GeV/fm^3. For relativistic matter, pressure and temperature are not independent variables, so the equation of state is a relation between the energy density and the pressure. This has been found through lattice computations, and compared to both perturbation theory and string theory. This is still a matter of active research. Response functions such as the specific heat and various quark number susceptibilities are currently being computed.

9.2.2   Flow

The equation of state is an important input into the flow equations. The speed of sound is currently under investigation in lattice computations. The mean free path of quarks and gluons has been computed using perturbation theory as well as string theory. Lattice computations have been slower here, although the first computations of transport coefficients have recently been concluded. These indicate that the mean free time of quarks and gluons in the QGP may be comparable to the average interparticle spacing: hence the QGP is a liquid as far as its flow properties go. This is very much an active field of research, and these conclusions may evolve rapidly. The incorporation of dissipative phenomena into hydrodynamics is another recent development that is still in an active stage.

9.2.3   Excitation spectrum

Does the QGP really contain (almost) free quarks and gluons? The study of thermodynamic and flow properties would indicate that this is an over-simplification. Many ideas are currently being evolved and will be put to test in the near future. It has been hypothesized recently that some mesons built from heavy quarks do not dissolve until the temperature reaches about 350 MeV. This has led to speculation that many other kinds of bound states may exist in the plasma. Some static properties of the plasma (similar to the Debye screening length) constrain the excitation spectrum.

9.2.4   Glasma hypothesis

Since 2008, there is a discussion about a hypothetical precursor state of the Quark–gluon plasma, the so-called "Glasma", where the dressed particles are condensed into some kind of glassy (or amorphous) state, below the genuine transition between the confined state and the plasma liquid. This would be analogous to the formation of metallic glasses, or amorphous alloys of them, below the genuine onset of the liquid metallic state.

9.3   Experimental situation

Those forms of the QGP that are easiest to compute are not those that are easiest to verify experimentally. While the balance of evidence points towards the QGP being the origin of the detailed properties of the fireball produced in the RHIC, this is the main barrier which prevents experimentalists from declaring a sighting of the QGP. For a summary see 2005 RHIC Assessment.

The important classes of experimental observations are

- Single particle spectra (photons and dileptons)
- Strangeness production
- Photon and muon rates (and J/ψ melting)
- Elliptic flow

- Jet quenching

- Fluctuations

- Hanbury Brown and Twiss effect and Bose–Einstein correlations

In short, a quark–gluon plasma flows like a splat of liquid, and because it's not "transparent" with respect to quarks, it can attenuate jets emitted by collisions. Furthermore, once formed, a ball of quark–gluon plasma, like any hot object, transfers heat internally by radiation. However, unlike in everyday objects, there is enough energy available that gluons (particles mediating the strong force) collide and produce an excess of the heavy (i.e. high-energy) strange quarks. Whereas, if the QGP didn't exist and there was a pure collision, the same energy would be converted into even heavier quarks such as charm quarks or bottom quarks.

9.4 Formation of quark matter

In April 2005, formation of quark matter was tentatively confirmed by results obtained at Brookhaven National Laboratory's Relativistic Heavy Ion Collider (RHIC). The consensus of the four RHIC research groups was that they had created a quark–gluon liquid of very low viscosity. However, contrary to what was at that time still the widespread assumption, it is yet unknown from theoretical predictions whether the QCD "plasma", especially close to the transition temperature, should behave like a gas or liquid. Authors favoring the weakly interacting interpretation derive their assumptions from the lattice QCD calculation, where the entropy density of quark–gluon plasma approaches the weakly interacting limit. However, since both energy density and correlation shows significant deviation from the weakly interacting limit, it has been pointed out by many authors that there is in fact no reason to assume a QCD "plasma" close to the transition point should be weakly interacting, like electromagnetic plasma (see, e.g.,[15]). That being said, systematically improvable perturbative QCD quasiparticle models do a very good job of reproducing the lattice data for thermodynamical observables (pressure, entropy, quark susceptibility), including the aforementioned "significant deviation from the weakly interacting limit", down to temperatures on the order of 2 to 3 times the critical temperature for the transition.[16][17][18]

9.5 See also

- Hadrons (that is mesons and baryons) and confinement

- Hadronization

- List of plasma (physics) articles

- Neutron stars

- Plasma physics

- QCD matter Quantum Chromodynamics matter

- Quantum electrodynamics

- Quantum chromodynamics

- Quantum hydrodynamics

- Relativistic plasma

- Relativistic nuclear collision

- Strangeness production

- Strange matter

- Color-glass condensate

9.6   References

[1] Bohr, Henrik; Nielsen, H. B. (1977). "Hadron production from a boiling quark soup: quark model predicting particle ratios in hadronic collisions". *Nuclear Physics B* **128** (2): 275. Bibcode:1977NuPhB.128..275B. doi:10.1016/0550-3213(77)90032-3.

[2] LHC creates liquid from Big Bang

[3] Quark-gluon plasma goes liquid - physicsworld.com

[4] A New State of Matter - Experiments

[5] Relativistic Heavy Ion Collider, RHIC

[6] http://www.bnl.gov/rhic/news2/news.asp?a=1074&t=pr 'Perfect' Liquid Hot Enough to be Quark Soup

[7] Alice Experiment: Welcome to ALICE Portal

[8] CERN Press Release November 4th 2010

[9] Hot stuff: CERN physicists create record-breaking subatomic soup : Nature News Blog

[10] The Indian Lattice Gauge Theory Initiative

[11] WA Zajc (2008). "The fluid nature of quark-gluon plasma".*Nuclear Physics A***805**: 283c–294c.arXiv:0802.3552.BiZ. doi:10.1016/j.nuclphysa.2008.02.285.

[12] Plümer, M.; Raha, S. & Weiner, R. M. (1984). "How free is the quark-gluon plasma". *Nucl. Phys. A* **418**: 549–557. Bibcode:1984NuPhA.418..549P. doi:10.1016/0375-9474(84)90575-X..

[13] Plümer, M.; Raha, S. & Weiner, R. M. (1984). "Effect of confinement on the sound velocity in a quark-gluon plasma". *Phys. Lett. B* **139** (3): 198–202. Bibcode:1984PhLB..139..198P. doi:10.1016/0370-2693(84)91244-9..

[14] Lattice-QCD calculations of the Quark-Gluon Plasma have been reviewed in and in

[15] Miklos Gyulassy (2004). "The QGP Discovered at RHIC". arXiv:nucl-th/0403032 [nucl-th].

[16] Andersen; Leganger; Strickland; Su (2011). "NNLO hard-thermal-loop thermodynamics for QCD". *Physics Letters B* **696** (5): 468. arXiv:1009.4644. Bibcode:2011PhLB..696..468A. doi:10.1016/j.physletb.2010.12.070.

[17] Andersen; Michael Strickland; Nan Su (2010). "Gluon Thermodynamics at Intermediate Coupling". *Physical Review Letters* **104** (12). arXiv:0911.0676. Bibcode:2010PhRvL.104l2003A. doi:10.1103/PhysRevLett.104.122003.

[18] Blaizot; Iancu; Rebhan (2003). "Thermodynamics of the high-temperature quark-gluon plasma". arXiv:hep-ph/0303185 [hep-ph].

9.7   External links

- The Relativistic Heavy Ion Collider at Brookhaven National Laboratory

- The Alice Experiment at CERN

- The Indian Lattice Gauge Theory Initiative

- Quark matter reviews: 2004 theory, 2004 experiment

- Quark-Gluon Plasma reviews: 2011 theory

- Lattice reviews: 2003, 2005

- BBC article mentioning Brookhaven results (2005)

- Physics News Update article on the quark-gluon liquid, with links to preprints

- Read for free : "Hadrons and Quark-Gluon Plasma" by Jean Letessier and Johann Rafelski Cambridge University Press (2002) ISBN 0-521-38536-9, Cambridge, UK;

Chapter 10

Atom

For other uses, see Atom (disambiguation).

An **atom** is the smallest constituent unit of ordinary matter that has the properties of a chemical element.[1] Every solid, liquid, gas, and plasma is made up of neutral or ionized atoms. Atoms are very small; typical sizes are around 100 pm (a ten-billionth of a meter, in the short scale).[2] However, atoms do not have well defined boundaries, and there are different ways to define their size which give different but close values.

Atoms are small enough that classical physics give noticeably incorrect results. Through the development of physics, atomic models have incorporated quantum principles to better explain and predict the behavior.

Every atom is composed of a nucleus and one or more electrons bound to the nucleus. The nucleus is made of one or more protons and typically a similar number of neutrons (none in hydrogen-1). Protons and neutrons are called nucleons. Over 99.94% of the atom's mass is in the nucleus. The protons have a positive electric charge, the electrons have a negative electric charge, and the neutrons have no electric charge. If the number of protons and electrons are equal, that atom is electrically neutral. If an atom has more or fewer electrons than protons, then it has an overall negative or positive charge, respectively, and it is called an ion.

Electrons of an atom are attracted to the protons in an atomic nucleus by this electromagnetic force. The protons and neutrons in the nucleus are attracted to each other by a different force, the nuclear force, which is usually stronger than the electromagnetic force repelling the positively charged protons from one another. Under certain circumstances the repelling electromagnetic force becomes stronger than the nuclear force, and nucleons can be ejected from the nucleus, leaving behind a different element: nuclear decay resulting in nuclear transmutation.

The number of protons in the nucleus defines to what chemical element the atom belongs: for example, all copper atoms contain 29 protons. The number of neutrons defines the isotope of the element.[3] The number of electrons influences the magnetic properties of an atom. Atoms can attach to one or more other atoms by chemical bonds to form chemical compounds such as molecules. The ability of atoms to associate and dissociate is responsible for most of the physical changes observed in nature, and is the subject of the discipline of chemistry.

Not all the matter of the universe is composed of atoms. Dark matter comprises more of the Universe than matter, and is composed not of atoms, but of particles of a currently unknown type.

10.1   History of atomic theory

Main article: Atomic theory

10.1.1 Atoms in philosophy

Main article: Atomism

The idea that matter is made up of discrete units is a very old one, appearing in many ancient cultures such as Greece and India. The word "atom", in fact, was coined by ancient Greek philosophers. However, these ideas were founded in philosophical and theological reasoning rather than evidence and experimentation. As a result, their views on what atoms look like and how they behave were incorrect. They also could not convince everybody, so atomism was but one of a number of competing theories on the nature of matter. It was not until the 19th century that the idea was embraced and refined by scientists, when the blossoming science of chemistry produced discoveries that only the concept of atoms could explain.

10.1.2 First evidence-based theory

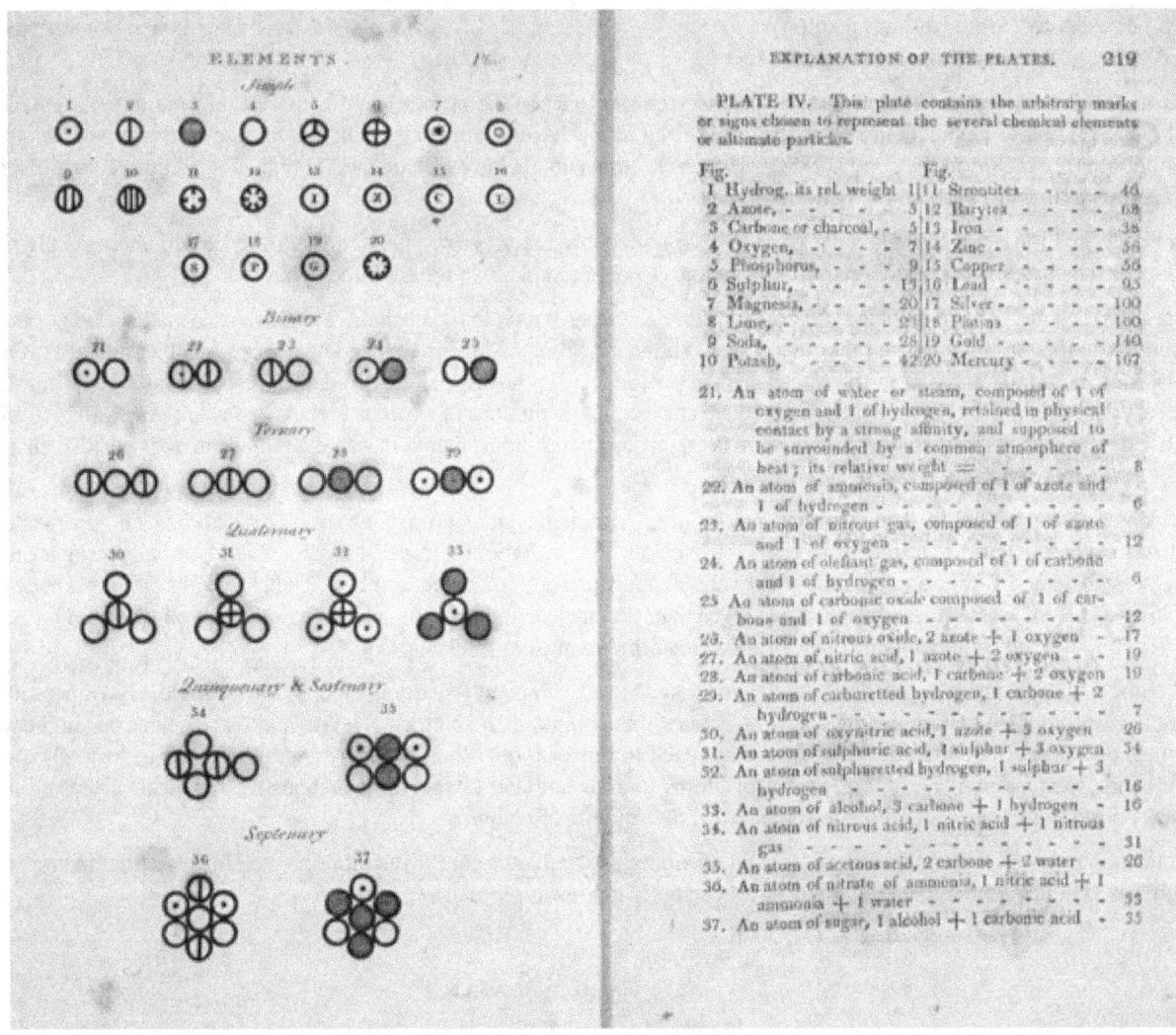

Various atoms and molecules as depicted in John Dalton's A New System of Chemical Philosophy *(1808).*

In the early 1800s, John Dalton used the concept of atoms to explain why elements always react in ratios of small whole numbers (the law of multiple proportions). For instance, there are two types of tin oxide: one is 88.1% tin and 11.9% oxygen and the other is 78.7% tin and 21.3% oxygen (tin(II) oxide and tin dioxide respectively). This means that 100g of tin will combine either with 13.5g or 27g of oxygen. 13.5 and 27 form a ratio of 1:2, a ratio of small whole numbers. This common pattern in chemistry suggested to Dalton that elements react in whole number multiples of discrete units—in other words, atoms. In the case of tin oxides, one tin atom will combine with either one or two oxygen atoms.[4]

Dalton also believed atomic theory could explain why water absorbs different gases in different proportions. For example, he found that water absorbs carbon dioxide far better than it absorbs nitrogen.[5] Dalton hypothesized this was due to the differences in mass and complexity of the gases' respective particles. Indeed, carbon dioxide molecules (CO_2) are heavier and larger than nitrogen molecules (N_2).

10.1.3 Brownian motion

In 1827, botanist Robert Brown used a microscope to look at dust grains floating in water and discovered that they moved about erratically, a phenomenon that became known as "Brownian motion". This was thought to be caused by water molecules knocking the grains about. In 1905 Albert Einstein produced the first mathematical analysis of the motion.[6][7][8] French physicist Jean Perrin used Einstein's work to experimentally determine the mass and dimensions of atoms, thereby conclusively verifying Dalton's atomic theory.[9]

10.1.4 Discovery of the electron

The physicist J. J. Thomson measured the mass of cathode rays, showing they were made of particles, but were around 1800 times lighter than the lightest atom, hydrogen. Therefore, they were not atoms, but a new particle, the first *subatomic* particle to be discovered, which he originally called "*corpuscle*" but was later named *electron*, after particles postulated by George Johnstone Stoney in 1874. He also showed they were identical to particles given off by photoelectric and radioactive materials.[10] It was quickly recognized that they are the particles that carry electric currents in metal wires, and carry the negative electric charge within atoms. Thomson was given the 1906 Nobel Prize in Physics for this work. Thus he overturned the belief that atoms are the indivisible, ultimate particles of matter.[11] Thomson also incorrectly postulated that the low mass, negatively charged electrons were distributed throughout the atom in a uniform sea of positive charge. This became known as the plum pudding model.

10.1.5 Discovery of the nucleus

Main article: Geiger-Marsden experiment

In 1909, Hans Geiger and Ernest Marsden, under the direction of Ernest Rutherford, bombarded a metal foil with alpha particles to observe how they scattered. They expected all the alpha particles to pass straight through with little deflection, because Thomson's model said that the charges in the atom are so diffuse that their electric fields could not affect the alpha particles much. However, Geiger and Marsden spotted alpha particles being deflected by angles greater than 90°, which was supposed to be impossible according to Thomson's model. To explain this, Rutherford proposed that the positive charge of the atom is concentrated in a tiny nucleus at the center of the atom.[12]

10.1.6 Discovery of isotopes

While experimenting with the products of radioactive decay, in 1913 radiochemist Frederick Soddy discovered that there appeared to be more than one type of atom at each position on the periodic table.[13] The term isotope was coined by Margaret Todd as a suitable name for different atoms that belong to the same element. J.J. Thomson created a technique for separating atom types through his work on ionized gases, which subsequently led to the discovery of stable isotopes.[14]

10.1.7 Bohr model

Main article: Bohr model

In 1913 the physicist Niels Bohr proposed a model in which the electrons of an atom were assumed to orbit the nucleus but could only do so in a finite set of orbits, and could jump between these orbits only in discrete changes of energy corresponding to absorption or radiation of a photon.[15] This quantization was used to explain why the electrons orbits are stable (given that normally, charges in acceleration, including circular motion, lose kinetic energy which is emitted as electromagnetic radiation, see *synchrotron radiation*) and why elements absorb and emit electromagnetic radiation in discrete spectra.[16]

Later in the same year Henry Moseley provided additional experimental evidence in favor of Niels Bohr's theory. These results refined Ernest Rutherford's and Antonius Van den Broek's model, which proposed that the atom contains in its nucleus a number of positive nuclear charges that is equal to its (atomic) number in the periodic table. Until these experiments, atomic number was not known to be a physical and experimental quantity. That it is equal to the atomic nuclear charge remains the accepted atomic model today.[17]

10.1.8   Chemical bonding explained

Chemical bonds between atoms were now explained, by Gilbert Newton Lewis in 1916, as the interactions between their constituent electrons.[18] As the chemical properties of the elements were known to largely repeat themselves according to the periodic law,[19] in 1919 the American chemist Irving Langmuir suggested that this could be explained if the electrons in an atom were connected or clustered in some manner. Groups of electrons were thought to occupy a set of electron shells about the nucleus.[20]

10.1.9   Further developments in quantum physics

The Stern–Gerlach experiment of 1922 provided further evidence of the quantum nature of the atom. When a beam of silver atoms was passed through a specially shaped magnetic field, the beam was split based on the direction of an atom's angular momentum, or spin. As this direction is random, the beam could be expected to spread into a line. Instead, the beam was split into two parts, depending on whether the atomic spin was oriented up or down.[21]

In 1924, Louis de Broglie proposed that all particles behave to an extent like waves. In 1926, Erwin Schrödinger used this idea to develop a mathematical model of the atom that described the electrons as three-dimensional waveforms rather than point particles. A consequence of using waveforms to describe particles is that it is mathematically impossible to obtain precise values for both the position and momentum of a particle at the same time; this became known as the uncertainty principle, formulated by Werner Heisenberg in 1926. In this concept, for a given accuracy in measuring a position one could only obtain a range of probable values for momentum, and vice versa. This model was able to explain observations of atomic behavior that previous models could not, such as certain structural and spectral patterns of atoms larger than hydrogen. Thus, the planetary model of the atom was discarded in favor of one that described atomic orbital zones around the nucleus where a given electron is most likely to be observed.[22][23]

10.1.10   Discovery of the neutron

The development of the mass spectrometer allowed the mass of atoms to be measured with increased accuracy. The device uses a magnet to bend the trajectory of a beam of ions, and the amount of deflection is determined by the ratio of an atom's mass to its charge. The chemist Francis William Aston used this instrument to show that isotopes had different masses. The atomic mass of these isotopes varied by integer amounts, called the whole number rule.[24] The explanation for these different isotopes awaited the discovery of the neutron, an uncharged particle with a mass similar to the proton, by the physicist James Chadwick in 1932. Isotopes were then explained as elements with the same number of protons, but different numbers of neutrons within the nucleus.[25]

10.1.11   Fission, high-energy physics and condensed matter

In 1938, the German chemist Otto Hahn, a student of Rutherford, directed neutrons onto uranium atoms expecting to get transuranium elements. Instead, his chemical experiments showed barium as a product.[26] A year later, Lise Meitner and her nephew Otto Frisch verified that Hahn's result were the first experimental *nuclear fission*.[27][28] In 1944, Hahn received the Nobel prize in chemistry. Despite Hahn's efforts, the contributions of Meitner and Frisch were not recognized.[29]

In the 1950s, the development of improved particle accelerators and particle detectors allowed scientists to study the impacts of atoms moving at high energies.[30] Neutrons and protons were found to be hadrons, or composites of smaller particles called quarks. The standard model of particle physics was developed that so far has successfully explained the properties of the nucleus in terms of these sub-atomic particles and the forces that govern their interactions.[31]

10.2 Structure

10.2.1 Subatomic particles

Main article: Subatomic particle

Though the word *atom* originally denoted a particle that cannot be cut into smaller particles, in modern scientific usage the atom is composed of various subatomic particles. The constituent particles of an atom are the electron, the proton and the neutron; all three are fermions. However, the hydrogen-1 atom has no neutrons and the hydron ion has no electrons.

The electron is by far the least massive of these particles at 9.11×10^{-31} kg, with a negative electrical charge and a size that is too small to be measured using available techniques.[32] It is the lightest particle with a positive rest mass measured. Under ordinary conditions, electrons are bound to the positively charged nucleus by the attraction created from opposite electric charges. If an atom has more or fewer electrons than its atomic number, then it becomes respectively negatively or positively charged as a whole; a charged atom is called an ion. Electrons have been known since the late 19th century, mostly thanks to J.J. Thomson; see history of subatomic physics for details.

Protons have a positive charge and a mass 1,836 times that of the electron, at 1.6726×10^{-27} kg. The number of protons in an atom is called its atomic number. Ernest Rutherford (1919) observed that nitrogen under alpha-particle bombardment ejects what appeared to be hydrogen nuclei. By 1920 he had accepted that the hydrogen nucleus is a distinct particle within the atom and named it proton.

Neutrons have no electrical charge and have a free mass of 1,839 times the mass of the electron,[33] or 1.6929×10^{-27} kg, the heaviest of the three constituent particles, but it can be reduced by the nuclear binding energy. Neutrons and protons (collectively known as nucleons) have comparable dimensions—on the order of 2.5×10^{-15} m—although the 'surface' of these particles is not sharply defined.[34] The neutron was discovered in 1932 by the English physicist James Chadwick.

In the Standard Model of physics, electrons are truly elementary particles with no internal structure. However, both protons and neutrons are composite particles composed of elementary particles called quarks. There are two types of quarks in atoms, each having a fractional electric charge. Protons are composed of two up quarks (each with charge $+\frac{2}{3}$) and one down quark (with a charge of $-\frac{1}{3}$). Neutrons consist of one up quark and two down quarks. This distinction accounts for the difference in mass and charge between the two particles.[35][36]

The quarks are held together by the strong interaction (or strong force), which is mediated by gluons. The protons and neutrons, in turn, are held to each other in the nucleus by the nuclear force, which is a residuum of the strong force that has somewhat different range-properties (see the article on the nuclear force for more). The gluon is a member of the family of gauge bosons, which are elementary particles that mediate physical forces.[35][36]

10.2.2 Nucleus

Main article: Atomic nucleus

All the bound protons and neutrons in an atom make up a tiny atomic nucleus, and are collectively called nucleons. The radius of a nucleus is approximately equal to $1.07 \sqrt[3]{A}$ fm, where A is the total number of nucleons.[37] This is much smaller than the radius of the atom, which is on the order of 10^5 fm. The nucleons are bound together by a short-ranged attractive potential called the residual strong force. At distances smaller than 2.5 fm this force is much more powerful than the electrostatic force that causes positively charged protons to repel each other.[38]

Atoms of the same element have the same number of protons, called the atomic number. Within a single element, the number of neutrons may vary, determining the isotope of that element. The total number of protons and neutrons determine the nuclide. The number of neutrons relative to the protons determines the stability of the nucleus, with certain isotopes undergoing radioactive decay.[39]

The proton, the electron, and the neutron are classified as fermions. Fermions obey the Pauli exclusion principle which prohibits *identical* fermions, such as multiple protons, from occupying the same quantum state at the same time. Thus, every proton in the nucleus must occupy a quantum state different from all other protons, and the same applies to all neutrons of the nucleus and to all electrons of the electron cloud. However, a proton and a neutron are allowed to occupy the same quantum state.[40]

For atoms with low atomic numbers, a nucleus that has more neutrons than protons tends to drop to a lower energy state through radioactive decay so that the neutron–proton ratio is closer to one. However, as the atomic number increases, a higher proportion of neutrons is required to offset the mutual repulsion of the protons. Thus, there are no stable nuclei with equal proton and neutron numbers above atomic number $Z = 20$ (calcium) and as Z increases, the neutron–proton ratio of stable isotopes increases.[40] The stable isotope with the highest proton–neutron ratio is lead-208 (about 1.5).

The number of protons and neutrons in the atomic nucleus can be modified, although this can require very high energies because of the strong force. Nuclear fusion occurs when multiple atomic particles join to form a heavier nucleus, such as through the energetic collision of two nuclei. For example, at the core of the Sun protons require energies of 3–10 keV to overcome their mutual repulsion—the coulomb barrier—and fuse together into a single nucleus.[41] Nuclear fission is the opposite process, causing a nucleus to split into two smaller nuclei—usually through radioactive decay. The nucleus can also be modified through bombardment by high energy subatomic particles or photons. If this modifies the number of protons in a nucleus, the atom changes to a different chemical element.[42][43]

If the mass of the nucleus following a fusion reaction is less than the sum of the masses of the separate particles, then the difference between these two values can be emitted as a type of usable energy (such as a gamma ray, or the kinetic energy of a beta particle), as described by Albert Einstein's mass–energy equivalence formula, $E = mc^2$, where m is the mass loss and c is the speed of light. This deficit is part of the binding energy of the new nucleus, and it is the non-recoverable loss of the energy that causes the fused particles to remain together in a state that requires this energy to separate.[44]

The fusion of two nuclei that create larger nuclei with lower atomic numbers than iron and nickel—a total nucleon number of about 60—is usually an exothermic process that releases more energy than is required to bring them together.[45] It is this energy-releasing process that makes nuclear fusion in stars a self-sustaining reaction. For heavier nuclei, the binding energy per nucleon in the nucleus begins to decrease. That means fusion processes producing nuclei that have atomic numbers higher than about 26, and atomic masses higher than about 60, is an endothermic process. These more massive nuclei can not undergo an energy-producing fusion reaction that can sustain the hydrostatic equilibrium of a star.[40]

10.2.3   Electron cloud

Main articles: Atomic orbital and Electron configuration

The electrons in an atom are attracted to the protons in the nucleus by the electromagnetic force. This force binds the electrons inside an electrostatic potential well surrounding the smaller nucleus, which means that an external source of energy is needed for the electron to escape. The closer an electron is to the nucleus, the greater the attractive force. Hence electrons bound near the center of the potential well require more energy to escape than those at greater separations.

Electrons, like other particles, have properties of both a particle and a wave. The electron cloud is a region inside the potential well where each electron forms a type of three-dimensional standing wave—a wave form that does not move relative to the nucleus. This behavior is defined by an atomic orbital, a mathematical function that characterises the probability that an electron appears to be at a particular location when its position is measured.[46] Only a discrete (or quantized) set of these orbitals exist around the nucleus, as other possible wave patterns rapidly decay into a more stable form.[47] Orbitals can have one or more ring or node structures, and they differ from each other in size, shape and orientation.[48]

Each atomic orbital corresponds to a particular energy level of the electron. The electron can change its state to a higher energy level by absorbing a photon with sufficient energy to boost it into the new quantum state. Likewise, through spontaneous emission, an electron in a higher energy state can drop to a lower energy state while radiating the excess energy as a photon. These characteristic energy values, defined by the differences in the energies of the quantum states, are responsible for atomic spectral lines.[47]

The amount of energy needed to remove or add an electron—the electron binding energy—is far less than the binding energy of nucleons. For example, it requires only 13.6 eV to strip a ground-state electron from a hydrogen atom,[49] compared to 2.23 *million* eV for splitting a deuterium nucleus.[50] Atoms are electrically neutral if they have an equal number of protons and electrons. Atoms that have either a deficit or a surplus of electrons are called ions. Electrons that are farthest from the nucleus may be transferred to other nearby atoms or shared between atoms. By this mechanism, atoms are able to bond into molecules and other types of chemical compounds like ionic and covalent network crystals.[51]

10.3 Properties

10.3.1 Nuclear properties

Main articles: Isotope, Stable isotope, List of nuclides and List of elements by stability of isotopes

By definition, any two atoms with an identical number of *protons* in their nuclei belong to the same chemical element. Atoms with equal numbers of protons but a different number of *neutrons* are different isotopes of the same element. For example, all hydrogen atoms admit exactly one proton, but isotopes exist with no neutrons (hydrogen-1, by far the most common form,[52] also called protium), one neutron (deuterium), two neutrons (tritium) and more than two neutrons. The known elements form a set of atomic numbers, from the single proton element hydrogen up to the 118-proton element ununoctium.[53] All known isotopes of elements with atomic numbers greater than 82 are radioactive.[54][55]

About 339 nuclides occur naturally on Earth,[56] of which 254 (about 75%) have not been observed to decay, and are referred to as "stable isotopes". However, only 90 of these nuclides are stable to all decay, even in theory. Another 164 (bringing the total to 254) have not been observed to decay, even though in theory it is energetically possible. These are also formally classified as "stable". An additional 34 radioactive nuclides have half-lives longer than 80 million years, and are long-lived enough to be present from the birth of the solar system. This collection of 288 nuclides are known as primordial nuclides. Finally, an additional 51 short-lived nuclides are known to occur naturally, as daughter products of primordial nuclide decay (such as radium from uranium), or else as products of natural energetic processes on Earth, such as cosmic ray bombardment (for example, carbon-14).[57][note 1]

For 80 of the chemical elements, at least one stable isotope exists. As a rule, there is only a handful of stable isotopes for each of these elements, the average being 3.2 stable isotopes per element. Twenty-six elements have only a single stable isotope, while the largest number of stable isotopes observed for any element is ten, for the element tin. Elements 43, 61, and all elements numbered 83 or higher have no stable isotopes.[58]

Stability of isotopes is affected by the ratio of protons to neutrons, and also by the presence of certain "magic numbers" of neutrons or protons that represent closed and filled quantum shells. These quantum shells correspond to a set of energy levels within the shell model of the nucleus; filled shells, such as the filled shell of 50 protons for tin, confers unusual stability on the nuclide. Of the 254 known stable nuclides, only four have both an odd number of protons *and* odd number of neutrons: hydrogen-2 (deuterium), lithium-6, boron-10 and nitrogen-14. Also, only four naturally occurring, radioactive odd–odd nuclides have a half-life over a billion years: potassium-40, vanadium-50, lanthanum-138 and tantalum-180m. Most odd–odd nuclei are highly unstable with respect to beta decay, because the decay products are even–even, and are therefore more strongly bound, due to nuclear pairing effects.[58]

10.3.2 Mass

Main articles: Atomic mass and mass number

The large majority of an atom's mass comes from the protons and neutrons that make it up. The total number of these particles (called "nucleons") in a given atom is called the mass number. It is a positive integer and dimensionless (instead of having dimension of mass), because it expresses a count. An example of use of a mass number is "carbon-12," which has 12 nucleons (six protons and six neutrons).

The actual mass of an atom at rest is often expressed using the unified atomic mass unit (u), also called dalton (Da). This unit is defined as a twelfth of the mass of a free neutral atom of carbon-12, which is approximately 1.66×10^{-27} kg.[59] Hydrogen-1 (the lightest isotope of hydrogen which is also the nuclide with the lowest mass) has an atomic weight of 1.007825 u.[60] The value of this number is called the atomic mass. A given atom has an atomic mass approximately equal (within 1%) to its mass number times the atomic mass unit (for example the mass of a nitrogen-14 is roughly 14 u). However, this number will not be exactly an integer except in the case of carbon-12 (see below).[61] The heaviest stable atom is lead-208,[54] with a mass of 207.9766521 u.[62]

As even the most massive atoms are far too light to work with directly, chemists instead use the unit of moles. One mole of atoms of any element always has the same number of atoms (about 6.022×10^{23}). This number was chosen so that if an element has an atomic mass of 1 u, a mole of atoms of that element has a mass close to one gram. Because of the definition of the unified atomic mass unit, each carbon-12 atom has an atomic mass of exactly 12 u, and so a mole of carbon-12 atoms weighs exactly 0.012 kg.[59]

10.3.3   Shape and size

Main article: Atomic radius

Atoms lack a well-defined outer boundary, so their dimensions are usually described in terms of an atomic radius. This is a measure of the distance out to which the electron cloud extends from the nucleus.[2] However, this assumes the atom to exhibit a spherical shape, which is only obeyed for atoms in vacuum or free space. Atomic radii may be derived from the distances between two nuclei when the two atoms are joined in a chemical bond. The radius varies with the location of an atom on the atomic chart, the type of chemical bond, the number of neighboring atoms (coordination number) and a quantum mechanical property known as spin.[63] On the periodic table of the elements, atom size tends to increase when moving down columns, but decrease when moving across rows (left to right).[64] Consequently, the smallest atom is helium with a radius of 32 pm, while one of the largest is caesium at 225 pm.[65]

When subjected to external forces, like electrical fields, the shape of an atom may deviate from spherical symmetry. The deformation depends on the field magnitude and the orbital type of outer shell electrons, as shown by group-theoretical considerations. Aspherical deviations might be elicited for instance in crystals, where large crystal-electrical fields may occur at low-symmetry lattice sites. Significant ellipsoidal deformations have recently been shown to occur for sulfur ions[66] and chalcogen ions[67] in pyrite-type compounds.

Atomic dimensions are thousands of times smaller than the wavelengths of light (400–700 nm) so they cannot be viewed using an optical microscope. However, individual atoms can be observed using a scanning tunneling microscope. To visualize the minuteness of the atom, consider that a typical human hair is about 1 million carbon atoms in width.[68] A single drop of water contains about 2 sextillion (2×10^{21}) atoms of oxygen, and twice the number of hydrogen atoms.[69] A single carat diamond with a mass of 2×10^{-4} kg contains about 10 sextillion (10^{22}) atoms of carbon.[note 2] If an apple were magnified to the size of the Earth, then the atoms in the apple would be approximately the size of the original apple.[70]

10.3.4   Radioactive decay

Main article: Radioactive decay

Every element has one or more isotopes that have unstable nuclei that are subject to radioactive decay, causing the nucleus to emit particles or electromagnetic radiation. Radioactivity can occur when the radius of a nucleus is large compared with the radius of the strong force, which only acts over distances on the order of 1 fm.[71]

The most common forms of radioactive decay are:[72][73]

- Alpha decay: this process is caused when the nucleus emits an alpha particle, which is a helium nucleus consisting of two protons and two neutrons. The result of the emission is a new element with a lower atomic number.

- Beta decay (and electron capture): these processes are regulated by the weak force, and result from a transformation of a neutron into a proton, or a proton into a neutron. The neutron to proton transition is accompanied by the emission of an electron and an antineutrino, while proton to neutron transition (except in electron capture) causes the emission of a positron and a neutrino. The electron or positron emissions are called beta particles. Beta decay either increases or decreases the atomic number of the nucleus by one. Electron capture is more common than positron emission, because it requires less energy. In this type of decay, an electron is absorbed by the nucleus, rather than a positron emitted from the nucleus. A neutrino is still emitted in this process, and a proton changes to a neutron.

- Gamma decay: this process results from a change in the energy level of the nucleus to a lower state, resulting in the emission of electromagnetic radiation. The excited state of a nucleus which results in gamma emission usually occurs following the emission of an alpha or a beta particle. Thus, gamma decay usually follows alpha or beta decay.

Other more rare types of radioactive decay include ejection of neutrons or protons or clusters of nucleons from a nucleus, or more than one beta particle. An analog of gamma emission which allows excited nuclei to lose energy in a different way, is internal conversion— a process that produces high-speed electrons that are not beta rays, followed by production of high-energy photons that are not gamma rays. A few large nuclei explode into two or more charged fragments of varying masses plus several neutrons, in a decay called spontaneous nuclear fission.

Each radioactive isotope has a characteristic decay time period—the half-life—that is determined by the amount of time needed for half of a sample to decay. This is an exponential decay process that steadily decreases the proportion of the remaining isotope by 50% every half-life. Hence after two half-lives have passed only 25% of the isotope is present, and so forth.[71]

10.3.5 Magnetic moment

Main articles: Electron magnetic moment and Nuclear magnetic moment

Elementary particles possess an intrinsic quantum mechanical property known as spin. This is analogous to the angular momentum of an object that is spinning around its center of mass, although strictly speaking these particles are believed to be point-like and cannot be said to be rotating. Spin is measured in units of the reduced Planck constant ($\hbar$), with electrons, protons and neutrons all having spin $\frac{1}{2}$ $\hbar$, or "spin-$\frac{1}{2}$". In an atom, electrons in motion around the nucleus possess orbital angular momentum in addition to their spin, while the nucleus itself possesses angular momentum due to its nuclear spin.[74]

The magnetic field produced by an atom—its magnetic moment—is determined by these various forms of angular momentum, just as a rotating charged object classically produces a magnetic field. However, the most dominant contribution comes from electron spin. Due to the nature of electrons to obey the Pauli exclusion principle, in which no two electrons may be found in the same quantum state, bound electrons pair up with each other, with one member of each pair in a spin up state and the other in the opposite, spin down state. Thus these spins cancel each other out, reducing the total magnetic dipole moment to zero in some atoms with even number of electrons.[75]

In ferromagnetic elements such as iron, cobalt and nickel, an odd number of electrons leads to an unpaired electron and a net overall magnetic moment. The orbitals of neighboring atoms overlap and a lower energy state is achieved when the spins of unpaired electrons are aligned with each other, a spontaneous process known as an exchange interaction. When the magnetic moments of ferromagnetic atoms are lined up, the material can produce a measurable macroscopic field. Paramagnetic materials have atoms with magnetic moments that line up in random directions when no magnetic field is present, but the magnetic moments of the individual atoms line up in the presence of a field.[75][76]

The nucleus of an atom will have no spin when it has even numbers of both neutrons and protons, but for other cases of odd numbers, the nucleus may have a spin. Normally nuclei with spin are aligned in random directions because of thermal equilibrium. However, for certain elements (such as xenon-129) it is possible to polarize a significant proportion of the nuclear spin states so that they are aligned in the same direction—a condition called hyperpolarization. This has important applications in magnetic resonance imaging.[77][78]

10.3.6   Energy levels

The potential energy of an electron in an atom is negative, its dependence of its position reaches the minimum (the most absolute value) inside the nucleus, and vanishes when the distance from the nucleus goes to infinity, roughly in an inverse proportion to the distance. In the quantum-mechanical model, a bound electron can only occupy a set of states centered on the nucleus, and each state corresponds to a specific energy level; see time-independent Schrödinger equation for theoretical explanation. An energy level can be measured by the amount of energy needed to unbind the electron from the atom, and is usually given in units of electronvolts (eV). The lowest energy state of a bound electron is called the ground state, i.e. stationary state, while an electron transition to a higher level results in an excited state.[79] The electron's energy raises when n increases because the (average) distance to the nucleus increases. Dependence of the energy on ℓ is caused not by electrostatic potential of the nucleus, but by interaction between electrons.

For an electron to transition between two different states, e.g. grounded state to first excited level (ionization), it must absorb or emit a photon at an energy matching the difference in the potential energy of those levels, according to Niels Bohr model, what can be precisely calculated by the Schrödinger equation. Electrons jump between orbitals in a particle-like fashion. For example, if a single photon strikes the electrons, only a single electron changes states in response to the photon; see Electron properties.

The energy of an emitted photon is proportional to its frequency, so these specific energy levels appear as distinct bands in the electromagnetic spectrum.[80] Each element has a characteristic spectrum that can depend on the nuclear charge, subshells filled by electrons, the electromagnetic interactions between the electrons and other factors.[81]

When a continuous spectrum of energy is passed through a gas or plasma, some of the photons are absorbed by atoms, causing electrons to change their energy level. Those excited electrons that remain bound to their atom spontaneously emit this energy as a photon, traveling in a random direction, and so drop back to lower energy levels. Thus the atoms behave like a filter that forms a series of dark absorption bands in the energy output. (An observer viewing the atoms from a view that does not include the continuous spectrum in the background, instead sees a series of emission lines from the photons emitted by the atoms.) Spectroscopic measurements of the strength and width of atomic spectral lines allow the composition and physical properties of a substance to be determined.[82]

Close examination of the spectral lines reveals that some display a fine structure splitting. This occurs because of spin–orbit coupling, which is an interaction between the spin and motion of the outermost electron.[83] When an atom is in an external magnetic field, spectral lines become split into three or more components; a phenomenon called the Zeeman effect. This is caused by the interaction of the magnetic field with the magnetic moment of the atom and its electrons. Some atoms can have multiple electron configurations with the same energy level, which thus appear as a single spectral line. The interaction of the magnetic field with the atom shifts these electron configurations to slightly different energy levels, resulting in multiple spectral lines.[84] The presence of an external electric field can cause a comparable splitting and shifting of spectral lines by modifying the electron energy levels, a phenomenon called the Stark effect.[85]

If a bound electron is in an excited state, an interacting photon with the proper energy can cause stimulated emission of a photon with a matching energy level. For this to occur, the electron must drop to a lower energy state that has an energy difference matching the energy of the interacting photon. The emitted photon and the interacting photon then move off in parallel and with matching phases. That is, the wave patterns of the two photons are synchronized. This physical property is used to make lasers, which can emit a coherent beam of light energy in a narrow frequency band.[86]

10.3.7   Valence and bonding behavior

Main articles: Valence (chemistry) and Chemical bond

Valency is the combining power of an element. It is equal to number of hydrogen atoms that atom can combine or displace in forming compounds.[87] The outermost electron shell of an atom in its uncombined state is known as the valence shell, and the electrons in that shell are called valence electrons. The number of valence electrons determines the bonding behavior with other atoms. Atoms tend to chemically react with each other in a manner that fills (or empties) their outer valence shells.[88] For example, a transfer of a single electron between atoms is a useful approximation for bonds that form between atoms with one-electron more than a filled shell, and others that are one-electron short of a full shell, such as occurs in the compound sodium chloride and other chemical ionic salts. However, many elements display multiple valences, or tendencies to share differing numbers of electrons in different compounds. Thus, chemical bonding between these elements takes many forms of electron-sharing that are more than simple electron transfers. Examples include the element carbon and the organic compounds.[89]

The chemical elements are often displayed in a periodic table that is laid out to display recurring chemical properties, and elements with the same number of valence electrons form a group that is aligned in the same column of the table. (The horizontal rows correspond to the filling of a quantum shell of electrons.) The elements at the far right of the table have their outer shell completely filled with electrons, which results in chemically inert elements known as the noble gases.[90][91]

10.3.8 States

Main articles: State of matter and Phase (matter)

Quantities of atoms are found in different states of matter that depend on the physical conditions, such as temperature and pressure. By varying the conditions, materials can transition between solids, liquids, gases and plasmas.[92] Within a state, a material can also exist in different allotropes. An example of this is solid carbon, which can exist as graphite or diamond.[93] Gaseous allotropes exist as well, such as dioxygen and ozone.

At temperatures close to absolute zero, atoms can form a Bose–Einstein condensate, at which point quantum mechanical effects, which are normally only observed at the atomic scale, become apparent on a macroscopic scale.[94][95] This super-cooled collection of atoms then behaves as a single super atom, which may allow fundamental checks of quantum mechanical behavior.[96]

10.4 Identification

The scanning tunneling microscope is a device for viewing surfaces at the atomic level. It uses the quantum tunneling phenomenon, which allows particles to pass through a barrier that would normally be insurmountable. Electrons tunnel through the vacuum between two planar metal electrodes, on each of which is an adsorbed atom, providing a tunneling-current density that can be measured. Scanning one atom (taken as the tip) as it moves past the other (the sample) permits plotting of tip displacement versus lateral separation for a constant current. The calculation shows the extent to which scanning-tunneling-microscope images of an individual atom are visible. It confirms that for low bias, the microscope images the space-averaged dimensions of the electron orbitals across closely packed energy levels—the Fermi level local density of states.[97][98]

An atom can be ionized by removing one of its electrons. The electric charge causes the trajectory of an atom to bend when it passes through a magnetic field. The radius by which the trajectory of a moving ion is turned by the magnetic field is determined by the mass of the atom. The mass spectrometer uses this principle to measure the mass-to-charge ratio of ions. If a sample contains multiple isotopes, the mass spectrometer can determine the proportion of each isotope in the sample by measuring the intensity of the different beams of ions. Techniques to vaporize atoms include inductively coupled plasma atomic emission spectroscopy and inductively coupled plasma mass spectrometry, both of which use a plasma to vaporize samples for analysis.[99]

A more area-selective method is electron energy loss spectroscopy, which measures the energy loss of an electron beam within a transmission electron microscope when it interacts with a portion of a sample. The atom-probe tomograph has sub-nanometer resolution in 3-D and can chemically identify individual atoms using time-of-flight mass spectrometry.[100]

Spectra of excited states can be used to analyze the atomic composition of distant stars. Specific light wavelengths contained in the observed light from stars can be separated out and related to the quantized transitions in free gas atoms. These colors can be replicated using a gas-discharge lamp containing the same element.[101] Helium was discovered in this way in the spectrum of the Sun 23 years before it was found on Earth.[102]

10.5   Origin and current state

Atoms form about 4% of the total energy density of the observable Universe, with an average density of about 0.25 atoms/m^3.[103] Within a galaxy such as the Milky Way, atoms have a much higher concentration, with the density of matter in the interstellar medium (ISM) ranging from 10^5 to 10^9 atoms/m^3.[104] The Sun is believed to be inside the Local Bubble, a region of highly ionized gas, so the density in the solar neighborhood is only about 10^3 atoms/m^3.[105] Stars form from dense clouds in the ISM, and the evolutionary processes of stars result in the steady enrichment of the ISM with elements more massive than hydrogen and helium. Up to 95% of the Milky Way's atoms are concentrated inside stars and the total mass of atoms forms about 10% of the mass of the galaxy.[106] (The remainder of the mass is an unknown dark matter.)[107]

10.5.1   Formation

Electrons are thought to exist in the Universe since early stages of the Big Bang. Atomic nuclei forms in nucleosynthesis reactions. In about three minutes Big Bang nucleosynthesis produced most of the helium, lithium, and deuterium in the Universe, and perhaps some of the beryllium and boron.[108][109][110]

Ubiquitousness and stability of atoms relies on their binding energy, which means that an atom has a lower energy than an unbound system of the nucleus and electrons. Where the temperature is much higher than ionization potential, the matter exists in the form of plasma—a gas of positively charged ions (possibly, bare nuclei) and electrons. When the temperature drops below the ionization potential, atoms become statistically favorable. Atoms (complete with bound electrons) became to dominate over charged particles 380,000 years after the Big Bang—an epoch called recombination, when the expanding Universe cooled enough to allow electrons to become attached to nuclei.[111]

Since the Big Bang, which produced no carbon or heavier elements, atomic nuclei have been combined in stars through the process of nuclear fusion to produce more of the element helium, and (via the triple alpha process) the sequence of elements from carbon up to iron;[112] see stellar nucleosynthesis for details.

Isotopes such as lithium-6, as well as some beryllium and boron are generated in space through cosmic ray spallation.[113] This occurs when a high-energy proton strikes an atomic nucleus, causing large numbers of nucleons to be ejected.

Elements heavier than iron were produced in supernovae through the r-process and in AGB stars through the s-process, both of which involve the capture of neutrons by atomic nuclei.[114] Elements such as lead formed largely through the radioactive decay of heavier elements.[115]

10.5.2   Earth

Most of the atoms that make up the Earth and its inhabitants were present in their current form in the nebula that collapsed out of a molecular cloud to form the Solar System. The rest are the result of radioactive decay, and their relative proportion can be used to determine the age of the Earth through radiometric dating.[116][117] Most of the helium in the crust of the Earth (about 99% of the helium from gas wells, as shown by its lower abundance of helium-3) is a product of alpha decay.[118]

There are a few trace atoms on Earth that were not present at the beginning (i.e., not "primordial"), nor are results of radioactive decay. Carbon-14 is continuously generated by cosmic rays in the atmosphere.[119] Some atoms on Earth have been artificially generated either deliberately or as by-products of nuclear reactors or explosions.[120][121] Of the transuranic elements—those with atomic numbers greater than 92—only plutonium and neptunium occur naturally on Earth.[122][123] Transuranic elements have radioactive lifetimes shorter than the current age of the Earth[124] and thus identifiable quantities of these elements have long since decayed, with the exception of traces of plutonium-244 possibly deposited by cosmic dust.[125] Natural deposits of plutonium and neptunium are produced by neutron capture in uranium ore.[126]

The Earth contains approximately 1.33×10^{50} atoms.[127] Although small numbers of independent atoms of noble gases exist, such as argon, neon, and helium, 99% of the atmosphere is bound in the form of molecules, including carbon dioxide and diatomic oxygen and nitrogen. At the surface of the Earth, an overwhelming majority of atoms combine to form various compounds, including water, salt, silicates and oxides. Atoms can also combine to create materials that do not consist of discrete molecules, including crystals and liquid or solid metals.[128][129] This atomic matter forms networked arrangements that lack the particular type of small-scale interrupted order associated with molecular matter.[130]

10.5.3 Rare and theoretical forms

Superheavy elements

Main article: Transuranium element

While isotopes with atomic numbers higher than lead (82) are known to be radioactive, an "island of stability" has been proposed for some elements with atomic numbers above 103. These superheavy elements may have a nucleus that is relatively stable against radioactive decay.[131] The most likely candidate for a stable superheavy atom, unbihexium, has 126 protons and 184 neutrons.[132]

Exotic matter

Main article: Exotic matter

Each particle of matter has a corresponding antimatter particle with the opposite electrical charge. Thus, the positron is a positively charged antielectron and the antiproton is a negatively charged equivalent of a proton. When a matter and corresponding antimatter particle meet, they annihilate each other. Because of this, along with an imbalance between the number of matter and antimatter particles, the latter are rare in the universe. The first causes of this imbalance are not yet fully understood, although theories of baryogenesis may offer an explanation. As a result, no antimatter atoms have been discovered in nature.[133][134] However, in 1996 the antimatter counterpart of the hydrogen atom (antihydrogen) was synthesized at the CERN laboratory in Geneva.[135][136]

Other exotic atoms have been created by replacing one of the protons, neutrons or electrons with other particles that have the same charge. For example, an electron can be replaced by a more massive muon, forming a muonic atom. These types of atoms can be used to test the fundamental predictions of physics.[137][138][139]

10.6 See also

- History of quantum mechanics
- Infinite divisibility
- List of basic chemistry topics
- Timeline of atomic and subatomic physics
- Vector model of the atom
- Nuclear model
- Radioactive isotope

10.7 Notes

[1] For more recent updates see Interactive Chart of Nuclides (Brookhaven National Laboratory).

[2] A carat is 200 milligrams. By definition, carbon-12 has 0.012 kg per mole. The Avogadro constant defines 6×10^{23} atoms per mole.

10.8   References

[1]   "Atom". *Compendium of Chemical Terminology (IUPAC Gold Book)* (2nd ed.). IUPAC. Retrieved 2015-04-25.

[2]   Ghosh, D. C.; Biswas, R. (2002). "Theoretical calculation of Absolute Radii of Atoms and Ions. Part 1. The Atomic Radii". *Int. J. Mol. Sci.* **3**: 87–113. doi:10.3390/i3020087.

[3]   Leigh, G. J., ed. (1990). *International Union of Pure and Applied Chemistry, Commission on the Nomenclature of Inorganic Chemistry, Nomenclature of Organic Chemistry – Recommendations 1990*. Oxford: Blackwell Scientific Publications. p. 35. ISBN 0-08-022369-9. An atom is the smallest unit quantity of an element that is capable of existence whether alone or in chemical combination with other atoms of the same or other elements.

[4]   Andrew G. van Melsen (1952). *From Atomos to Atom*. Mineola, N.Y.: Dover Publications. ISBN 0-486-49584-1.

[5]   Dalton, John. "On the Absorption of Gases by Water and Other Liquids", in *Memoirs of the Literary and Philosophical Society of Manchester*. 1803. Retrieved on August 29, 2007.

[6]   Einstein, Albert (1905). "Über die von der molekularkinetischen Theorie der Wärme geforderte Bewegung von in ruhenden Flüssigkeiten suspendierten Teilchen" (PDF). *Annalen der Physik* (in German) **322** (8): 549–560. Bibcode:1905AnP...322..549E. doi:10.1002/andp.19053220806. Retrieved 4 February 2007.

[7]   Mazo, Robert M. (2002). *Brownian Motion: Fluctuations, Dynamics, and Applications*. Oxford University Press. pp. 1–7. ISBN 0-19-851567-7. OCLC 48753074.

[8]   Lee, Y.K.; Hoon, K. (1995). "Brownian Motion". Imperial College. Archived from the original on 18 December 2007. Retrieved 18 December 2007.

[9]   Patterson, G. (2007). "Jean Perrin and the triumph of the atomic doctrine". *Endeavour* **31** (2): 50–53. doi: PMID 17602746.

[10]   Thomson, J. J. (August 1901). "On bodies smaller than atoms". *The Popular Science Monthly* (Bonnier Corp.): 323–335. Retrieved 2009-06-21.

[11]   "J.J. Thomson". Nobel Foundation. 1906. Retrieved 20 December 2007.

[12]   Rutherford, E. (1911). "The Scattering of α and β Particles by Matter and the Structure of the Atom" (PDF). *Philosophical Magazine* **21** (125): 669–88. doi:10.1080/14786440508637080.

[13]   "Frederick Soddy, The Nobel Prize in Chemistry 1921". Nobel Foundation. Retrieved 18 January 2008.

[14]   Thomson, Joseph John (1913). "Rays of positive electricity". *Proceedings of the Royal Society*. A**89**(607): 1–20. Bibcod. doi:10.1098/rspa.1913.0057.

[15]   Stern, David P. (16 May 2005). "The Atomic Nucleus and Bohr's Early Model of the Atom". NASA/Goddard Space Flight Center. Retrieved 20 December 2007.

[16]   Bohr, Niels (11 December 1922). "Niels Bohr, The Nobel Prize in Physics 1922, Nobel Lecture". Nobel Foundation. Retrieved 16 February 2008.

[17]   Pais, Abraham (1986). *Inward Bound: Of Matter and Forces in the Physical World*. New York: Oxford University Press. pp. 228–230. ISBN 0-19-851971-0.

[18]   Lewis, Gilbert N. (1916). "The Atom and the Molecule". *Journal of the American Chemical Society* **38** (4): 762–786. doi:10.1021/ja02261a002.

[19]   Scerri, Eric R. (2007). *The periodic table: its story and its significance*. Oxford University Press US. pp. 205–226. ISBN 0-19-530573-6.

[20]   Langmuir, Irving (1919). "The Arrangement of Electrons in Atoms and Molecules". *Journal of the American Chemical Society* **41** (6): 868–934. doi:10.1021/ja02227a002.

[21]   Scully, Marlan O.; Lamb, Willis E.; Barut, Asim (1987). "On the theory of the Stern-Gerlach apparatus". *Foundations of Physics* **17** (6): 575–583. Bibcode:1987FoPh...17..575S. doi:10.1007/BF01882788.

[22]   Brown, Kevin (2007). "The Hydrogen Atom". MathPages. Retrieved 21 December 2007.

[23] Harrison, David M. (2000). "The Development of Quantum Mechanics". University of Toronto. Archived from the original on 25 December 2007. Retrieved 21 December 2007.

[24] Aston, Francis W. (1920). "The constitution of atmospheric neon". *Philosophical Magazine* **39**(6): 449–55. doi:10.1080/148.

[25] Chadwick, James (12 December 1935). "Nobel Lecture: The Neutron and Its Properties". Nobel Foundation. Retrieved 21 December 2007.

[26] "Otto Hahn, Lise Meitner and Fritz Strassmann". *Chemical Achievers: The Human Face of the Chemical Sciences*. Chemical Heritage Foundation. Archived from the original on 24 October 2009. Retrieved 15 September 2009.

[27] Meitner, Lise; Frisch, Otto Robert (1939). "Disintegration of uranium by neutrons: a new type of nuclear reaction". *Nature* **143** (3615): 239–240. Bibcode:1939Natur.143..239M. doi:10.1038/143239a0.

[28] Schroeder, M. "Lise Meitner – Zur 125. Wiederkehr Ihres Geburtstages" (in German). Retrieved 4 June 2009.

[29] Crawford, E.; Sime, Ruth Lewin; Walker, Mark (1997). "A Nobel tale of postwar injustice". *Physics Today* **50** (9): 26–32. Bibcode:1997PhT....50i..26C. doi:10.1063/1.881933.

[30] Kullander, Sven (28 August 2001). "Accelerators and Nobel Laureates". Nobel Foundation. Retrieved 31 January 2008.

[31] "The Nobel Prize in Physics 1990". Nobel Foundation. 17 October 1990. Retrieved 31 January 2008.

[32] Demtröder, Wolfgang (2002). *Atoms, Molecules and Photons: An Introduction to Atomic- Molecular- and Quantum Physics* (1st ed.). Springer. pp. 39–42. ISBN 3-540-20631-0. OCLC 181435713.

[33] Woan, Graham (2000). *The Cambridge Handbook of Physics*. Cambridge University Press. p. 8. ISBN 0-521-57507-9. OCLC 224032426.

[34] MacGregor, Malcolm H. (1992). *The Enigmatic Electron*. Oxford University Press. pp. 33–37. ISBN 0-19-521833-7. OCLC 223372888.

[35] Particle Data Group (2002). "The Particle Adventure". Lawrence Berkeley Laboratory. Archived from the original on 4 January 2007. Retrieved 3 January 2007.

[36] Schombert, James (18 April 2006). "Elementary Particles". University of Oregon. Retrieved 3 January 2007.

[37] Jevremovic, Tatjana (2005). *Nuclear Principles in Engineering*. Springer. p. 63. ISBN 0-387-23284-2. OCLC 228384008.

[38] Pfeffer, Jeremy I.; Nir, Shlomo (2000). *Modern Physics: An Introductory Text*. Imperial College Press. pp. 330–336. ISBN 1-86094-250-4. OCLC 45900880.

[39] Wenner, Jennifer M. (10 October 2007). "How Does Radioactive Decay Work?". Carleton College. Retrieved 9 January 2008.

[40] Raymond, David (7 April 2006). "Nuclear Binding Energies". New Mexico Tech. Archived from the original on 11 December 2006. Retrieved 3 January 2007.

[41] Mihos, Chris (23 July 2002). "Overcoming the Coulomb Barrier". Case Western Reserve University. Retrieved 13 February 2008.

[42] Staff (30 March 2007). "ABC's of Nuclear Science". Lawrence Berkeley National Laboratory. Archived from the original on 5 December 2006. Retrieved 3 January 2007.

[43] Makhijani, Arjun; Saleska, Scott (2 March 2001). "Basics of Nuclear Physics and Fission". Institute for Energy and Environmental Research. Archived from the original on 16 January 2007. Retrieved 3 January 2007.

[44] Shultis, J. Kenneth; Faw, Richard E. (2002). *Fundamentals of Nuclear Science and Engineering*. CRC Press. pp. 10–17. ISBN 0-8247-0834-2. OCLC 123346507.

[45] Fewell, M. P. (1995). "The atomic nuclide with the highest mean binding energy". *American Journal of Physics* **63** (7): 653–658. Bibcode:1995AmJPh..63..653F. doi:10.1119/1.17828.

[46] Mulliken, Robert S. (1967). "Spectroscopy, Molecular Orbitals, and Chemical Bonding". *Science* **157**(3784): 13–24. BibcM. doi:10.1126/science.157.3784.13. PMID 5338306.

[47] Brucat, Philip J. (2008). "The Quantum Atom". University of Florida. Archived from the original on 7 December 2006. Retrieved 4 January 2007.

[48] Manthey, David (2001). "Atomic Orbitals". Orbital Central. Archived from the original on 10 January 2008. Retrieved 21 January 2008.

[49] Herter, Terry (2006). "Lecture 8: The Hydrogen Atom". Cornell University. Retrieved 14 February 2008.

[50] Bell, R. E.; Elliott, L. G. (1950). "Gamma-Rays from the Reaction $H^1(n,\gamma)D^2$ and the Binding Energy of the Deuteron". *Physical Review* **79** (2): 282–285. Bibcode:1950PhRv...79..282B. doi:10.1103/PhysRev.79.282.

[51] Smirnov, Boris M. (2003). *Physics of Atoms and Ions*. Springer. pp. 249–272. ISBN 0-387-95550-X.

[52] Matis, Howard S. (9 August 2000). "The Isotopes of Hydrogen". *Guide to the Nuclear Wall Chart*. Lawrence Berkeley National Lab. Archived from the original on 18 December 2007. Retrieved 21 December 2007.

[53] Weiss, Rick (17 October 2006). "Scientists Announce Creation of Atomic Element, the Heaviest Yet". Washington Post. Retrieved 21 December 2007.

[54] Sills, Alan D. (2003). *Earth Science the Easy Way*. Barron's Educational Series. pp. 131–134. ISBN 0-7641-2146-4. OCLC 51543743.

[55] Dumé, Belle (23 April 2003). "Bismuth breaks half-life record for alpha decay". Physics World. Archived from the original on 14 December 2007. Retrieved 21 December 2007.

[56] Lindsay, Don (30 July 2000). "Radioactives Missing From The Earth". Don Lindsay Archive. Archived from the original on 28 April 2007. Retrieved 23 May 2007.

[57] Tuli, Jagdish K. (April 2005). "Nuclear Wallet Cards". National Nuclear Data Center, Brookhaven National Laboratory. Retrieved 16 April 2011.

[58] CRC Handbook (2002).

[59] Mills, Ian; Cvitaš, Tomislav; Homann, Klaus; Kallay, Nikola; Kuchitsu, Kozo (1993). *Quantities, Units and Symbols in Physical Chemistry* (PDF) (2nd ed.). Oxford: International Union of Pure and Applied Chemistry, Commission on Physiochemical Symbols Terminology and Units, Blackwell Scientific Publications. p. 70. ISBN 0-632-03583-8. OCLC 27011505.

[60] Chieh, Chung (22 January 2001). "Nuclide Stability". University of Waterloo. Retrieved 4 January 2007.

[61] "Atomic Weights and Isotopic Compositions for All Elements". National Institute of Standards and Technology. Archived from the original on 31 December 2006. Retrieved 4 January 2007.

[62] Audi, G.; Wapstra, A.H.; Thibault, C. (2003). "The Ame2003 atomic mass evaluation (II)" (PDF). *Nuclear Physics A* **729** (1): 337–676. Bibcode:2003NuPhA.729..337A. doi:10.1016/j.nuclphysa.2003.11.003.

[63] Shannon, R. D. (1976). "Revised effective ionic radii and systematic studies of interatomic distances in halides and chalcogenides". *Acta Crystallographica A* **32** (5): 751–767. Bibcode:1976AcCrA..32..751S. doi:10.1107/S0567739476001551.

[64] Dong, Judy (1998). "Diameter of an Atom". The Physics Factbook. Archived from the original on 4 November 2007. Retrieved 19 November 2007.

[65] Zumdahl, Steven S. (2002). *Introductory Chemistry: A Foundation* (5th ed.). Houghton Mifflin. ISBN 0-618-34342-3. OCLC 173081482. Archived from the original on 4 March 2008. Retrieved 5 February 2008.

[66] Birkholz, M.; Rudert, R. (2008). "Interatomic distances in pyrite-structure disulfides – a case for ellipsoidal modeling of sulfur ions]" (PDF). *phys. stat. sol. b* **245**: 1858–1864. Bibcode:2008PSSBR.245.1858B. doi:10.1002/pssb.200879532.

[67] Birkholz, M. (2014). "Modeling the Shape of Ions in Pyrite-Type Crystals". *Crystals* **4**: 390–403. doi:10.3390/cryst4030390.

[68] Staff (2007). "Small Miracles: Harnessing nanotechnology". Oregon State University. Retrieved 7 January 2007.—describes the width of a human hair as 10^5 nm and 10 carbon atoms as spanning 1 nm.

[69] Padilla, Michael J.; Miaoulis, Ioannis; Cyr, Martha (2002). *Prentice Hall Science Explorer: Chemical Building Blocks*. Upper Saddle River, New Jersey USA: Prentice-Hall, Inc. p. 32. ISBN 0-13-054091-9. OCLC 47925884. There are 2,000,000 (that's2sextillion)atoms of oxygen in one drop of water—and twice as many atoms of hydrogen.

[70] Feynman, Richard (1995). *Six Easy Pieces*. The Penguin Group. p. 5. ISBN 978-0-14-027666-4. OCLC 40499574.

[71] "Radioactivity". Splung.com. Archived from the original on 4 December 2007. Retrieved 19 December 2007.

[72] L'Annunziata, Michael F. (2003). *Handbook of Radioactivity Analysis*. Academic Press. pp. 3–56. ISBN 0-12-436603-1. OCLC 16212955.

[73] Firestone, Richard B. (22 May 2000). "Radioactive Decay Modes". Berkeley Laboratory. Retrieved 7 January 2007.

[74] Hornak, J. P. (2006). "Chapter 3: Spin Physics". *The Basics of NMR*. Rochester Institute of Technology. Archived from the original on 3 February 2007. Retrieved 7 January 2007.

[75] Schroeder, Paul A. (25 February 2000). "Magnetic Properties". University of Georgia. Archived from the original on 29 April 2007. Retrieved 7 January 2007.

[76] Goebel, Greg (1 September 2007). "[4.3] Magnetic Properties of the Atom". *Elementary Quantum Physics*. In The Public Domain website. Retrieved 7 January 2007.

[77] Yarris, Lynn (Spring 1997). "Talking Pictures". *Berkeley Lab Research Review*. Archived from the original on 13 January 2008. Retrieved 9 January 2008.

[78] Liang, Z.-P.; Haacke, E. M. (1999). Webster, J. G., ed. *Encyclopedia of Electrical and Electronics Engineering: Magnetic Resonance Imaging*. vol. 2. John Wiley & Sons. pp. 412–426. ISBN 0-471-13946-7.

[79] Zeghbroeck, Bart J. Van (1998). "Energy levels". Shippensburg University. Archived from the original on 15 January 2005. Retrieved 23 December 2007.

[80] Fowles, Grant R. (1989). *Introduction to Modern Optics*. Courier Dover Publications. pp. 227–233. ISBN 0-486-65957-7. OCLC 18834711.

[81] Martin, W. C.; Wiese, W. L. (May 2007). "Atomic Spectroscopy: A Compendium of Basic Ideas, Notation, Data, and Formulas". National Institute of Standards and Technology. Archived from the original on 8 February 2007. Retrieved 8 January 2007.

[82] "Atomic Emission Spectra — Origin of Spectral Lines". Avogadro Web Site. Retrieved 10 August 2006.

[83] Fitzpatrick, Richard (16 February 2007). "Fine structure". University of Texas at Austin. Retrieved 14 February 2008.

[84] Weiss, Michael (2001). "The Zeeman Effect". University of California-Riverside. Archived from the original on 2 February 2008. Retrieved 6 February 2008.

[85] Beyer, H. F.; Shevelko, V. P. (2003). *Introduction to the Physics of Highly Charged Ions*. CRC Press. pp. 232–236. ISBN 0-7503-0481-2. OCLC 47150433.

[86] Watkins, Thayer. "Coherence in Stimulated Emission". San José State University. Archived from the original on 12 January 2008. Retrieved 23 December 2007.

[87] oxford dictionary – valency

[88] Reusch, William (16 July 2007). "Virtual Textbook of Organic Chemistry". Michigan State University. Retrieved 11 January 2008.

[89] "Covalent bonding – Single bonds". chemguide. 2000.

[90] Husted, Robert et al. (11 December 2003). "Periodic Table of the Elements". Los Alamos National Laboratory. Archived from the original on 10 January 2008. Retrieved 11 January 2008.

[91] Baum, Rudy (2003). "It's Elemental: The Periodic Table". Chemical & Engineering News. Retrieved 11 January 2008.

[92] Goodstein, David L. (2002). *States of Matter*. Courier Dover Publications. pp. 436–438. ISBN 0-13-843557-X.

[93] Brazhkin, Vadim V. (2006). "Metastable phases, phase transformations, and phase diagrams in physics and chemistry". *Physics-Uspekhi* **49** (7): 719–24. Bibcode:2006PhyU...49..719B. doi:10.1070/PU2006v049n07ABEH006013.

[94] Myers, Richard (2003). *The Basics of Chemistry*. Greenwood Press. p. 85. ISBN 0-313-31664-3. OCLC 50164580.

[95] Staff (9 October 2001). "Bose-Einstein Condensate: A New Form of Matter". National Institute of Standards and Technology. Archived from the original on 3 January 2008. Retrieved 16 January 2008.

[96] Colton, Imogen; Fyffe, Jeanette (3 February 1999). "Super Atoms from Bose-Einstein Condensation". The University of Melbourne. Archived from the original on 29 August 2007. Retrieved 6 February 2008.

[97] Jacox, Marilyn; Gadzuk, J. William (November 1997). "Scanning Tunneling Microscope". National Institute of Standards and Technology. Archived from the original on 7 January 2008. Retrieved 11 January 2008.

[98] "The Nobel Prize in Physics 1986". The Nobel Foundation. Retrieved 11 January 2008.—in particular, see the Nobel lecture by G. Binnig and H. Rohrer.

[99] Jakubowski, N.; Moens, Luc; Vanhaecke, Frank (1998). "Sector field mass spectrometers in ICP-MS". *Spectrochimica Acta Part B: Atomic Spectroscopy* **53** (13): 1739–63. Bibcode:1998AcSpe..53.1739J. doi:10.1016/S0584-8547(98)00222-5.

[100] Müller, Erwin W.; Panitz, John A.; McLane, S. Brooks (1968). "The Atom-Probe Field Ion Microscope". *Review of Scientific Instruments* **39** (1): 83–86. Bibcode:1968RScI...39...83M. doi:10.1063/1.1683116.

[101] Lochner, Jim; Gibb, Meredith; Newman, Phil (30 April 2007). "What Do Spectra Tell Us?". NASA/Goddard Space Flight Center. Archived from the original on 16 January 2008. Retrieved 3 January 2008.

[102] Winter, Mark (2007). "Helium". WebElements. Archived from the original on 30 December 2007. Retrieved 3 January 2008.

[103] Hinshaw, Gary (10 February 2006). "What is the Universe Made Of?". NASA/WMAP. Archived from the original on 31 December 2007. Retrieved 7 January 2008.

[104] Choppin, Gregory R.; Liljenzin, Jan-Olov; Rydberg, Jan (2001). *Radiochemistry and Nuclear Chemistry*. Elsevier. p. 441. ISBN 0-7506-7463-6. OCLC 162592180.

[105] Davidsen, Arthur F. (1993). "Far-Ultraviolet Astronomy on the Astro-1 Space Shuttle Mission". *Science* **259** (5093): 327–34. Bibcode:1993Sci...259..327D. doi:10.1126/science.259.5093.327. PMID 17832344.

[106] Lequeux, James (2005). *The Interstellar Medium*. Springer. p. 4. ISBN 3-540-21326-0. OCLC 133157789.

[107] Smith, Nigel (6 January 2000). "The search for dark matter". Physics World. Archived from the original on 16 February 2008. Retrieved 14 February 2008.

[108] Croswell, Ken (1991). "Boron, bumps and the Big Bang: Was matter spread evenly when the Universe began? Perhaps not; the clues lie in the creation of the lighter elements such as boron and beryllium". *New Scientist* (1794): 42. Archived from the original on 7 February 2008. Retrieved 14 January 2008.

[109] Copi, Craig J.; Schramm, DN; Turner, MS (1995). "Big-Bang Nucleosynthesis and the Baryon Density of the Universe". *Science* **267** (5195): 192–99. arXiv:astro-ph/9407006. Bibcode:1995Sci...267..192C. doi:10.1126/science.7809624. PMID 7809624.

[110] Hinshaw, Gary (15 December 2005). "Tests of the Big Bang: The Light Elements". NASA/WMAP. Archived from the original on 17 January 2008. Retrieved 13 January 2008.

[111] Abbott, Brian (30 May 2007). "Microwave (WMAP) All-Sky Survey". Hayden Planetarium. Retrieved 13 January 2008.

[112] Hoyle, F. (1946). "The synthesis of the elements from hydrogen". *Monthly Notices of the Royal Astronomical Society* **106**: 343–83. Bibcode:1946MNRAS.106..343H. doi:10.1093/mnras/106.5.343.

[113] Knauth, D. C.; Knauth, D. C.; Lambert, David L.; Crane, P. (2000). "Newly synthesized lithium in the interstellar medium". *Nature* **405** (6787): 656–58. doi:10.1038/35015028. PMID 10864316.

[114] Mashnik, Stepan G. (2000). "On Solar System and Cosmic Rays Nucleosynthesis and Spallation Processes". arXiv:astro-ph/0008382 [astro-ph].

[115] Kansas Geological Survey (4 May 2005). "Age of the Earth". University of Kansas. Retrieved 14 January 2008.

[116] Manuel 2001, pp. 407–430, 511–519.

[117] Dalrymple, G. Brent (2001). "The age of the Earth in the twentieth century: a problem (mostly) solved". *Geological Society, London, Special Publications* **190** (1): 205–21. Bibcode:2001GSLSP.190..205D. doi:10.1144/GSL.SP.2001.190.01.14. Retrieved 14 January 2008.

[118] Anderson, Don L.; Foulger, G. R.; Meibom, Anders (2 September 2006). "Helium: Fundamental models". MantlePlumes.org. Archived from the original on 8 February 2007. Retrieved 14 January 2007.

[119] Pennicott, Katie (10 May 2001). "Carbon clock could show the wrong time". PhysicsWeb. Archived from the original on 15 December 2007. Retrieved 14 January 2008.

[120] Yarris, Lynn (27 July 2001). "New Superheavy Elements 118 and 116 Discovered at Berkeley Lab". Berkeley Lab. Archived from the original on 9 January 2008. Retrieved 14 January 2008.

[121] Diamond, H et al. (1960). "Heavy Isotope Abundances in Mike Thermonuclear Device". *Physical Review* **119** (6): 2000–04. Bibcode:1960PhRv..119.2000D. doi:10.1103/PhysRev.119.2000.

[122] Poston Sr., John W. (23 March 1998). "Do transuranic elements such as plutonium ever occur naturally?". Scientific American.

[123] Keller, C. (1973). "Natural occurrence of lanthanides, actinides, and superheavy elements". *Chemiker Zeitung* **97** (10): 522–30. OSTI 4353086.

[124] Zaider, Marco; Rossi, Harald H. (2001). *Radiation Science for Physicians and Public Health Workers*. Springer. p. 17. ISBN 0-306-46403-9. OCLC 44110319.

[125] Manuel 2001, pp. 407–430,511–519.

[126] "Oklo Fossil Reactors". Curtin University of Technology. Archived from the original on 18 December 2007. Retrieved 15 January 2008.

[127] Weisenberger, Drew. "How many atoms are there in the world?". Jefferson Lab. Retrieved 16 January 2008.

[128] Pidwirny, Michael. "Fundamentals of Physical Geography". University of British Columbia Okanagan. Archived from the original on 21 January 2008. Retrieved 16 January 2008.

[129] Anderson, Don L. (2002). "The inner inner core of Earth". *Proceedings of the National Academy of Sciences* **99** (22): 13966–68. Bibcode:2002PNAS...9913966A. doi:10.1073/pnas.232565899. PMC 137819. PMID 12391308.

[130] Pauling, Linus (1960). *The Nature of the Chemical Bond*. Cornell University Press. pp. 5–10. ISBN 0-8014-0333-2. OCLC 17518275.

[131] Anonymous (2 October 2001). "Second postcard from the island of stability". *CERN Courier*. Archived from the original on 3 February 2008. Retrieved 14 January 2008.

[132] Jacoby, Mitch (2006). "As-yet-unsynthesized superheavy atom should form a stable diatomic molecule with fluorine". *Chemical & Engineering News* **84** (10): 19. doi:10.1021/cen-v084n010.p019a.

[133] Koppes, Steve (1 March 1999). "Fermilab Physicists Find New Matter-Antimatter Asymmetry". University of Chicago. Retrieved 14 January 2008.

[134] Cromie, William J. (16 August 2001). "A lifetime of trillionths of a second: Scientists explore antimatter". Harvard University Gazette. Retrieved 14 January 2008.

[135] Hijmans, Tom W. (2002). "Particle physics: Cold antihydrogen". *Nature* **419** (6906): 439–40. Bibcode:2002Natur.419..439H. doi:10.1038/419439a. PMID 12368837.

[136] Staff (30 October 2002). "Researchers 'look inside' antimatter". BBC News. Retrieved 14 January 2008.

[137] Barrett, Roger (1990). "The Strange World of the Exotic Atom". *New Scientist* (1728): 77–115. Archived from the original on 21 December 2007. Retrieved 4 January 2008.

[138] Indelicato, Paul (2004). "Exotic Atoms". *Physica Scripta* **T112** (1): 20–26. arXiv:physics/0409058.Bibcode:2004PhST..11. doi:10.1238/Physica.Topical.112a00020.

[139] Ripin, Barrett H. (July 1998). "Recent Experiments on Exotic Atoms". American Physical Society. Retrieved 15 February 2008.

10.9   Sources

- Manuel, Oliver (2001). *Origin of Elements in the Solar System: Implications of Post-1957 Observations.* Springer. ISBN 0-306-46562-0. OCLC 228374906.

10.10   Further reading

- Dalton, J. (1808). *A New System of Chemical Philosophy, Part 1.* London and Manchester: S. Russell.

- Gangopadhyaya, Mrinalkanti (1981). *Indian Atomism: History and Sources.* Atlantic Highlands, New Jersey: Humanities Press. ISBN 0-391-02177-X. OCLC 10916778.

- Harrison, Edward Robert (2003). *Masks of the Universe: Changing Ideas on the Nature of the Cosmos.* Cambridge University Press. ISBN 0-521-77351-2. OCLC 50441595.

- Iannone, A. Pablo (2001). *Dictionary of World Philosophy.* Routledge. ISBN 0-415-17995-5. OCLC 44541769.

- King, Richard (1999). *Indian philosophy: an introduction to Hindu and Buddhist thought.* Edinburgh University Press. ISBN 0-7486-0954-7.

- Levere, Trevor, H. (2001). *Transforming Matter – A History of Chemistry for Alchemy to the Buckyball.* The Johns Hopkins University Press. ISBN 0-8018-6610-3.

- Liddell, Henry George; Scott, Robert. "A Greek-English Lexicon". Perseus Digital Library.

- Liddell, Henry George; Scott, Robert. "ἄτομος". *A Greek-English Lexicon.* Perseus Digital Library. Retrieved 21 June 2010.

- McEvilley, Thomas (2002). *The shape of ancient thought: comparative studies in Greek and Indian philosophies.* Allworth Press. ISBN 1-58115-203-5.

- Moran, Bruce T. (2005). *Distilling Knowledge: Alchemy, Chemistry, and the Scientific Revolution.* Harvard University Press. ISBN 0-674-01495-2.

- Ponomarev, Leonid Ivanovich (1993). *The Quantum Dice.* CRC Press. ISBN 0-7503-0251-8. OCLC 26853108.

- Roscoe, Henry Enfield (1895). *John Dalton and the Rise of Modern Chemistry.* Century science series. New York: Macmillan. Retrieved 3 April 2011.

- Siegfried, Robert (2002). *From Elements to Atoms: A History of Chemical Composition.* DIANE. ISBN 0-87169-924-9. OCLC 186607849.

- Teresi, Dick (2003). *Lost Discoveries: The Ancient Roots of Modern Science.* Simon & Schuster. pp. 213–214. ISBN 0-7432-4379-X.

- Various (2002). Lide, David R., ed. *Handbook of Chemistry & Physics* (88th ed.). CRC. ISBN 0-8493-0486-5. OCLC 179976746. Archived from the original on 23 May 2008. Retrieved 23 May 2008.

- Wurtz, Charles Adolphe (1881). *The Atomic Theory.* New York: D. Appleton and company. ISBN 0-559-43636-X.

10.11 External links

- "Quantum Mechanics and the Structure of Atoms" on YouTube

- Freudenrich, Craig C. "How Atoms Work". How Stuff Works. Archived from the original on 8 January 2007. Retrieved 9 January 2007.

- "The Atom". *Free High School Science Texts: Physics*. Wikibooks. Retrieved 10 July 2010.

- Anonymous (2007). "The atom". Science aid+. Retrieved 10 July 2010.—a guide to the atom for teens.

- Anonymous (3 January 2006). "Atoms and Atomic Structure". BBC. Archived from the original on 2 January 2007. Retrieved 11 January 2007.

- Various (3 January 2006). "Physics 2000, Table of Contents". University of Colorado. Archived from the original on 14 January 2008. Retrieved 11 January 2008.

- Various (3 February 2006). "What does an atom look like?". University of Karlsruhe. Retrieved 12 May 2008.

THOMSON

RUTHERFORD

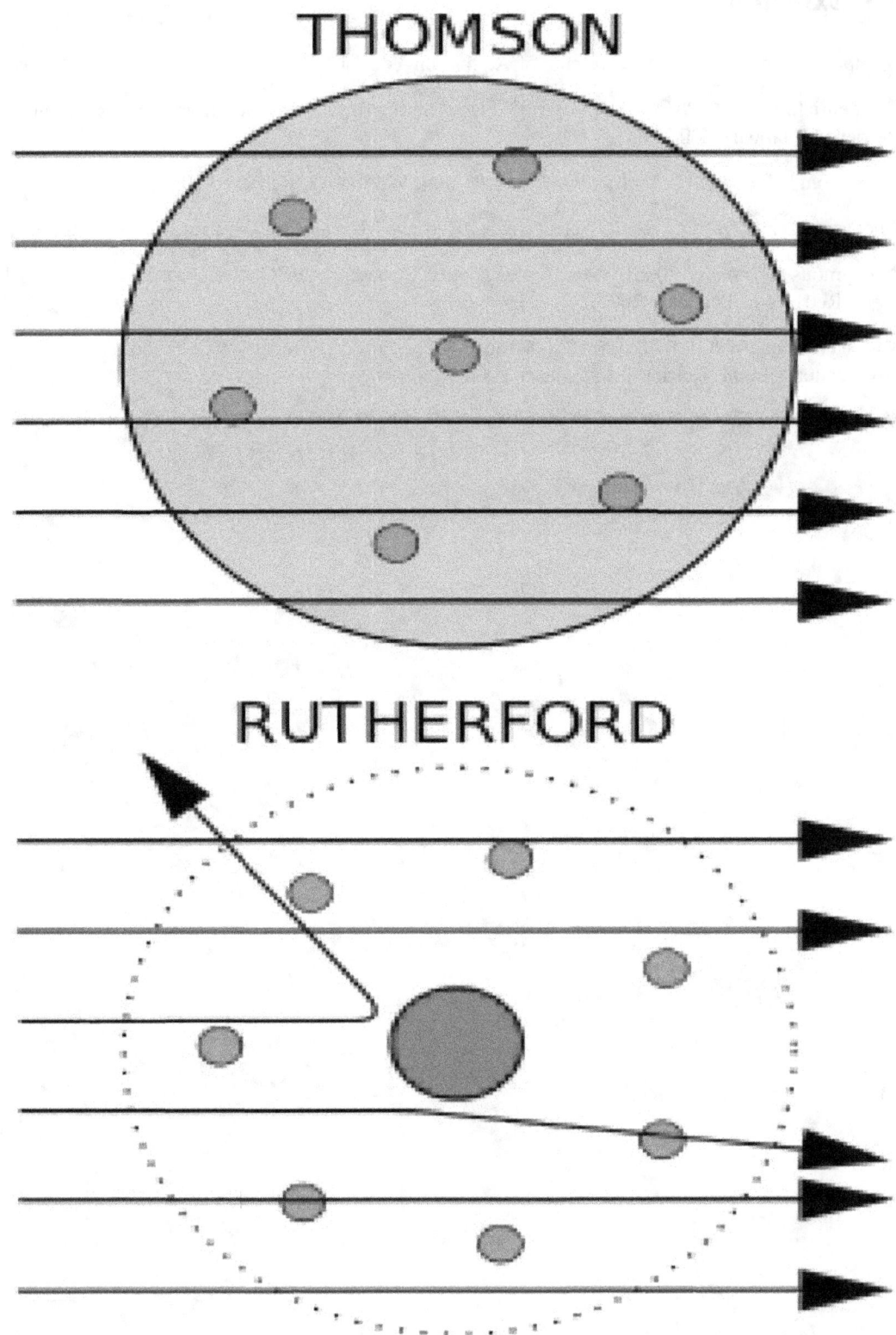

The Geiger– Marsden experiment Top: Expected results: alpha particles passing through the plum pudding model of the atom with negligible deflection. Bottom: Observed results: a small portion of the particles were deflected by the concentrated positive charge of the nucleus.

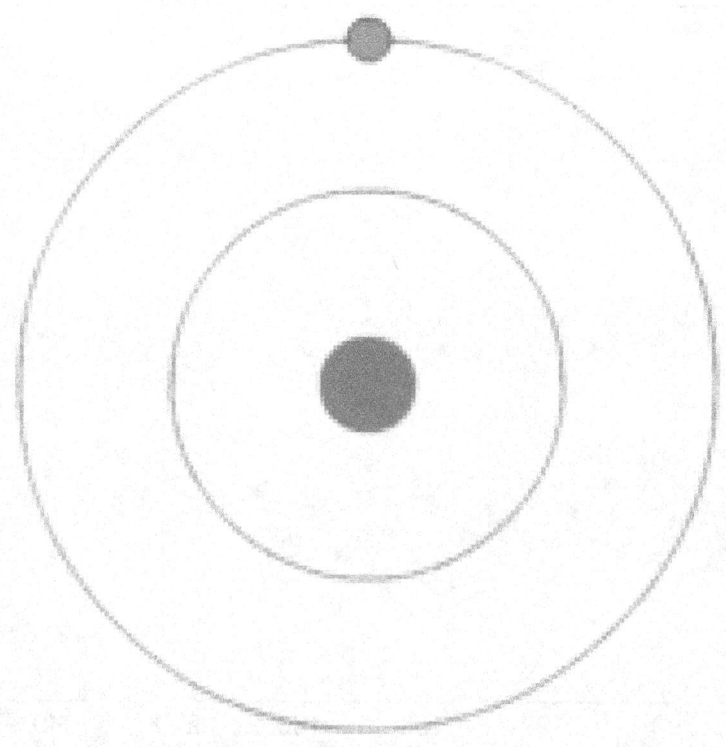

The Bohr model of the atom, with an electron making instantaneous "quantum leaps" from one orbit to another. This model is obsolete.

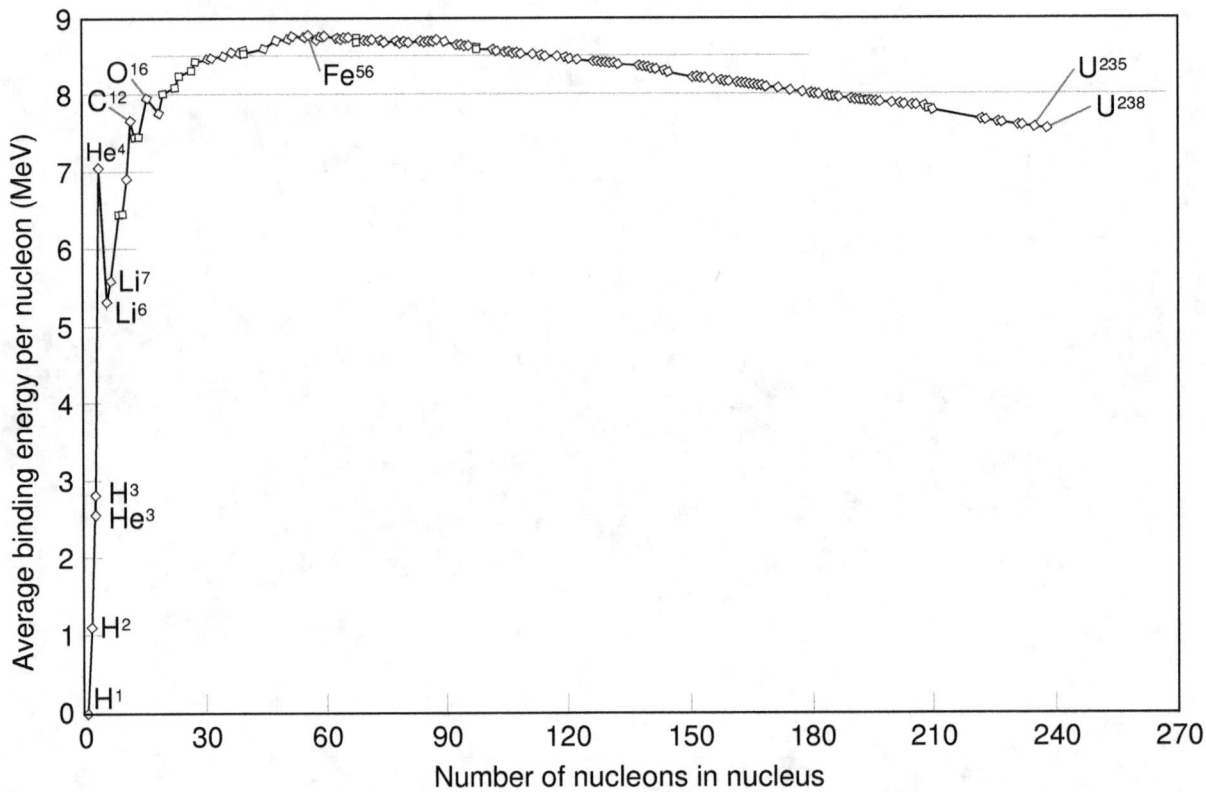

The binding energy needed for a nucleon to escape the nucleus, for various isotopes

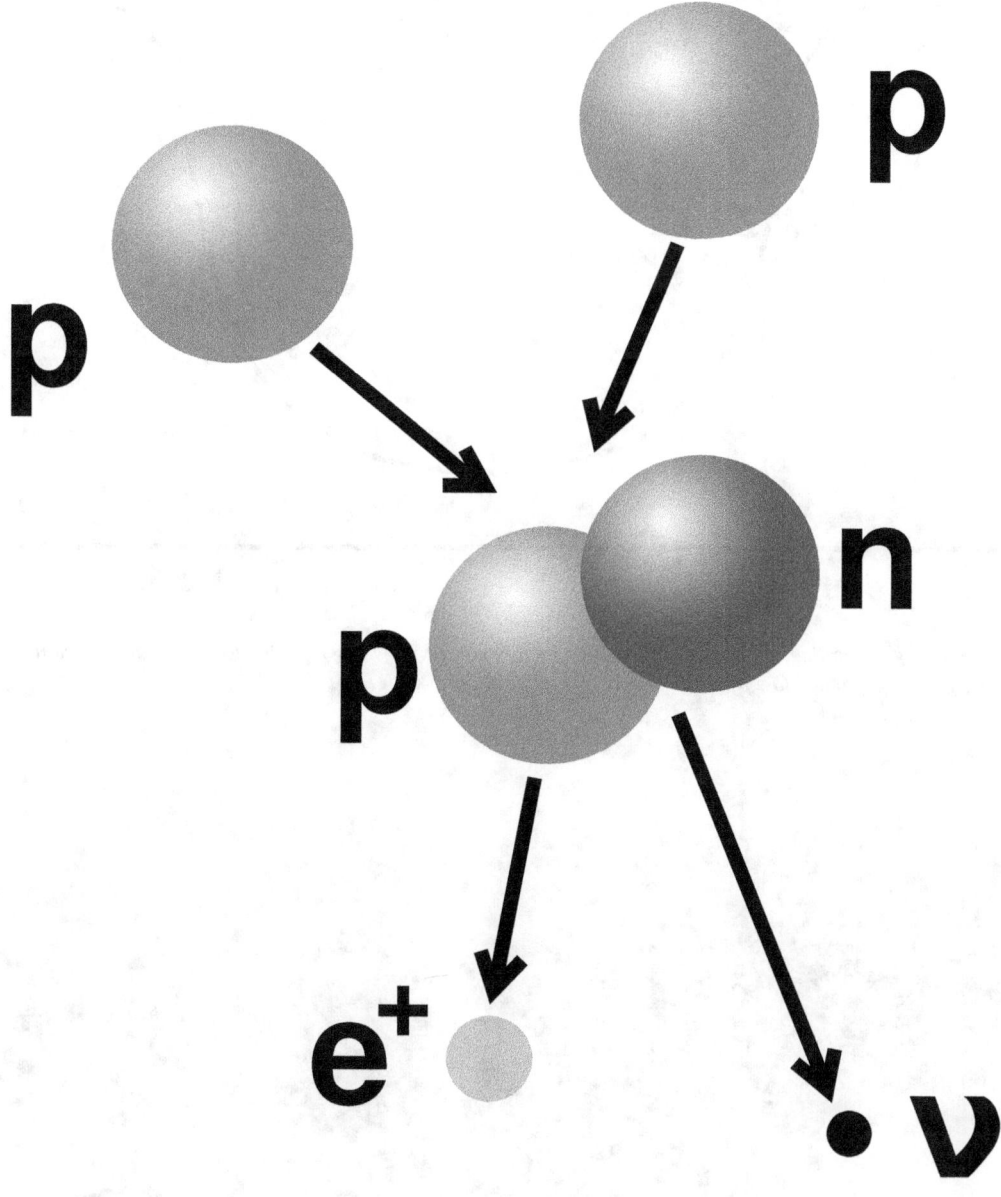

Illustration of a nuclear fusion process that forms a deuterium nucleus, consisting of a proton and a neutron, from two protons. A positron (e⁺)—an antimatter electron—is emitted along with an electron neutrino.

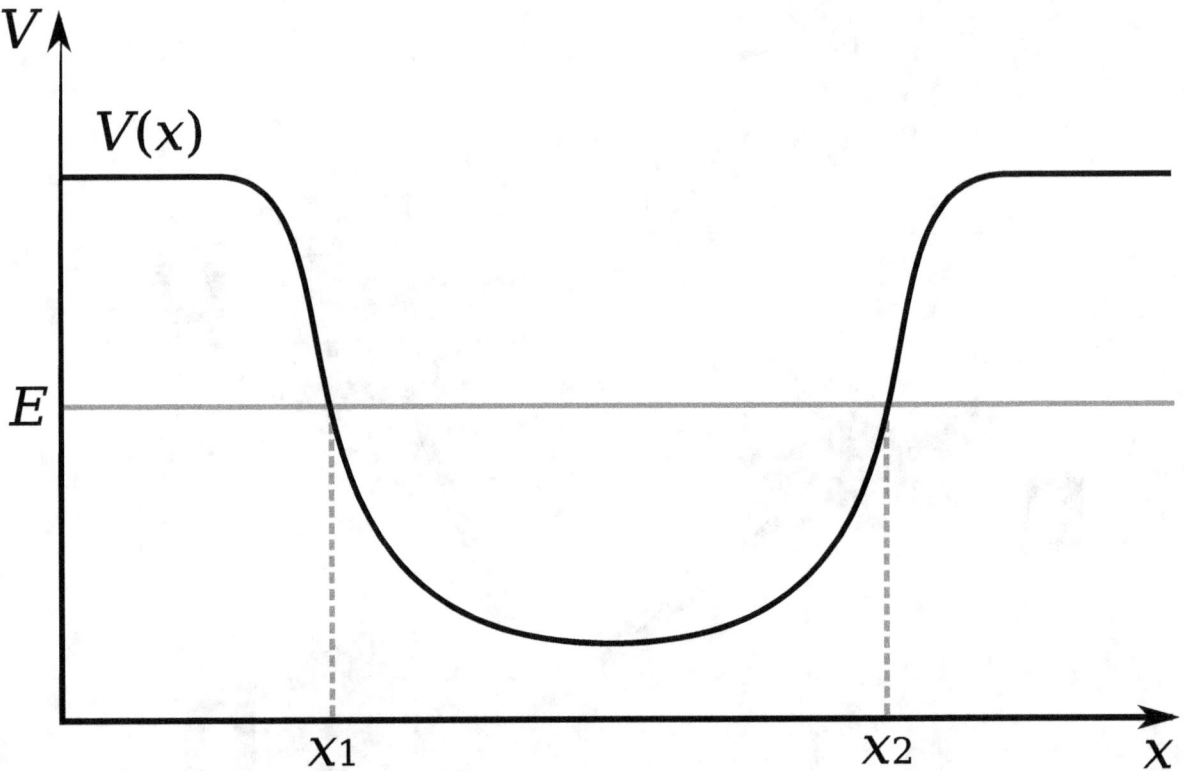

A potential well, showing, according to classical mechanics, the minimum energy $V(x)$ needed to reach each position x. Classically, a particle with energy E is constrained to a range of positions between x_1 and x_2.

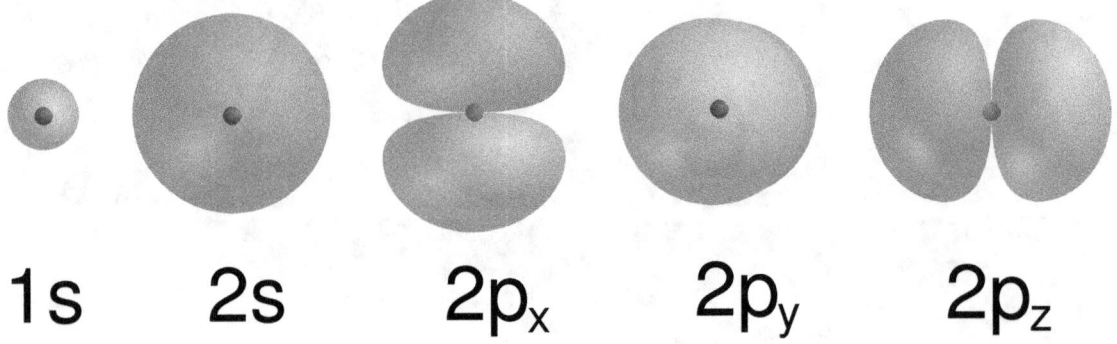

Wave functions of the first five atomic orbitals. The three 2p orbitals each display a single angular node that has an orientation and a minimum at the center.

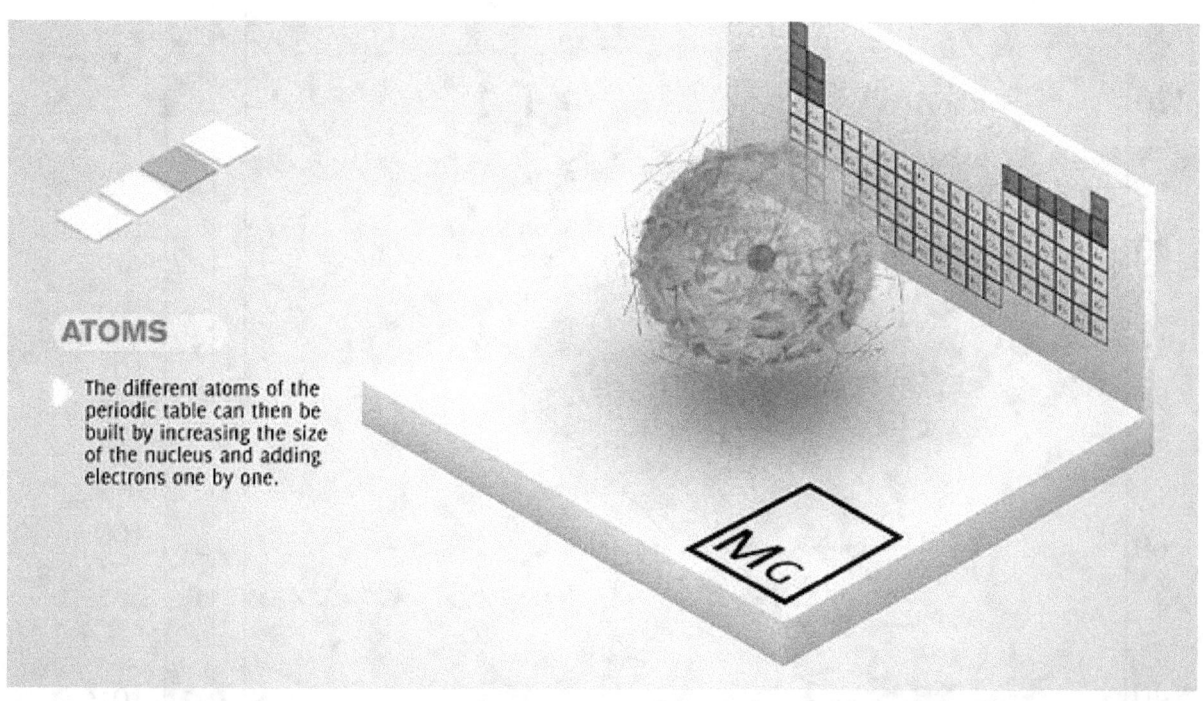

How atoms are constructed from electron orbitals and link to the periodic table

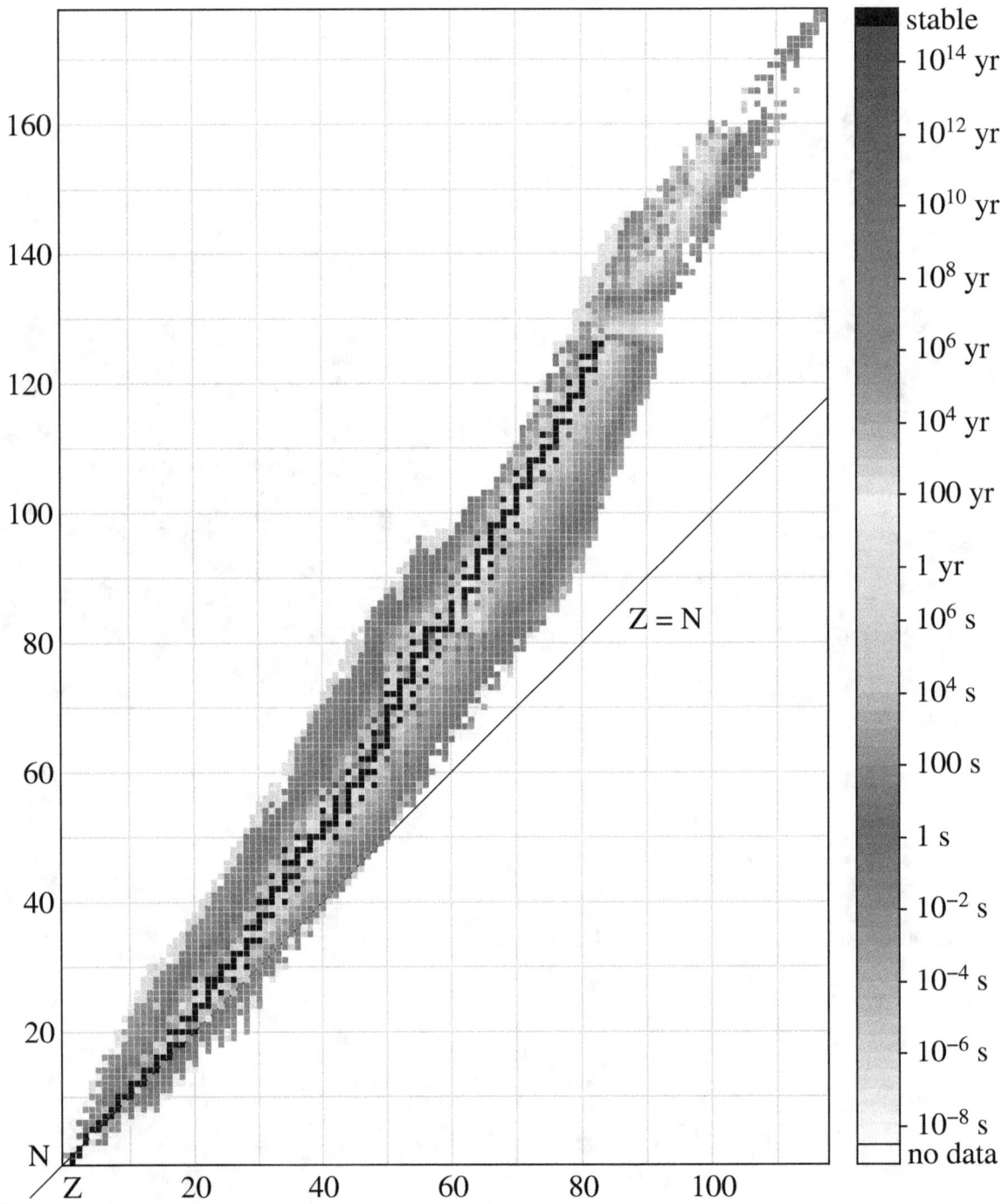

This diagram shows the half-life (T½) of various isotopes with Z protons and N neutrons.

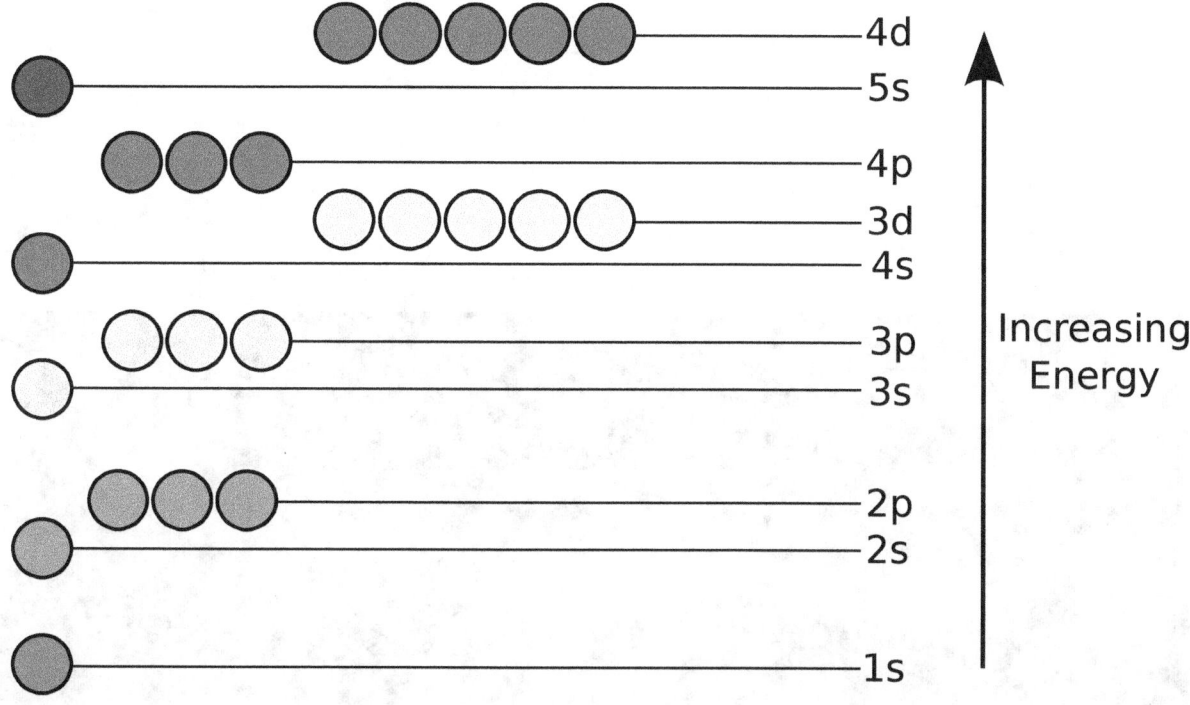

These electron's energy levels (not to scale) are sufficient for ground states of atoms up to cadmium ($5s^2\,4d^{10}$) inclusively. Do not forget that even the top of the diagram is lower than an unbound electron state.

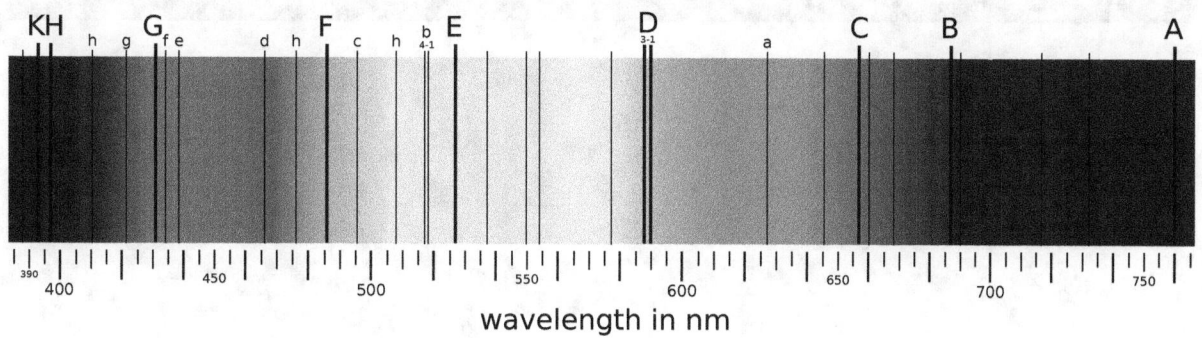

An example of absorption lines in a spectrum

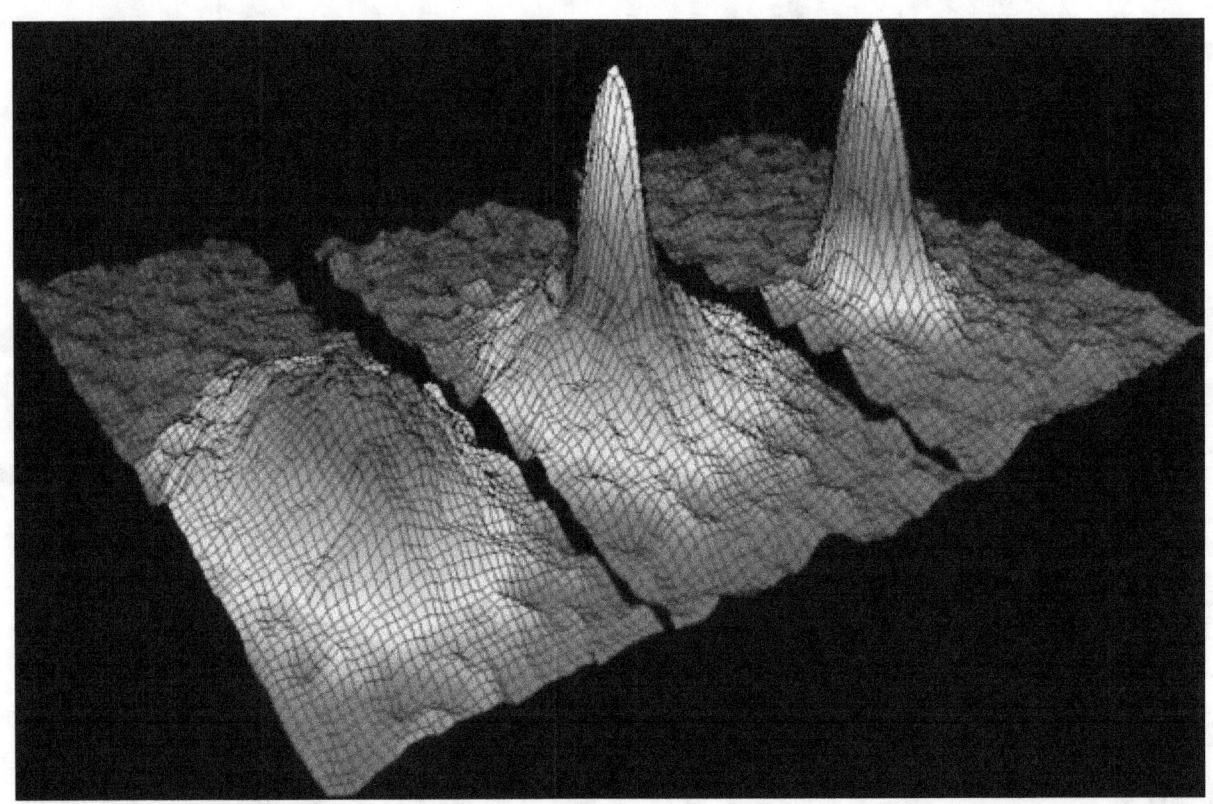

Snapshots illustrating the formation of a Bose–Einstein condensate

Scanning tunneling microscope image showing the individual atoms making up this gold (100) surface. The surface atoms deviate from the bulk crystal structure and arrange in columns several atoms wide with pits between them (See surface reconstruction).

Chapter 11

Degenerate matter

Degenerate matter[1][2] in physics is a collection of free, non-interacting particles with a pressure and other physical characteristics determined by quantum mechanical effects. It is the analogue of an ideal gas in classical mechanics. The degenerate state of matter, in the sense of deviant from an ideal gas, arises at extraordinarily high density (in compact stars) or at extremely low temperatures in laboratories.[3][4] It occurs for matter particles such as electrons, neutrons, protons, and fermions in general and is referred to as **electron-degenerate matter**, **neutron-degenerate matter**, etc. In a mixture of particles, such as ions and electrons in white dwarfs or metals, the electrons may be degenerate, while the ions are not.

In a quantum mechanical description, free particles limited to a finite volume may take only a discrete set of energies, called quantum states. The Pauli exclusion principle prevents identical fermions from occupying the same quantum state. At lowest total energy (when the thermal energy of the particles is negligible), all the lowest energy quantum states are filled. This state is referred to as full degeneracy. The pressure (called degeneracy pressure or Fermi pressure) remains nonzero even near absolute zero temperature.[3][4] Adding particles or reducing the volume forces the particles into higher-energy quantum states. This requires a compression force, and is made manifest as a resisting pressure. The key feature is that this degeneracy pressure does not depend on the temperature and only on the density of the fermions. It keeps dense stars in equilibrium independent of the thermal structure of the star.

Degenerate matter is also called a **Fermi gas** or a **degenerate gas**. A degenerate state with velocities of the fermions close to the speed of light (particle energy larger than its rest mass energy) is called **relativistic degenerate matter**.

Degenerate matter was first described for a mixture of ions and electrons in 1926 by Ralph H. Fowler,[5] showing that at densities observed in white dwarfs the electrons (obeying Fermi–Dirac statistics, the term degenerate was not yet in use) have a pressure much higher than the partial pressure of the ions.

11.1 Concept

Imagine that a plasma is cooled and compressed repeatedly. Eventually, it will not be possible to compress the plasma any further, because the Pauli exclusion principle states that two fermions cannot share the same quantum state. When in this state, since there is no extra space for any particles, we can also say that a particle's location is extremely defined. Therefore, since (according to the Heisenberg uncertainty principle) $\Delta p \Delta x \geq \hbar/2$ where Δp is the uncertainty in the particle's momentum and Δx is the uncertainty in position, then we must say that their momentum is extremely uncertain since the particles are located in a very confined space. Therefore, *even though the plasma is cold*, the particles must be moving very fast on average. This leads to the conclusion that in order to compress an object into a very small space, tremendous force is required to control its particles' momentum.

Unlike a classical ideal gas, whose pressure is proportional to its temperature ($P = nkT/V$, where P is pressure, V is the volume, n is the number of particles—typically atoms or molecules—k is Boltzmann's constant, and T is temperature), the pressure exerted by degenerate matter depends only weakly on its temperature. In particular, the pressure remains nonzero even at absolute zero temperature. At relatively low densities, the pressure of a fully degenerate gas is given by

$P = K(n/V)5/3$
, where K depends on the properties of the particles making up the gas. At very high densities, where most of the particles are forced into quantum states with relativistic energies, the pressure is given by $P = K'(n/V)4/3$
, where K' again depends on the properties of the particles making up the gas.[6]

All matter experiences both normal thermal pressure and degeneracy pressure, but in commonly encountered gases, thermal pressure dominates so much that degeneracy pressure can be ignored. Likewise, degenerate matter still has normal thermal pressure, but at extremely high densities the degeneracy pressure usually dominates.

Exotic examples of degenerate matter include neutronium, strange matter, metallic hydrogen and white dwarf matter. Degeneracy pressure contributes to the pressure of conventional solids, but these are not usually considered to be degenerate matter because a significant contribution to their pressure is provided by electrical repulsion of atomic nuclei and the screening of nuclei from each other by electrons. In metals it is useful to treat the conduction electrons alone as a degenerate, free electron gas while the majority of the electrons are regarded as occupying bound quantum states. This contrasts with degenerate matter that forms the body of a white dwarf, where all the electrons would be treated as occupying free particle momentum states.

11.2 Degenerate gases

Degenerate gases are gases composed of fermions that have a particular configuration that usually forms at high densities. Fermions are particles with half-integer spin. Their behavior is regulated by a set of quantum mechanical rules called the Fermi–Dirac statistics. One particular rule is the Pauli exclusion principle, which states that there can be only one fermion occupying each quantum state, which also applies to electrons that are not bound to a nucleus but merely confined to a fixed volume, such as in the deep interior of a star. Such particles as electrons, protons, neutrons, and neutrinos are all fermions and obey Fermi–Dirac statistics.

A fermion gas in which all energy states below some energy level are filled is called a fully degenerate fermion gas. The difference between this energy level and the lowest energy level is known as the Fermi energy. The electron gas in ordinary metals and in the interior of white dwarf stars constitute two examples of a degenerate electron gas. Most stars are supported against their own gravitation by normal thermal gas pressure. White dwarf stars are supported by the degeneracy pressure of the electron gas in their interior, while for neutron stars the degenerate particles are neutrons.

11.2.1 Electron degeneracy

In an ordinary fermion gas in which thermal effects dominate, most of the available electron energy levels are unfilled and the electrons are free to move to these states. As particle density is increased, electrons progressively fill the lower energy states and additional electrons are forced to occupy states of higher energy even at low temperatures. Degenerate gases strongly resist further compression because the electrons cannot move to already filled lower energy levels due to the Pauli exclusion principle. Since electrons cannot give up energy by moving to lower energy states, no thermal energy can be extracted. The momentum of the fermions in the fermion gas nevertheless generates pressure, termed *degeneracy pressure*.

Under high densities the matter becomes a degenerate gas when the electrons are all stripped from their parent atoms. In the core of a star, once hydrogen burning in nuclear fusion reactions stops, it becomes a collection of positively charged ions, largely helium and carbon nuclei, floating in a sea of electrons, which have been stripped from the nuclei. Degenerate gas is an almost perfect conductor of heat and does not obey the ordinary gas laws. White dwarfs are luminous not because they are generating any energy but rather because they have trapped a large amount of heat which is gradually radiated away. Normal gas exerts higher pressure when it is heated and expands, but the pressure in a degenerate gas does not depend on the temperature. When gas becomes super-compressed, particles position right up against each other to produce degenerate gas that behaves more like a solid. In degenerate gases the kinetic energies of electrons are quite high and the rate of collision between electrons and other particles is quite low, therefore degenerate electrons can travel great distances at velocities that approach the speed of light. Instead of temperature, the pressure in a degenerate gas depends only on the speed of the degenerate particles; however, adding heat does not increase the speed. Pressure is only increased by the mass of the particles, which increases the gravitational force pulling the particles closer together. Therefore, the phenomenon is the opposite of that normally found in matter where if the mass of the matter is increased, the object becomes bigger. In degenerate gas, when the mass is increased, the pressure is increased, and the particles become spaced closer together, so the object becomes smaller. Degenerate gas can be compressed to very high densities, typical values being in the range of 10,000 kilograms per cubic centimeter.

There is an upper limit to the mass of an electron-degenerate object, the Chandrasekhar limit, beyond which electron degeneracy pressure cannot support the object against collapse. The limit is approximately 1.44[7]solar masses for objects with compositions similar to the sun. The mass cutoff changes with the chemical composition of the object, as this affects the ratio of mass to number of electrons present. Celestial objects below this limit are white dwarf stars, formed by the collapse of the cores of stars that run out of fuel. During collapse, an electron-degenerate gas forms in the core, providing sufficient degeneracy pressure as it is compressed to resist further collapse. Above this mass limit, a neutron star (supported by neutron degeneracy pressure) or a black hole may be formed instead.

11.2.2   Proton degeneracy

Sufficiently dense matter containing protons experiences proton degeneracy pressure, in a manner similar to the electron degeneracy pressure in electron-degenerate matter: protons confined to a sufficiently small volume have a large uncertainty in their momentum due to the Heisenberg uncertainty principle. Because protons are much more massive than electrons, the same momentum represents a much smaller velocity for protons than for electrons. As a result, in matter with approximately equal numbers of protons and electrons, proton degeneracy pressure is much smaller than electron degeneracy pressure, and proton degeneracy is usually modeled as a correction to the equations of state of electron-degenerate matter.

11.2.3   Neutron degeneracy

Neutron degeneracy is analogous to electron degeneracy and is demonstrated in neutron stars, which are primarily supported by the pressure from a degenerate neutron gas.[8] This happens when the core of a white dwarf star above the vicinity of 1.4 solar masses, the Chandrasekhar limit, collapses and is not halted by the degenerate electrons. As the star collapses, the Fermi energy of the electrons increases to the point where it is energetically favorable for them to combine with protons to produce neutrons (via inverse beta decay, also termed electron capture and "neutralization"). The result of this collapse is an extremely compact star composed of nuclear matter, which is predominantly a degenerate neutron gas, sometimes called neutronium, with a small admixture of degenerate proton and electron gases.

Neutrons in a degenerate neutron gas are spaced much more closely than electrons in an electron-degenerate gas, because the more massive neutron has a much shorter wavelength at a given energy. In the case of neutron stars and white dwarf stars, this is compounded by the fact that the pressures within neutron stars are much higher than those in white dwarfs. The pressure increase is caused by the fact that the compactness of a neutron star causes gravitational forces to be much higher than in a less compact body with similar mass. This results in a star with a diameter on the order of a thousandth that of a white dwarf.

There is an upper limit to the mass of a neutron-degenerate object, the Tolman–Oppenheimer–Volkoff limit, which is analogous to the Chandrasekhar limit for electron-degenerate objects. The precise limit is unknown, as it depends on the equations of state of nuclear matter, for which a highly accurate model is not yet available. Above this limit, a neutron star may collapse into a black hole, or into other, denser forms of degenerate matter (such as quark matter) if these forms exist and have suitable properties (mainly related to degree of compressibility, or "stiffness", described by the equations of state).

11.2.4   Quark degeneracy

At densities greater than those supported by neutron degeneracy, quark matter is expected to occur. Several variations of this have been proposed that represent quark-degenerate states. Strange matter is a degenerate gas of quarks that is often assumed to contain strange quarks in addition to the usual up and down quarks. Color superconductor materials are degenerate gases of quarks in which quarks pair up in a manner similar to Cooper pairing in electrical superconductors. The equations of state for the various proposed forms of quark-degenerate matter vary widely, and are usually also poorly defined, due to the difficulty of modeling strong force interactions.

Quark-degenerate matter may occur in the cores of neutron stars, depending on the equations of state of neutron-degenerate matter. It may also occur in hypothetical quark stars, formed by the collapse of objects above the Tolman–Oppenheimer–Volkoff mass limit for neutron-degenerate objects. Whether quark-degenerate matter forms at all in these situations depends on the equations of state of both neutron-degenerate matter and quark-degenerate matter, both of which are poorly known.

11.2.5   Preon degeneracy hypothesis

Preons are subatomic particles proposed to be the constituents of quarks, which become composite particles in preon-based models. If preons exist, preon-degenerate matter might occur at densities greater than that which can be supported by quark-degenerate matter. The expected properties of preon-degenerate matter depend very strongly on the model chosen to describe preons, and the existence of preons is not assumed by the majority of the scientific community, due to conflicts between the preon models originally proposed and experimental data from particle accelerators.

11.2.6   Singularity

At densities greater than those supported by any degeneracy, gravity overwhelms all other forces. To the best of our current understanding, the body collapses to form a black hole. At the same time, the material must be converted from fermions, subject to degeneracy pressure, to bosons, which are not. Physics cannot currently predict what sort of bosons those might be.

In the frame of reference that is co-moving with the collapsing matter, general relativity models without quantum mechanics have all the matter ending up in an infinitely dense singularity at the center of the event horizon. It is a general result of quantum mechanics that no object can be confined in a space smaller than its own wavelength, making such a singularity impossible, but we do not have a theory that combines GR and QM sufficiently to tell us what the structure inside a black hole might be.

In the frame of reference of an observer at infinity, the collapse viewed in electromagnetic radiation asymptotically appears to approach the event horizon. Observations in gravitational waves, which are planned at LIGO, would let distant observers detect effects from matter falling inside the event horizon.

As a consequence of relativity, the extreme gravitational field and orbital velocity experienced by infalling matter around a black hole would "slow" time for that matter relative to a distant observer. In the slowed proper time experienced by the infalling matter, the fall to the center of the black hole would be quite short in duration.

11.3   See also

- Compact star
- White dwarf
- Neutron star
- Quark star—QCD matter
- Preon star—preon
- Pauli exclusion principle
- Uncertainty principle
- Neutronium
- Electron degeneracy pressure
- Nuclear matter

- Gravitational time dilation

- List of plasma (physics) articles

11.4 Notes

[1] H.S. Goldberg, M.D. Scadron (1987). *Physics of Stellar Evolution and Cosmology.* Taylor & Francis. p. 202. ISBN 0-677-05540-4.

[2] An Introduction to Modern Astrophysics §16.3 "The Physics of Degenerate Matter – Carroll & Ostlie, 2007, second edition. ISBN 0-8053-0402-9

[3] see http://apod.nasa.gov/apod/ap100228.html

[4] Andrew G. Truscott, Kevin E. Strecker, William I. McAlexander, Guthrie Partridge, and Randall G. Hulet, "Observation of Fermi Pressure in a Gas of Trapped Atoms", Science, 2 March 2001

[5] On Dense Matter, R. H. Fowler, *Monthly Notices of the Royal Astronomical Society* **87** (1926), pp. 114–122.

[6] *Stellar Structure and Evolution* section 15.3 – R Kippenhahn & A. Weigert, 1990, 3rd printing 1994. ISBN 0-387-58013-1

[7] ENCYCLOPAEDIA BRITANNICA

[8] Potekhin, A. Y. (2011). "The Physics of Neutron Stars". arXiv:1102.5735. Bibcode:2010PhyU...53.1235Y.doi:10.379.

11.5 References

- Cohen-Tanoudji, Claude (2011). *Advances in Atomic Physics.* World Scientific. p. 791. ISBN 978-981-277-496-5.

11.6 External links

- Detailed mathematical explanation of degenerate gases

- Mass-radius diagram of degenerate star types

Chapter 12

QCD matter

Quark matter or **QCD matter** refers to any of a number of theorized phases of matter whose degrees of freedom include quarks and gluons. These theoretical phases would occur at extremely high temperatures and densities, billions of times higher than can be produced in equilibrium in laboratories. Under such extreme conditions, the familiar structure of matter, where the basic constituents are nuclei (consisting of nucleons which are bound states of quarks) and electrons, is disrupted. In quark matter it is more appropriate to treat the quarks themselves as the basic degrees of freedom.

In the standard model of particle physics, the strong force is described by the theory of quantum chromodynamics (QCD). At ordinary temperatures or densities this force just confines the quarks into composite particles (hadrons) of size around 10^{-15} m = 1 femtometer = 1 fm (corresponding to the QCD energy scale $\Lambda QCD \approx 200$ MeV) and its effects are not noticeable at longer distances. However, when the temperature reaches the QCD energy scale (T of order 10^{12} kelvins) or the density rises to the point where the average inter-quark separation is less than 1 fm (quark chemical potential μ around 400 MeV), the hadrons are melted into their constituent quarks, and the strong interaction becomes the dominant feature of the physics. Such phases are called quark matter or QCD matter.

The strength of the color force makes the properties of quark matter unlike gas or plasma, instead leading to a state of matter more reminiscent of a liquid. At high densities, quark matter is a Fermi liquid, but is predicted to exhibit color superconductivity at high densities and temperatures below 10^{12} K.

12.1 Occurrence

12.1.1 Natural occurrence

- In the early universe, at high temperatures according to the Big Bang theory, when the universe was only a few tens of microseconds old, the phase of matter took the form of a hot phase of quark matter called the quark–gluon plasma (QGP).

- Compact stars (neutron stars). A neutron star is much cooler than 10^{12} K, but it is compressed by its own mass to such high densities, that it is reasonable to surmise that quark matter may exist in the core.[1] Compact stars composed mostly or entirely of quark matter are called quark stars or strange stars, yet at this time no star with properties expected of these objects has been observed.

- Strangelets. These are theoretically postulated (but as yet unobserved) lumps of strange matter comprising nearly equal amounts of up, down and strange quarks.

- Cosmic ray impacts. Cosmic rays comprise also high energy atomic nuclei, particularly that of iron. Laboratory experiments suggest that interaction with heavy noble gas in the upper atmosphere would lead to quark–gluon plasma formation.

12.1.2 Laboratory experiments

- Heavy-ion collisions at very high energies can produce small short-lived regions of space whose energy density is comparable to that of the 20-micro-second-old universe. This has been achieved by colliding heavy nuclei at high speeds, and a first time claim of formation of quark–gluon plasma came from the SPS accelerator at CERN in February 2000.[2] This work has been continued at more powerful accelerators, such as RHIC at Brookhaven National Laboratory in the USA, and as of 2010 at the LHC at CERN located in the border area of Switzerland & France. There is good evidence that the quark–gluon plasma has also been produced at RHIC.[3]

12.2 Thermodynamics

The context for understanding the thermodynamics of quark matter is the standard model of particle physics, which contains six different flavors of quarks, as well as leptons like electrons and neutrinos. These interact via the strong interaction, electromagnetism, and also the weak interaction which allows one flavor of quark to turn into another. Electromagnetic interactions occur between particles that carry electrical charge; strong interactions occur between particles that carry color charge.

The correct thermodynamic treatment of quark matter depends on the physical context. For large quantities that exist for long periods of time (the "thermodynamic limit"), we must take into account the fact that the only conserved charges in the standard model are quark number (equivalent to baryon number), electric charge, the eight color charges, and lepton number. Each of these can have an associated chemical potential. However, large volumes of matter must be electrically and color-neutral, which determines the electric and color charge chemical potentials. This leaves a three-dimensional phase space, parameterized by quark chemical potential, lepton chemical potential, and temperature.

In compact stars quark matter would occupy cubic kilometers and exist for millions of years, so the thermodynamic limit is appropriate. However, the neutrinos escape, violating lepton number, so the phase space for quark matter in compact stars only has two dimensions, temperature (T) and quark number chemical potential μ. A strangelet is not in the thermodynamic limit of large volume, so it is like an exotic nucleus: it may carry electric charge.

A heavy-ion collision is in neither the thermodynamic limit of large volumes nor long times. Putting aside questions of whether it is sufficiently equilibrated for thermodynamics to be applicable, there is certainly not enough time for weak interactions to occur, so flavor is conserved, and there are independent chemical potentials for all six quark flavors. The initial conditions (the impact parameter of the collision, the number of up and down quarks in the colliding nuclei, and the fact that they contain no quarks of other flavors) determine the chemical potentials. (Reference for this section:,[4][5]).

12.3 Phase diagram

The phase diagram of quark matter is not well known, either experimentally or theoretically. A commonly conjectured form of the phase diagram is shown in the figure.[4] It is applicable to matter in a compact star, where the only relevant thermodynamic potentials are quark chemical potential μ and temperature T. For guidance it also shows the typical values of μ and T in heavy-ion collisions and in the early universe. For readers who are not familiar with the concept of a chemical potential, it is helpful to think of μ as a measure of the imbalance between quarks and antiquarks in the system. Higher μ means a stronger bias favoring quarks over antiquarks. At low temperatures there are no antiquarks, and then higher μ generally means a higher density of quarks.

Ordinary atomic matter as we know it is really a mixed phase, droplets of nuclear matter (nuclei) surrounded by vacuum, which exists at the low-temperature phase boundary between vacuum and nuclear matter, at $\mu = 310$ MeV and T close to zero. If we increase the quark density (i.e. increase μ) keeping the temperature low, we move into a phase of more and more compressed nuclear matter. Following this path corresponds to burrowing more and more deeply into a neutron star. Eventually, at an unknown critical value of μ, there is a transition to quark matter. At ultra-high densities we expect to find the color-flavor-locked (CFL) phase of color-superconducting quark matter. At intermediate densities we expect some other phases (labelled "non-CFL quark liquid" in the figure) whose nature is presently unknown,.[4][5] They might be other forms of color-superconducting quark matter, or something different.

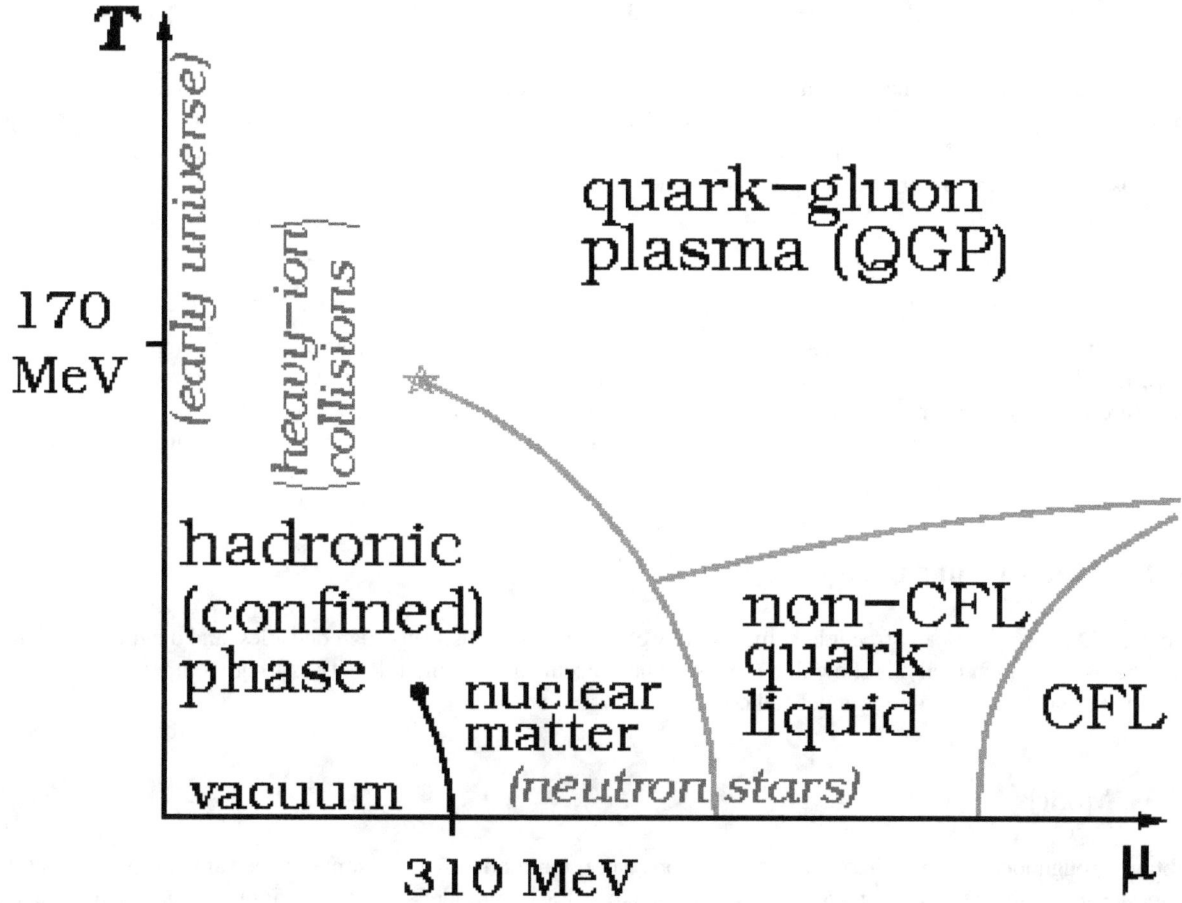

Conjectured form of the phase diagram of QCD matter, with temperature as ordinate (in mega-electron volts) and quark chemical potential as abscissa.[4]

Now, imagine starting at the bottom left corner of the phase diagram, in the vacuum where $\mu = T = 0$. If we heat up the system without introducing any preference for quarks over antiquarks, this corresponds to moving vertically upwards along the T axis. At first, quarks are still confined and we create a gas of hadrons (pions, mostly). Then around $T = 150$ MeV there is a crossover to the quark gluon plasma: thermal fluctuations break up the pions, and we find a gas of quarks, antiquarks, and gluons, as well as lighter particles such as photons, electrons, positrons, etc. Following this path corresponds to travelling far back in time (so to say), to the state of the universe shortly after the big bang (where there was a very tiny preference for quarks over antiquarks).

The line that rises up from the nuclear/quark matter transition and then bends back towards the T axis, with its end marked by a star, is the conjectured boundary between confined and unconfined phases. Until recently it was also believed to be a boundary between phases where chiral symmetry is broken (low temperature and density) and phases where it is unbroken (high temperature and density). It is now known that the CFL phase exhibits chiral symmetry breaking, and other quark matter phases may also break chiral symmetry, so it is not clear whether this is really a chiral transition line. The line ends at the "chiral critical point", marked by a star in this figure, which is a special temperature and density at which striking physical phenomena, analogous to critical opalescence, are expected. (Reference for this section:[4][5][6]).

For a complete description of phase diagram it is required that one must have complete understanding of dense, strongly interacting hadronic matter and strongly interacting quark matter from some underlying theory e.g. quantum chromodynamics (QCD). However because such a description requires the proper understanding of QCD in its non-perturbative regime, which is still far from being completely understood, any theoretical advance remains very challenging.

12.4   Theoretical challenges: calculation techniques

The phase structure of quark matter remains mostly conjectural because it is difficult to perform calculations predicting the properties of quark matter. The reason is that QCD, the theory describing the dominant interaction between quarks, is strongly coupled at the densities and temperatures of greatest physical interest, and hence it is very hard to obtain any predictions from it. Here are brief descriptions of some of the standard approaches.

12.4.1   Lattice gauge theory

The only first-principles calculational tool currently available is lattice QCD, i.e. brute-force computer calculations. Because of a technical obstacle known as the fermion sign problem, this method can only be used at low density and high temperature ($\mu < T$), and it predicts that the crossover to the quark–gluon plasma will occur around $T = 150$ MeV [7] However, it cannot be used to investigate the interesting color-superconducting phase structure at high density and low temperature.[8]

12.4.2   Weak coupling theory

Because QCD is asymptotically free it becomes weakly coupled at unrealistically high densities, and diagrammatic methods can be used.[5] Such methods show that the CFL phase occurs at very high density. At high temperatures, however, diagrammatic methods are still not under full control.

12.4.3   Models

To obtain a rough idea of what phases might occur, one can use a model that has some of the same properties as QCD, but is easier to manipulate. Many physicists use Nambu-Jona-Lasinio models, which contain no gluons, and replace the strong interaction with a four-fermion interaction. Mean-field methods are commonly used to analyse the phases. Another approach is the bag model, in which the effects of confinement are simulated by an additive energy density that penalizes unconfined quark matter.

12.4.4   Effective theories

Many physicists simply give up on a microscopic approach, and make informed guesses of the expected phases (perhaps based on NJL model results). For each phase, they then write down an effective theory for the low-energy excitations, in terms of a small number of parameters, and use it to make predictions that could allow those parameters to be fixed by experimental observations.[6]

12.4.5   Other approaches

Main article: AdS/QCD

There are other methods that are sometimes used to shed light on QCD, but for various reasons have not yet yielded useful results in studying quark matter.

1/N expansion

Treat the number of colors N, which is actually 3, as a large number, and expand in powers of $1/N$. It turns out that at high density the higher-order corrections are large, and the expansion gives misleading results.[4]

Supersymmetry

Adding scalar quarks (squarks) and fermionic gluons (gluinos) to the theory makes it more tractable, but the thermodynamics of quark matter depends crucially on the fact that only fermions can carry quark number, and on the number of degrees of freedom in general.

12.5 Experimental challenges

Experimentally, it is hard to map the phase diagram of quark matter because it has been rather difficult to learn how to tune to high enough temperatures and density in the laboratory experiment using collisions of relativistic heavy ions as experimental tools. However, these collisions ultimately will provide information about the crossover from hadronic matter to QGP. It has been suggested that the observations of compact stars may also constrain the information about the high-density low-temperature region. Models of the cooling, spin-down, and precession of these stars offer information about the relevant properties of their interior. As observations become more precise, physicists hope to learn more.[4]

One of the natural subjects for future research is the search for the exact location of the chiral critical point. Some ambitious lattice QCD calculations may have found evidence for it, and future calculations will clarify the situation. Heavy-ion collisions might be able to measure its position experimentally, but this will require scanning across a range of values of μ and T.[9]

12.6 See also

- Color–flavor locking

- Lattice QCD

- Quantum chromodynamics

- Quark–gluon plasma

- Quark star

- Strange matter

- Strangeness production

- 1/N expansion

12.7 References

[1] Shapiro and Teukolsky: *Black Holes, White Dwarfs and Neutron Stars: The Physics of Compact Objects*, Wiley 2008

[2] Ulrich Heinz; Maurice Jacob (2000). "Evidence for a New State of Matter: an Assessment of the Results from the CERN Lead Beam Programme". arXiv:nucl-th/0002042 [nucl-th].

[3] Berndt Müller (2005). "Quark Matter 2005 -- Theoretical Summary". arXiv:nucl-th/0508062 [nucl-th].

[4] Alford, Mark G.; Schmitt, Andreas; Rajagopal, Krishna; Schäfer, Thomas (2008). "Color superconductivity in dense quark matter". *Review of Modern Physics* **80** (4): 1455–1515. arXiv:0709.4635. Bibcode:2008RvMP...80.1455A. doi:

[5] Rischke, D (2004). "The quark–gluon plasma in equilibrium". *Progress in Particle and Nuclear Physics* **52**: 197. arXiv:nucl-th/0305030. Bibcode:2004PrPNP..52..197R. doi:10.1016/j.ppnp.2003.09.002.

[6] T. Schäfer (2004). "Quark matter". In A. B. Santra. *Mesons and Quarks*. 14th National Nuclear Physics Summer School. Alpha Science International. arXiv:hep-ph/0304281. ISBN 978-81-7319-589-1.

[7]P. Petreczky (2012). "Lattice QCD at non-zero temperature".*J.Phys. G***39**(9): 093002.arXiv:1203.5320.Bibcode:2012J. doi:10.1088/0954-3899/39/9/093002.

[8] Christian Schmidt (2006). "Lattice QCD at Finite Density". *PoS LAT2006*: 021. arXiv:hep-lat/0610116.Bibcode:2.

[9] Rajagopal, K (1999). "Mapping the QCD phase diagram". *Nuclear Physics A* **661**: 150–161. arXiv:hep-ph/99050R. doi:10.1016/S0375-9474(99)85017-9.

12.8   Further reading

- S. Hands (2001). "The phase diagram of QCD". *Contemporary Physics* **42** (4): 209. arXiv:physics/0105022. Bibcode:2001ConPh..42..209H. doi:10.1080/00107510110063843.

- K. Rajagopal (2001). "Free the quarks" (PDF). *Beam Line* **32** (2): 9–15.

12.9   External links

- Virtual Journal on QCD Matter

- RHIC finds Exotic Antimatter

Chapter 13

Strange matter

This article is about physics. For the book series, see Strange Matter.

Strange matter is a particular form of quark matter, usually thought of as a "liquid" of up, down and strange quarks. It is to be contrasted with nuclear matter, which is a liquid of neutrons and protons (which themselves are built out of up and down quarks), and with non-strange quark matter, which is a quark liquid containing only up and down quarks. At high enough density, strange matter is expected to be color superconducting. Strange matter is hypothesized to occur in the core of neutron stars, or, more speculatively, as isolated droplets that may vary in size from femtometers (strangelets) to kilometers (quark stars).

13.1 Two meanings of the term "strange matter"

In particle physics and astrophysics, the term is used in two ways, one broader and the other more specific[1][2]

1. The broader meaning is just quark matter that contains three flavors of quarks: up, down, and strange. In this definition, there is a critical pressure and an associated critical density, and when nuclear matter (made of protons and neutrons) is compressed beyond this density, the protons and neutrons dissociate into quarks, yielding quark matter (probably strange matter).

2. The narrower meaning is that quark matter is *more stable than nuclear matter*, i.e. that the true ground state of matter is quark matter. The idea that this could happen is the "strange matter hypothesis" of Bodmer[3] and Witten.[4] In this definition, the critical pressure is zero. The nuclei that we see in the matter around us, which are droplets of nuclear matter, are actually metastable, and given enough time (or the right external stimulus) would decay into droplets of strange matter, i.e. strangelets.

13.2 Strange matter that is only stable at high pressure

Under the broader definition, strange matter might occur inside neutron stars, if the pressure at their core is high enough (i.e. above the critical pressure). At the sort of densities we expect in the center of a neutron star, the quark matter would probably be strange matter. It could conceivably be non-strange quark matter, if the effective mass of the strange quark were too high. Charm and heavier quarks would only occur at much higher densities.

A neutron star with a quark matter core is often[1][2] called a hybrid star. However, it is hard to know whether hybrid stars really exist in nature because physicists currently have little idea of the likely value of the critical pressure or density. It seems plausible that the transition to quark matter will already have occurred when the separation between the nucleons becomes much smaller than their size, so the critical density must be less than about 100 times nuclear saturation density. But a more precise estimate is not yet available, because the strong interaction that governs the behavior of quarks is mathematically intractable, and numerical calculations using lattice QCD are currently blocked by the fermion sign problem.

One major area of activity in neutron star physics is the attempt to find observable signatures by which we could tell, from earth based observations of neutron stars, whether they have quark matter (probably strange matter) in their core.

13.3   Strange matter that is stable at zero pressure

If the "strange matter hypothesis" is true then nuclear matter is metastable against decaying into strange matter. The lifetime for spontaneous decay is very long, so we do not see this decay process happening around us.[4] However, under this hypothesis there should be strange matter in the universe:

1. Quark stars (often called "strange stars") consist of quark matter from their core to their surface. They would be several kilometers across, and may have a very thin crust of nuclear matter.[2]

2. Strangelets are small pieces of strange matter, perhaps as small as nuclei. They would be produced when strange stars are formed or collide, or when a nucleus decays.[1]

13.4   See also

- Exotic matter

- Negative matter

- Quark matter

- Quark star

- Strangeness production

- Strangelet

- Quark

- QCD matter

13.5   References

[1] J. Madsen, "Physics and astrophysics of strange quark matter" arXiv:astro-ph/9809032, Lect. Notes Phys. 516:162-203 (1999)

[2] F. Weber, "Strange quark matter and compact stars", arXiv:astro-ph/0407155, Prog.Part.Nucl.Phys.54:193-288,2005.

[3] A. Bodmer "Collapsed Nuclei" Phys. Rev. D4, 1601 (1971)

[4] Edward Witten, "Cosmic Separation Of Phases" Phys. Rev. D30, 272 (1984)

Chapter 14

Phase diagram

For the use of this term in mathematics and physics, see phase space.

A **phase diagram** in physical chemistry, engineering, mineralogy, and materials science is a type of chart used to show conditions at which thermodynamically distinct phases can occur at equilibrium.

14.1 Overview

Common components of a phase diagram are *lines of equilibrium* or *phase boundaries*, which refer to lines that mark conditions under which multiple phases can coexist at equilibrium. Phase transitions occur along lines of equilibrium.

Triple points are points on phase diagrams where lines of equilibrium intersect. Triple points mark conditions at which three different phases can coexist. For example, the water phase diagram has a triple point corresponding to the single temperature and pressure at which solid, liquid, and gaseous water can coexist in a stable equilibrium.

The solidus is the temperature below which the substance is stable in the solid state. The liquidus is the temperature above which the substance is stable in a liquid state. There may be a gap between the solidus and liquidus; within the gap, the substance consists of a mixture of crystals and liquid (like a "slurry").[1]

14.2 Types of phase diagrams

14.2.1 2D phase diagrams

The simplest phase diagrams are pressure–temperature diagrams of a single simple substance, such as water. The axes correspond to the pressure and temperature. The phase diagram shows, in pressure–temperature space, the lines of equilibrium or phase boundaries between the three phases of solid, liquid, and gas.

The curves on the phase diagram show the points where the free energy (and other derived properties) becomes non-analytic: their derivatives with respect to the coordinates (temperature and pressure in this example) change discontinuously (abruptly). For example, the heat capacity of a container filled with ice will change abruptly as the container is heated past the melting point. The open spaces, where the free energy is analytic, correspond to single phase regions. Single phase regions are separated by lines of non-analytical behavior, where phase transitions occur, which are called **phase boundaries**.

In the diagram on the left, the phase boundary between liquid and gas does not continue indefinitely. Instead, it terminates at a point on the phase diagram called the critical point. This reflects the fact that, at extremely high temperatures and pressures, the liquid and gaseous phases become indistinguishable,[2] in what is known as a supercritical fluid. In water, the critical point occurs at around $T_c = 647.096$ K (373.946 °C), $p_c = 22.064$ MPa (217.75 atm) and $\rho_c = 356$ kg/m^3.[3]

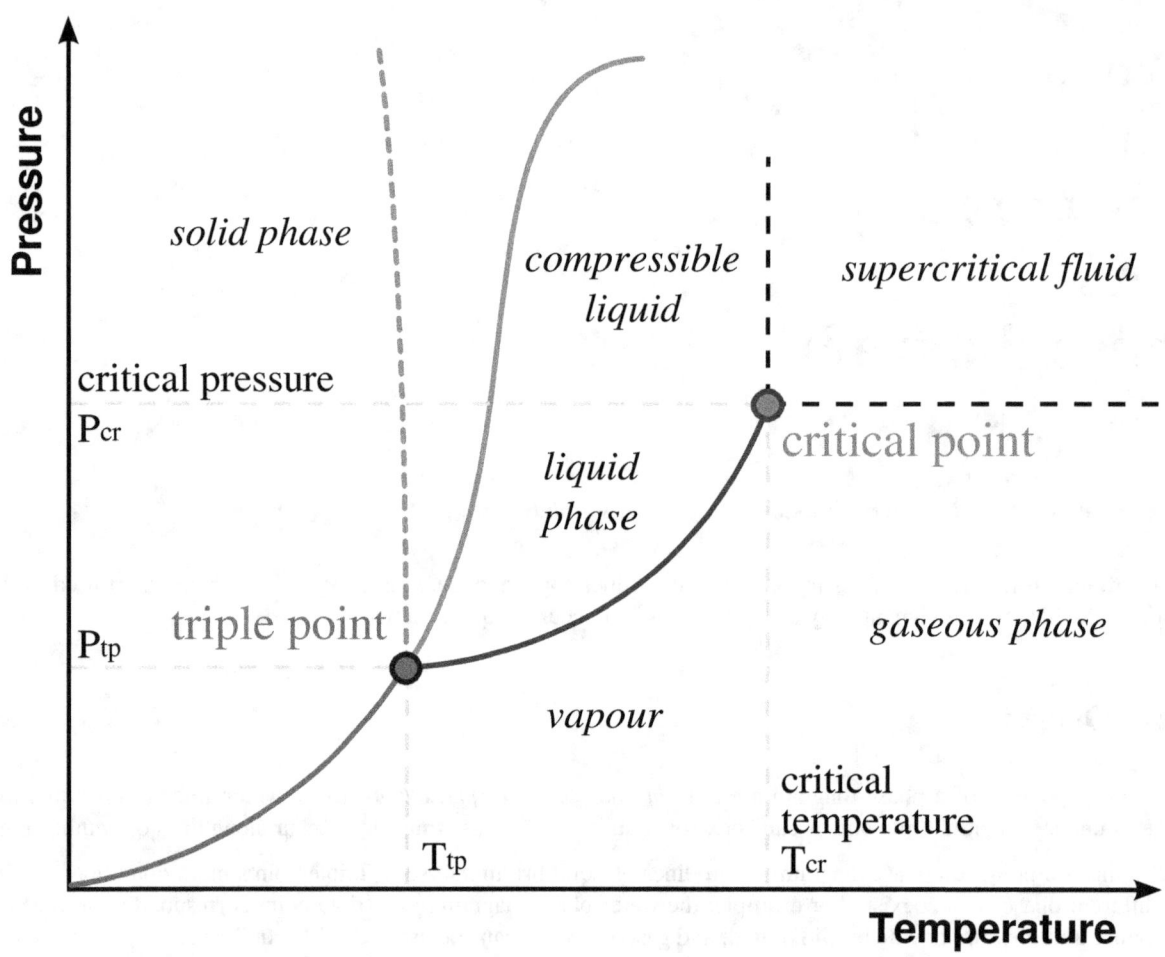

A typical phase diagram. The solid green line applies to most substances; the dotted green line gives the anomalous behavior of water. The green lines mark the freezing point and the blue line the boiling point, showing how they vary with pressure.

The existence of the liquid–gas critical point reveals a slight ambiguity in labelling the single phase regions. When going from the liquid to the gaseous phase, one usually crosses the phase boundary, but it is possible to choose a path that never crosses the boundary by going to the right of the critical point. Thus, the liquid and gaseous phases can blend continuously into each other. The solid–liquid phase boundary can only end in a critical point if the solid and liquid phases have the same symmetry group.

The solid–liquid phase boundary in the phase diagram of most substances has a positive slope; the greater the pressure on a given substance, the closer together the molecules of the substance are brought to each other, which increases the effect of the substance's intermolecular forces. Thus, the substance requires a higher temperature for its molecules to have enough energy to break out of the fixed pattern of the solid phase and enter the liquid phase. A similar concept applies to liquid–gas phase changes.[4] Water, because of its particular properties, is one of the several exceptions to the rule.

Other thermodynamic properties

In addition to just temperature or pressure, other thermodynamic properties may be graphed in phase diagrams. Examples of such thermodynamic properties include specific volume, specific enthalpy, or specific entropy. For example, single-component graphs of temperature vs. specific entropy (T vs. s) for water/steam or for a refrigerant are commonly used to illustrate thermodynamic cycles such as a Carnot cycle, Rankine cycle, or vapor-compression refrigeration cycle.

In a two-dimensional graph, two of the thermodynamic quantities may be shown on the horizontal and vertical axes. Additional thermodynamic quantities may each be illustrated in increments as a series of lines - curved, straight, or a combination of curved and straight. Each of these **iso**-lines represents the thermodynamic quantity at a certain constant value.

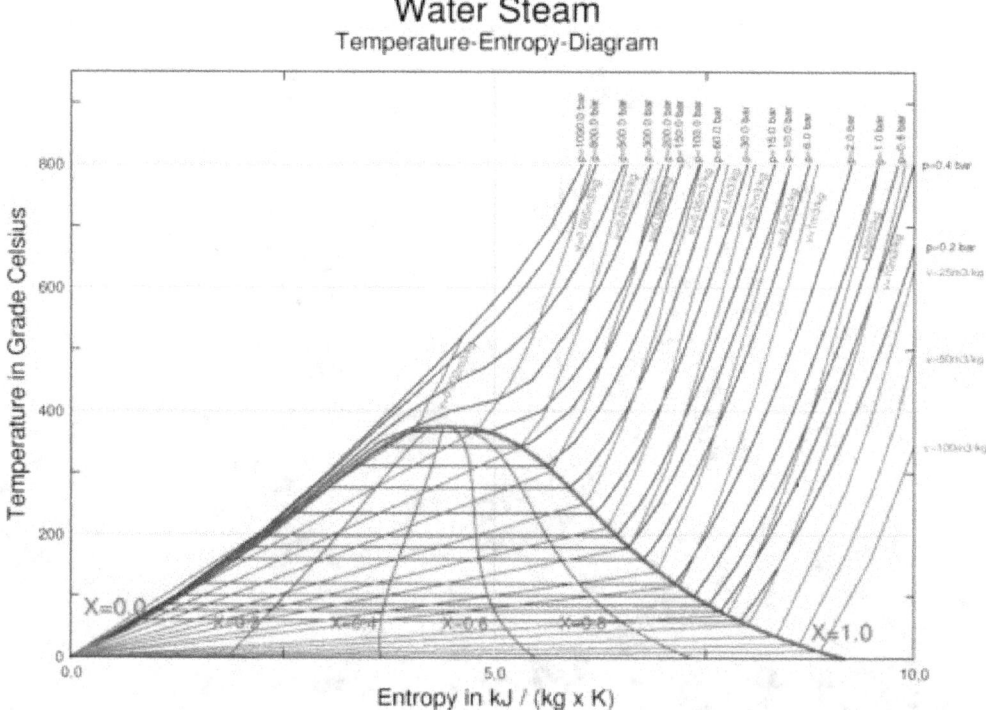

*Temperature vs. specific entropy phase diagram for water/steam. In the area under the red dome, liquid water and steam coexist in equilibrium. The critical point is at the top of the dome. Liquid water is to the left of the dome. Steam is to the right of the dome. The blue lines/curves are **isobars** showing constant pressure. The green lines/curves are **isochors** showing constant specific volume. The red curves show constant quality.*

14.2.2   3D phase diagrams

It is possible to envision three-dimensional (3D) graphs showing three thermodynamic quantities.[5][6] For example for a single component, a 3D Cartesian coordinate type graph can show temperature (T) on one axis, pressure (p) on a second axis, and specific volume (v) on a third. Such a 3D graph is sometimes called a p–v–T diagram. The equilibrium conditions are shown as curves on a curved surface in 3D with areas for solid, liquid, and vapor phases and areas where solid and liquid, solid and vapor, or liquid and vapor coexist in equilibrium. A line on the surface called a **triple line** is where solid, liquid and vapor can all coexist in equilibrium. The critical point remains a point on the surface even on a 3D phase diagram.

An orthographic projection of the 3D p–v–T graph showing pressure and temperature as the vertical and horizontal axes collapses the 3D plot into the standard 2D pressure–temperature diagram. When this is done, the solid–vapor, solid–liquid, and liquid–vapor surfaces collapse into three corresponding curved lines meeting at the triple point, which is the collapsed orthographic projection of the triple line.

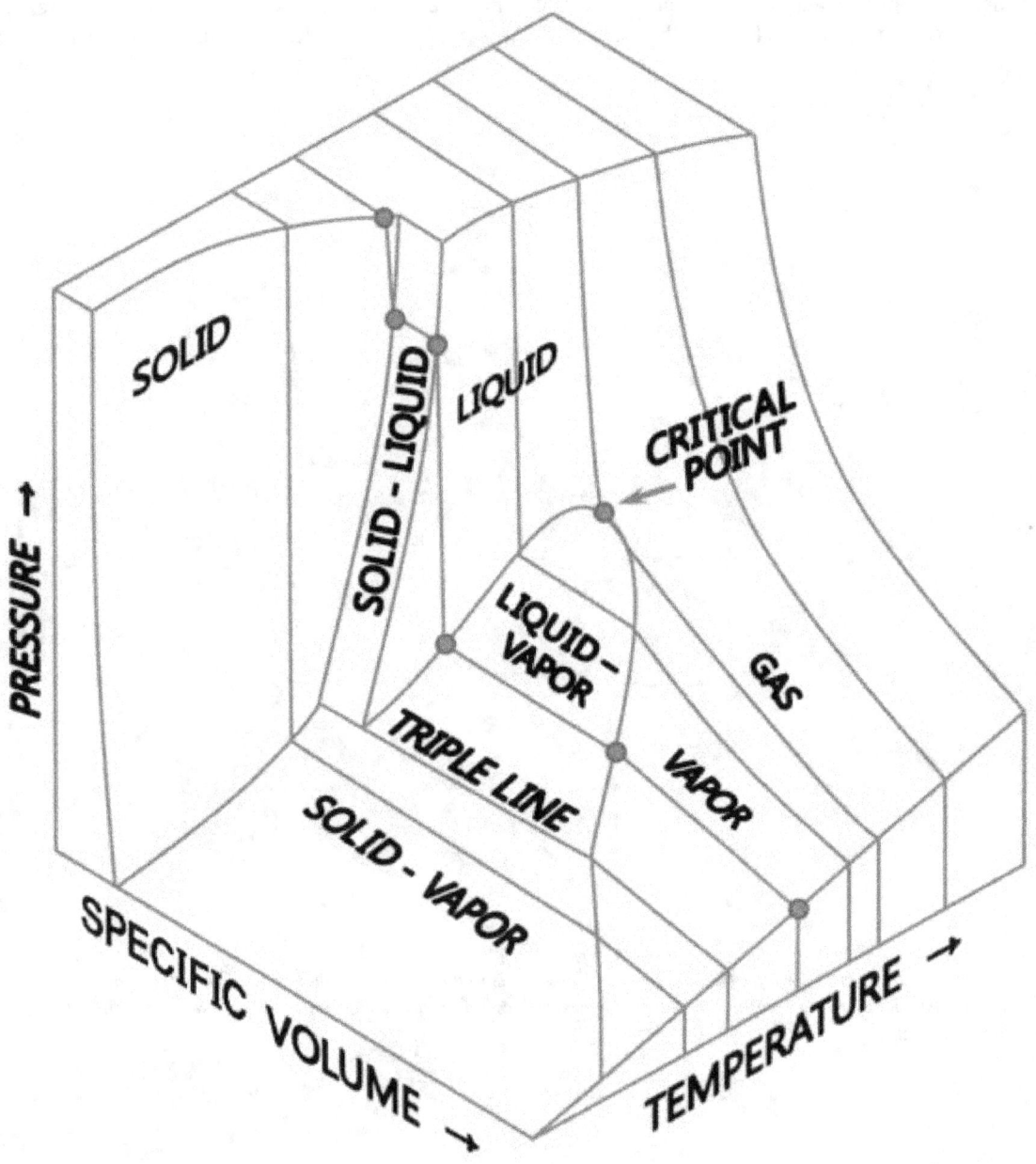

p–v–T 3D diagram for fixed amount of pure material

14.2.3   Binary phase diagrams

Other much more complex types of phase diagrams can be constructed, particularly when more than one pure component is present. In that case, concentration becomes an important variable. Phase diagrams with more than two dimensions can be constructed that show the effect of more than two variables on the phase of a substance. Phase diagrams can use other variables in addition to or in place of temperature, pressure and composition, for example the strength of an applied electrical or magnetic field, and they can also involve substances that take on more than just three states of matter.

One type of phase diagram plots temperature against the relative concentrations of two substances in a binary mixture

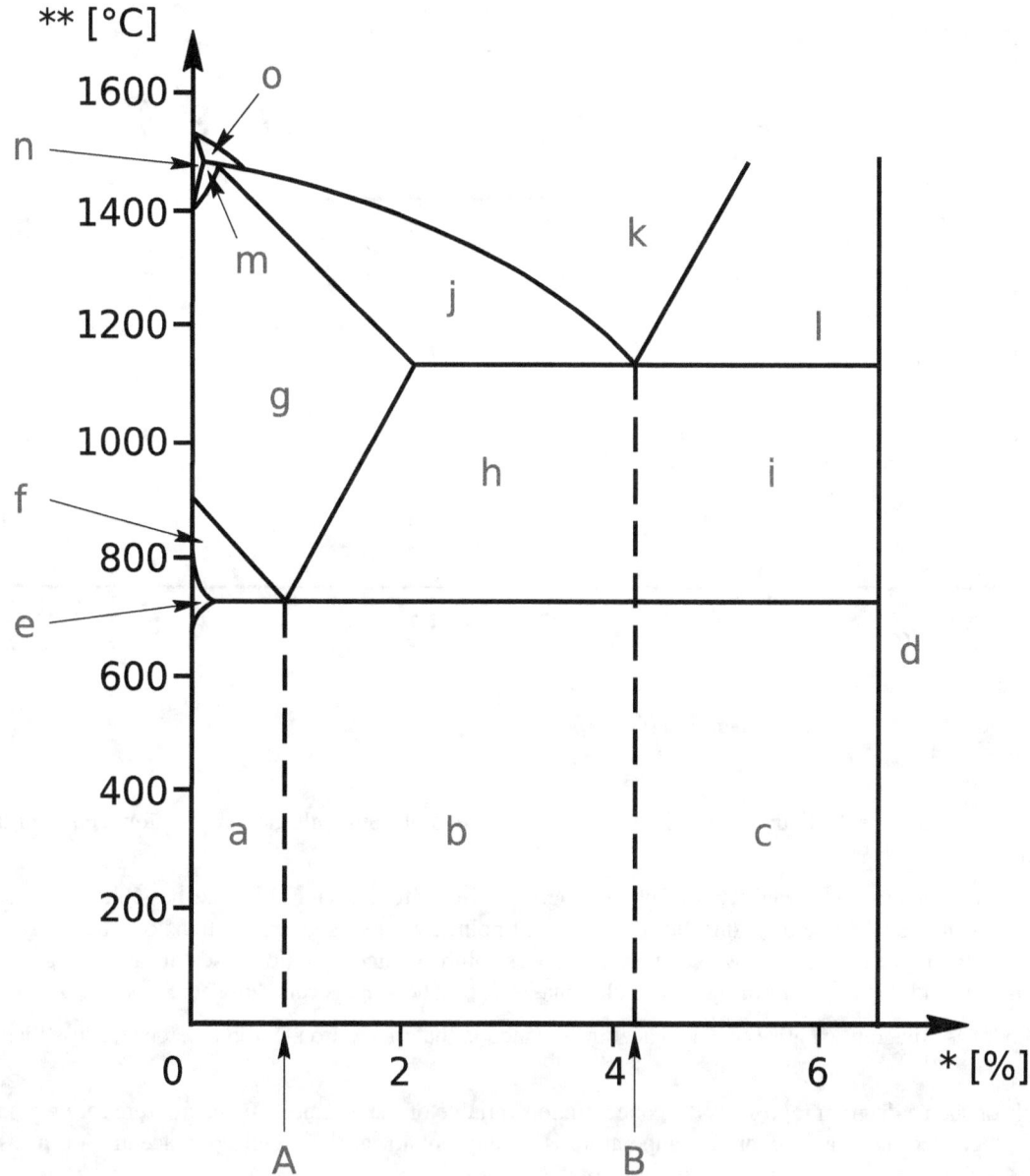

The iron–iron carbide (Fe–Fe₃C) phase diagram. The percentage of carbon present and the temperature define the phase of the iron carbon alloy and therefore its physical characteristics and mechanical properties. The percentage of carbon determines the type of the ferrous alloy: iron, steel or cast iron

called a *binary phase diagram*, as shown at right. Such a mixture can be either a solid solution, eutectic or peritectic, among others. These two types of mixtures result in very different graphs. Another type of binary phase diagram is a *boiling-point diagram* for a mixture of two components, i. e. chemical compounds. For two particular volatile components at a certain pressure such as atmospheric pressure, a boiling-point diagram shows what vapor (gas) compositions are in equilibrium with given liquid compositions depending on temperature. In a typical binary boiling-point diagram, temperature is plotted on a vertical axis and mixture composition on a horizontal axis.

A simple example diagram with hypothetical components 1 and 2 in a non-azeotropic mixture is shown at right. The fact that there are two separate curved lines joining the boiling points of the pure components means that the vapor composition

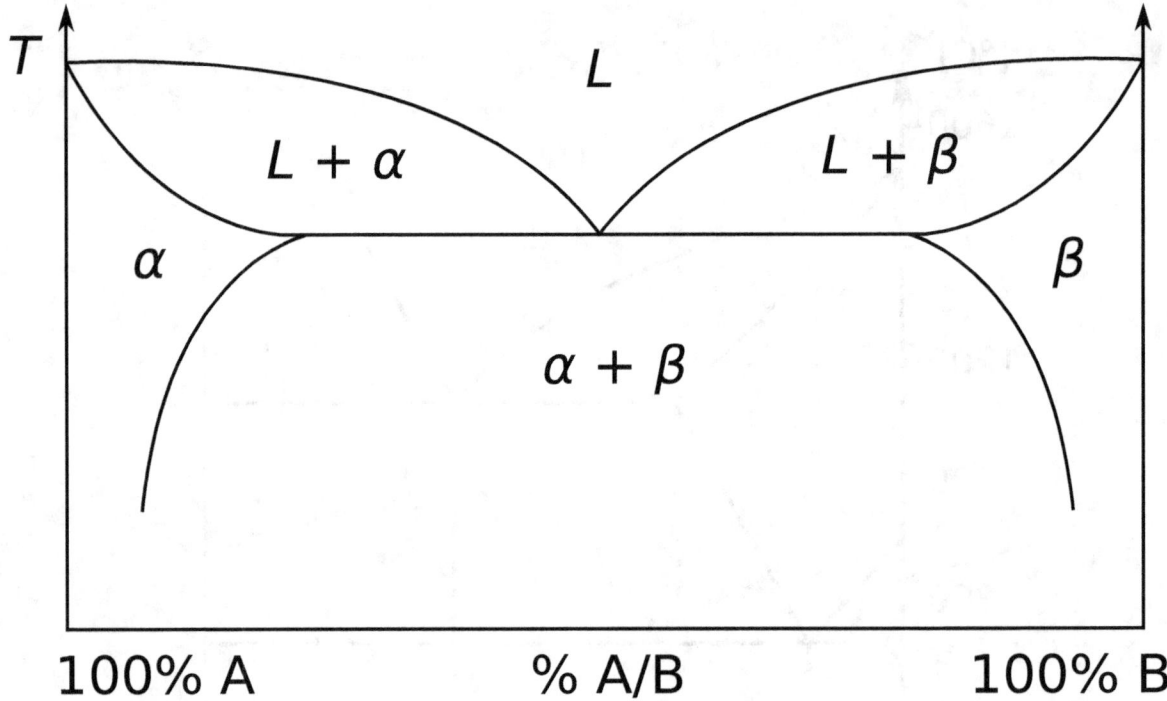

A phase diagram for a binary system displaying a eutectic point.

is usually not the same as the liquid composition the vapor is in equilibrium with. See Vapor–liquid equilibrium for more information.

In addition to the above-mentioned types of phase diagrams, there are thousands of other possible combinations. Some of the major features of phase diagrams include congruent points, where a solid phase transforms directly into a liquid. There is also the peritectoid, a point where two solid phases combine into one solid phase during cooling. The inverse of this, when one solid phase transforms into two solid phases during heating, is called the eutectoid.

A complex phase diagram of great technological importance is that of the iron–carbon system for less than 7% carbon (see steel).

The x-axis of such a diagram represents the concentration variable of the mixture. As the mixtures are typically far from dilute and their density as a function of temperature is usually unknown, the preferred concentration measure is mole fraction. A volume-based measure like molarity would be inadvisable.

14.2.4   Crystal phase diagrams

Polymorphic and polyamorphic substances have multiple crystal or amorphous phases, which can be graphed in a similar fashion to solid, liquid, and gas phases.

14.2.5   Mesophase diagrams

Some organic materials pass through intermediate states between solid and liquid; these states are called mesophases. Attention has been directed to mesophases because they enable display devices and have become commercially important through the so-called liquid-crystal technology. Phase diagrams are used to describe the occurrence of mesophases.[8]

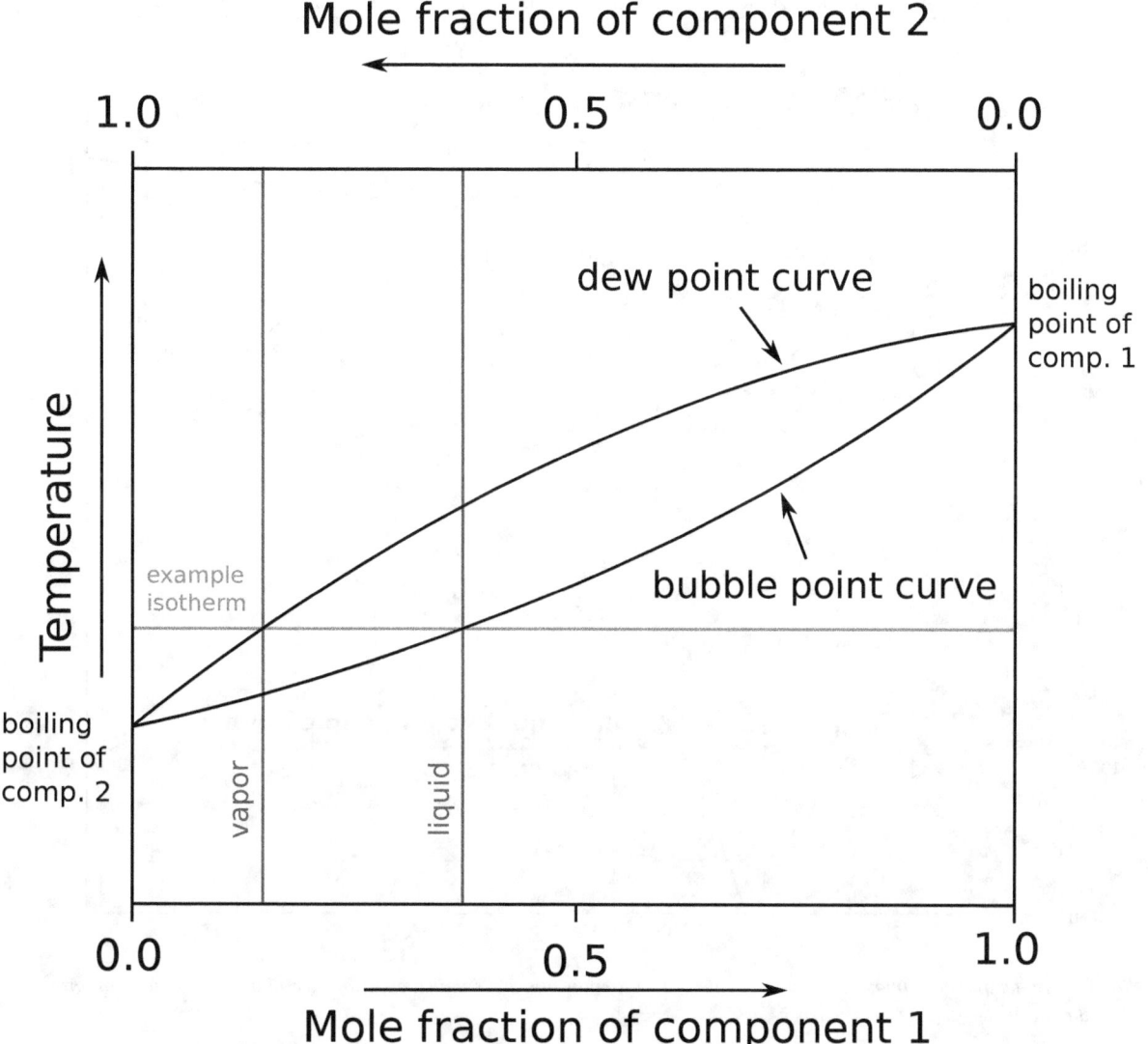

Boiling-point diagram

14.3 See also

- CALPHAD (method)

- Congruent melting and incongruent melting

- Gibbs phase rule

- Glass databases

- Hamiltonian mechanics

- Phase separation

- Schreinemaker's analysis

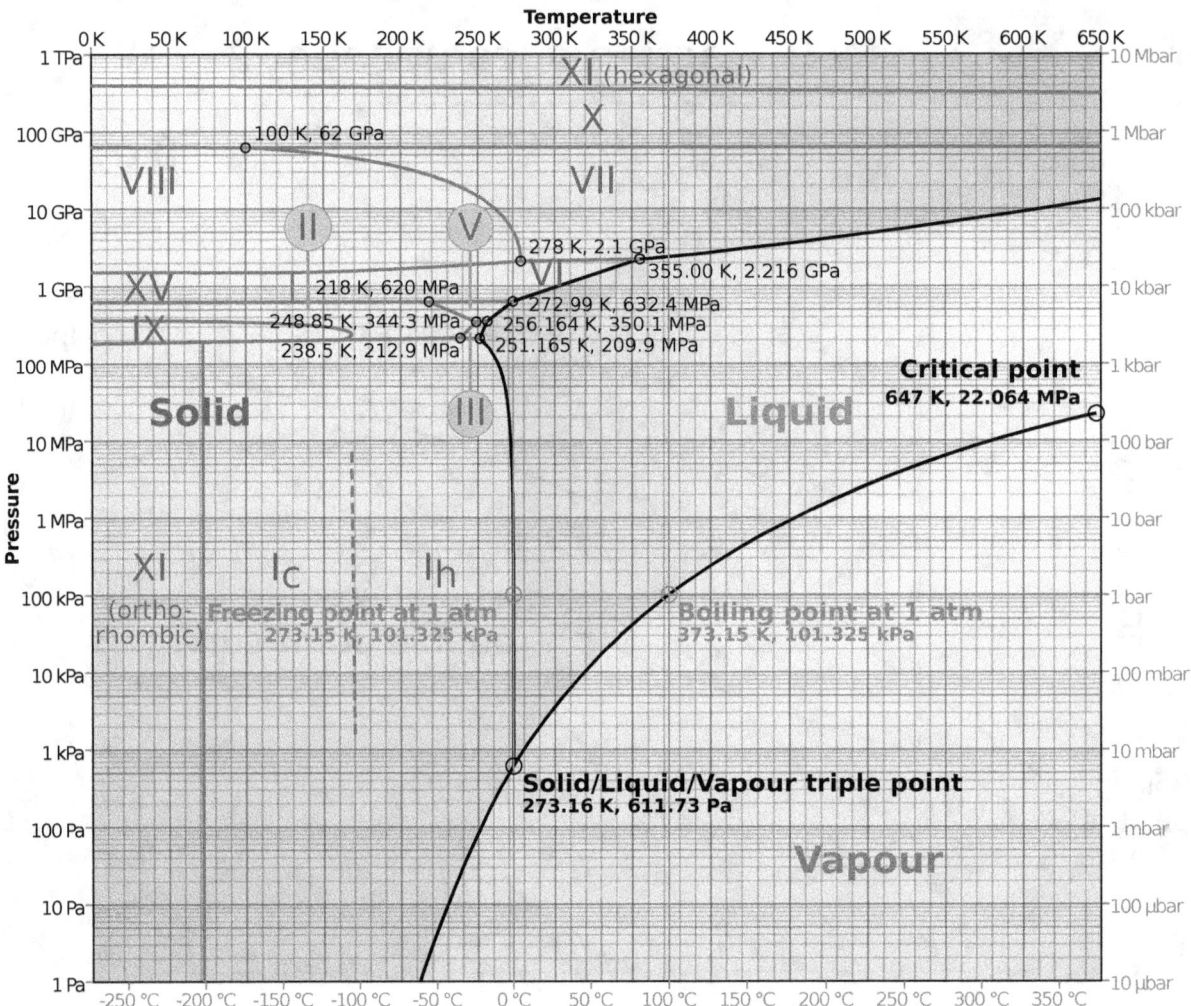

Log-lin pressure–temperature phase diagram of water. The Roman numerals indicate various ice phases. A similar diagram may be found on the site Water structure and science.[7]

14.4   References

[1] Predel, Bruno; Hoch, Michael J. R; Pool, Monte. *Phase Diagrams and Heterogeneous Equilibria : A Practical Introduction.* Springer. ISBN 3-540-14011-5.

[2] P. Papon, J. Leblond, and P.H.E. Meijer, *The Physics of Phase Transition - Concepts and Applications* Springer 1999.

[3] The International Association for the Properties of Water and Steam "Guideline on the Use of Fundamental Physical Constants and Basic Constants of Water", 2001, p. 5

[4] *Chemistry: The Study of Matter, Fourth Edition.* Prentice Hall. pp. 266–273. ISBN 0-13-127333-7.

[5] *Heat and Thermodynamics*, Mark W. Zemansky, Richard H. Dittman, McGraw-Hill, 6th ed., 1981, Figures 2-3, 2-4, 2-5, 10-10, P10-1, ISBN 0-07-072808-9.

[6] Web applet: 3D Phase Diagrams for Water, Carbon Dioxide and Ammonia. Described in A. Herráez, R.M. Hanson, and L. Glasser *J. Chem. Educ.* **86** (5), 566 (May 2009).

[7] Water structure and science Site by Martin Chaplin, accessed 2 July 2015.

[8] Sivaramakrishna Chandrasekhar (1992) *Liquid Crystals*, 2nd edition, pages 27–9, 356, Cambridge University Press ISBN 0-521-41747-3 .

14.5 External links

- Iron-Iron Carbide Phase Diagram Example

- How to build a phase diagram

- Phase Changes: Phase Diagrams: Part 1

- Equilibrium Fe-C phase diagram

- Phase diagrams for lead free solders

- DoITPoMS Phase Diagram Library

- DoITPoMS Teaching and Learning Package- "Phase Diagrams and Solidification"

Chapter 15

Antimatter

For other uses, see Antimatter (disambiguation).

In particle physics, **antimatter** is material composed of antiparticles, which have the same mass as particles of ordinary matter but opposite charges, as well as other particle properties such as lepton and baryon numbers and quantum spin. Collisions between particles and antiparticles lead to the annihilation of both, giving rise to variable proportions of intense photons (gamma rays), neutrinos, and less massive particle–antiparticle pairs. The total consequence of annihilation is a release of energy available for work, proportional to the total matter and antimatter mass, in accord with the mass–energy equivalence equation, $E = mc^2$.[1]

Antiparticles bind with each other to form antimatter, just as ordinary particles bind to form normal matter. For example, a positron (the antiparticle of the electron) and an antiproton (the antiparticle of the proton) can form an antihydrogen atom. Physical principles indicate that complex antimatter atomic nuclei are possible, as well as anti-atoms corresponding to the known chemical elements. Studies of cosmic rays have identified both positrons and antiprotons, presumably produced by collisions between particles of ordinary matter. Satellite-based searches of cosmic rays for antideuteron and antihelium particles have yielded nothing. [2] [3]

There is considerable speculation as to why the observable universe is composed almost entirely of ordinary matter, as opposed to a more even mixture of matter and antimatter. This asymmetry of matter and antimatter in the visible universe is one of the great unsolved problems in physics.[4] The process by which this inequality between particles and antiparticles developed is called baryogenesis.

Antimatter in the form of anti-atoms is one of the most difficult materials to produce. Antimatter in the form of individual anti-particles, however, is commonly produced by particle accelerators and in some types of radioactive decay. The nuclei of antihelium (both helium-3 and helium-4) have been artificially produced with difficulty. These are the most complex anti-nuclei so far observed. [5]

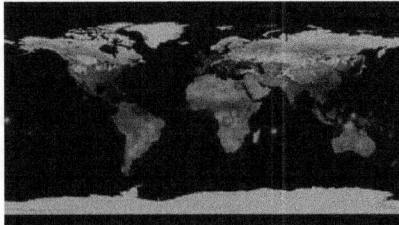

There are some 500 terrestrial gamma-ray flashes daily. The red dots show those the Fermi Gamma-ray Space Telescope spotted through 2010.

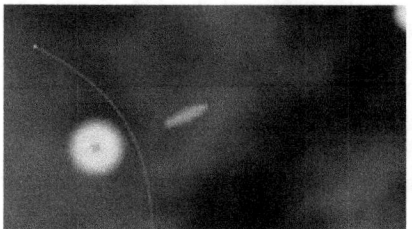

A video showing how scientists used the Fermi Gamma-ray Space Telescope's gamma-ray detector to uncover bursts of antimatter from thunderstorms

15.1 History of the concept

The idea of negative matter appears in past theories of matter that have now been abandoned. Using the once popular vortex theory of gravity, the possibility of matter with negative gravity was discussed by William Hicks in the 1880s. Between the 1880s and the 1890s, Karl Pearson proposed the existence of "squirts" [6] and sinks of the flow of aether. The squirts represented normal matter and the sinks represented negative matter. Pearson's theory required a fourth dimension for the aether to flow from and into.[7]

The term antimatter was first used by Arthur Schuster in two rather whimsical letters to *Nature* in 1898,[8] in which he coined the term. He hypothesized antiatoms, as well as whole antimatter solar systems, and discussed the possibility of matter and antimatter annihilating each other. Schuster's ideas were not a serious theoretical proposal, merely speculation, and like the previous ideas, differed from the modern concept of antimatter in that it possessed negative gravity.[9]

The modern theory of antimatter began in 1928, with a paper[10] by Paul Dirac. Dirac realised that his relativistic version of the Schrödinger wave equation for electrons predicted the possibility of antielectrons. These were discovered by Carl D. Anderson in 1932 and named positrons (a portmanteau of "positive electron"). Although Dirac did not himself use the term antimatter, its use follows on naturally enough from antielectrons, antiprotons, etc.[11] A complete periodic table of antimatter was envisaged by Charles Janet in 1929.[12]

The Feynman–Stueckelberg interpretation states that antimatter and antiparticles are regular particles traveling backward in time.[13]

15.2 Notation

One way to denote an antiparticle is by adding a bar over the particle's symbol. For example, the proton and antiproton are denoted as p and p, respectively. The same rule applies if one were to address a particle by its constituent components. A proton is made up of uud quarks, so an antiproton must therefore be formed from uud antiquarks. Another convention is to distinguish particles by their electric charge. Thus, the electron and positron are denoted simply as e− and e+ respectively. However, to prevent confusion, the two conventions are never mixed.

15.3 Origin and asymmetry

See also: Baryogenesis

Almost all matter observable from the Earth seems to be made of matter rather than antimatter. If antimatter-dominated regions of space existed, the gamma rays produced in annihilation reactions along the boundary between matter and antimatter regions would be detectable.[14]

Antiparticles are created everywhere in the universe where high-energy particle collisions take place. High-energy cosmic rays impacting Earth's atmosphere (or any other matter in the Solar System) produce minute quantities of antiparticles in the resulting particle jets, which are immediately annihilated by contact with nearby matter. They may similarly be produced in regions like the center of the Milky Way and other galaxies, where very energetic celestial events occur (principally the interaction of relativistic jets with the interstellar medium). The presence of the resulting antimatter is detectable by the two gamma rays produced every time positrons annihilate with nearby matter. The frequency and wavelength of the gamma rays indicate that each carries 511 keV of energy (i.e., the rest mass of an electron multiplied by c^2).

Recent observations by the European Space Agency's INTEGRAL satellite may explain the origin of a giant antimatter cloud surrounding the galactic center. The observations show that the cloud is asymmetrical and matches the pattern of X-ray binaries (binary star systems containing black holes or neutron stars), mostly on one side of the galactic center. While the mechanism is not fully understood, it is likely to involve the production of electron–positron pairs, as ordinary matter gains kinetic energy while falling into a stellar remnant.[15][16]

Antimatter may exist in relatively large amounts in far-away galaxies due to cosmic inflation in the primordial time of the universe. Antimatter galaxies, if they exist, are expected to have the same chemistry and absorption and emission spectra as normal-matter galaxies, and their astronomical objects would be observationally identical, making them difficult to distinguish.[17] NASA is trying to determine if such galaxies exist by looking for X-ray and gamma-ray signatures of annihilation events in colliding superclusters.[18]

15.4 Natural production

Main article: Positron emission

Positrons are produced naturally in β^+ decays of naturally occurring radioactive isotopes (for example, potassium-40) and in interactions of gamma quanta (emitted by radioactive nuclei) with matter. Antineutrinos are another kind of antiparticle created by natural radioactivity (β^- decay). Many different kinds of antiparticles are also produced by (and contained in) cosmic rays. In January 2011, research by the American Astronomical Society discovered antimatter (positrons) originating above thunderstorm clouds; positrons are produced in gamma-ray flashes created by electrons accelerated by strong electric fields in the clouds.[19][20] Antiprotons have also been found to exist in the Van Allen Belts around the Earth by the PAMELA module.[21][22]

Antiparticles are also produced in any environment with a sufficiently high temperature (mean particle energy greater than the pair production threshold). During the period of baryogenesis, when the universe was extremely hot and dense, matter and antimatter were continually produced and annihilated. The presence of remaining matter, and absence of detectable remaining antimatter,[23] also called baryon asymmetry, is attributed to CP-violation: a violation of the CP-symmetry relating matter to antimatter. The exact mechanism of this violation during baryogenesis remains a mystery.

15.4.1 Observation in cosmic rays

Main article: Cosmic ray

Satellite experiments have found evidence of positrons and a few antiprotons in primary cosmic rays, amounting to less than 1% of the particles in primary cosmic rays. This antimatter cannot all have been created in the Big Bang, but is instead attributed to have been produced by cyclic processes at high energies. For instance, electron-positron pairs may be formed in pulsars, as a magnetized neutron star rotation cycle shears electron-positron pairs from the star surface. Therein the antimatter forms a wind which crashes upon the ejecta of the progenitor supernovae. This weathering takes place as "the cold, magnetized relativistic wind launched by the star hits the non-relativistically expanding ejecta, a shock wave system forms in the impact: the outer one propagates in the ejecta, while a reverse shock propagates back towards the star." [24] The former ejection of matter in the outer shock wave and the latter production of antimatter in the reverse shock wave are steps in a space weather cycle.

Preliminary results from the presently operating Alpha Magnetic Spectrometer (*AMS-02*) on board the International Space Station show that positrons in the cosmic rays arrive with no directionality, and with energies that range from 10 GeV to

250 GeV. In September, 2014, new results with almost twice as much data were presented in a talk at CERN and published in Physical Review Letters.[25][26] A new measurement of positron fraction up to 500 GeV was reported, showing that positron fraction peaks at a maximum of about 16% of total electron+positron events, around an energy of 275 ± 32 GeV. At higher energies, up to 500 GeV, the ratio of positrons to electrons begins to fall again. The absolute flux of positrons also begins to fall before 500 GeV, but peaks at energies far higher than electron energies, which peak about 10 GeV.[27] These results on interpretation have been suggested to be due to positron production in annihilation events of massive dark matter particles.[28]

Cosmic ray antiprotons also have a much higher energy than their normal-matter counterparts (protons). They arrive at Earth with a characteristic energy maximum of 2 GeV, indicating their production in a fundamentally different process from cosmic ray protons, which on average have only one-sixth of the energy.[29]

There is no evidence of complex antimatter atomic nuclei, such as antihelium nuclei (i.e., anti-alpha particles), in cosmic rays. These are actively being searched for. A prototype of the *AMS-02* designated *AMS-01*, was flown into space aboard the Space Shuttle *Discovery* on STS-91 in June 1998. By not detecting any antihelium at all, the *AMS-01* established an upper limit of 1.1×10^{-6} for the antihelium to helium flux ratio.[30]

15.5 Artificial production

15.5.1 Positrons

Main article: Positron

Positrons were reported[31] in November 2008 to have been generated by Lawrence Livermore National Laboratory in larger numbers than by any previous synthetic process. A laser drove electrons through a millimeter-radius gold target's nuclei, which caused the incoming electrons to emit energy quanta that decayed into both matter and antimatter. Positrons were detected at a higher rate and in greater density than ever previously detected in a laboratory. Previous experiments made smaller quantities of positrons using lasers and paper-thin targets; however, new simulations showed that short, ultra-intense lasers and millimeter-thick gold are a far more effective source.[32]

15.5.2 Antiprotons, antineutrons, and antinuclei

Main articles: Antiproton and Antineutron

The existence of the antiproton was experimentally confirmed in 1955 by University of California, Berkeley physicists Emilio Segrè and Owen Chamberlain, for which they were awarded the 1959 Nobel Prize in Physics.[33] An antiproton consists of two up antiquarks and one down antiquark (uud). The properties of the antiproton that have been measured all match the corresponding properties of the proton, with the exception of the antiproton having opposite electric charge and magnetic moment from the proton. Shortly afterwards, in 1956, the antineutron was discovered in proton–proton collisions at the Bevatron (Lawrence Berkeley National Laboratory) by Bruce Cork and colleagues.[34]

In addition to antibaryons, anti-nuclei consisting of multiple bound antiprotons and antineutrons have been created. These are typically produced at energies far too high to form antimatter atoms (with bound positrons in place of electrons). In 1965, a group of researchers led by Antonino Zichichi reported production of nuclei of antideuterium at the Proton Synchrotron at CERN.[35] At roughly the same time, observations of antideuterium nuclei were reported by a group of American physicists at the Alternating Gradient Synchrotron at Brookhaven National Laboratory.[36]

15.5.3 Antihydrogen atoms

Main article: Antihydrogen

In 1995, CERN announced that it had successfully brought into existence nine hot antihydrogen atoms by implementing the SLAC/Fermilab concept during the PS210 experiment. The experiment was performed using the Low Energy Antiproton Ring (LEAR), and was led by Walter Oelert and Mario Macri.[37] Fermilab soon confirmed the CERN findings by producing approximately 100 antihydrogen atoms at their facilities. The antihydrogen atoms created during PS210 and subsequent experiments (at both CERN and Fermilab) were extremely energetic and were not well suited to study. To resolve this hurdle, and to gain a better understanding of antihydrogen, two collaborations were formed in the late 1990s, namely, ATHENA and ATRAP. In 2005, ATHENA disbanded and some of the former members (along with others) formed the ALPHA Collaboration, which is also based at CERN. The primary goal of these collaborations is the creation of less energetic ("cold") antihydrogen, better suited to study.

In 1999, CERN activated the Antiproton Decelerator, a device capable of decelerating antiprotons from 3500 MeV to 5.3 MeV — still too "hot" to produce study-effective antihydrogen, but a huge leap forward. In late 2002 the ATHENA project announced that they had created the world's first "cold" antihydrogen.[38] The ATRAP project released similar results very shortly thereafter.[39] The antiprotons used in these experiments were cooled by decelerating them with the Antiproton Decelerator, passing them through a thin sheet of foil, and finally capturing them in a Penning–Malmberg trap.[40] The overall cooling process is workable, but highly inefficient; approximately 25 million antiprotons leave the Antiproton Decelerator and roughly 25,000 make it to the Penning–Malmberg trap, which is about $\frac{1}{1000}$ or 0.1% of the original amount.

The antiprotons are still hot when initially trapped. To cool them further, they are mixed into an electron plasma. The electrons in this plasma cool via cyclotron radiation, and then sympathetically cool the antiprotons via Coulomb collisions. Eventually, the electrons are removed by the application of short-duration electric fields, leaving the antiprotons with energies less than 100 meV.[41] While the antiprotons are being cooled in the first trap, a small cloud of positrons is captured from radioactive sodium in a Surko-style positron accumulator.[42] This cloud is then recaptured in a second trap near the antiprotons. Manipulations of the trap electrodes then tip the antiprotons into the positron plasma, where some combine with antiprotons to form antihydrogen. This neutral antihydrogen is unaffected by the electric and magnetic fields used to trap the charged positrons and antiprotons, and within a few microseconds the antihydrogen hits the trap walls, where it annihilates. Some hundreds of millions of antihydrogen atoms have been made in this fashion.

Most of the sought-after high-precision tests of the properties of antihydrogen could only be performed if the antihydrogen were trapped, that is, held in place for a relatively long time. While antihydrogen atoms are electrically neutral, the spins of their component particles produce a magnetic moment. These magnetic moments can interact with an inhomogeneous magnetic field; some of the antihydrogen atoms can be attracted to a magnetic minimum. Such a minimum can be created by a combination of mirror and multipole fields.[43] Antihydrogen can be trapped in such a magnetic minimum (minimum-B) trap; in November 2010, the ALPHA collaboration announced that they had so trapped 38 antihydrogen atoms for about a sixth of a second.[44][45] This was the first time that neutral antimatter had been trapped.

On 26 April 2011, ALPHA announced that they had trapped 309 antihydrogen atoms, some for as long as 1,000 seconds (about 17 minutes). This was longer than neutral antimatter had ever been trapped before.[46][47] ALPHA has used these trapped atoms to initiate research into the spectral properties of the antihydrogen.[48]

The biggest limiting factor in the large-scale production of antimatter is the availability of antiprotons. Recent data released by CERN states that, when fully operational, their facilities are capable of producing ten million antiprotons per minute.[49] Assuming a 100% conversion of antiprotons to antihydrogen, it would take 100 billion years to produce 1 gram or 1 mole of antihydrogen (approximately 6.02×10^{23} atoms of antihydrogen).

15.5.4   Antihelium

Antihelium-3 nuclei (3He) were first observed in the 1970s in proton–nucleus collision experiments at Institute for High Energy Physics by Y. Prockoshkin's group (Protvino near Moscow, USSR)[50] and later created in nucleus–nucleus collision experiments.[51] Nucleus–nucleus collisions produce antinuclei through the coalescense of antiprotons and antineutrons created in these reactions. In 2011, the STAR detector reported the observation of artificially created antihelium-4 nuclei (anti-alpha particles) (4He) from such collisions.[52]

15.5.5 Preservation

Antimatter cannot be stored in a container made of ordinary matter because antimatter reacts with any matter it touches, annihilating itself and an equal amount of the container. Antimatter in the form of charged particles can be contained by a combination of electric and magnetic fields, in a device called a Penning trap. This device cannot, however, contain antimatter that consists of uncharged particles, for which atomic traps are used. In particular, such a trap may use the dipole moment (electric or magnetic) of the trapped particles. At high vacuum, the matter or antimatter particles can be trapped and cooled with slightly off-resonant laser radiation using a magneto-optical trap or magnetic trap. Small particles can also be suspended with optical tweezers, using a highly focused laser beam.[53]

In 2011, CERN scientists were able to preserve antihydrogen for approximately 17 minutes.[54]

15.5.6 Cost

Scientists claim that antimatter is the costliest material to make.[55] In 2006, Gerald Smith estimated $250 million could produce 10 milligrams of positrons[56] (equivalent to $25 billion per gram); in 1999, NASA gave a figure of $62.5 trillion per gram of antihydrogen.[55] This is because production is difficult (only very few antiprotons are produced in reactions in particle accelerators), and because there is higher demand for other uses of particle accelerators. According to CERN, it has cost a few hundred million Swiss francs to produce about 1 billionth of a gram (the amount used so far for particle/antiparticle collisions).[57] In comparison, to produce the first atomic weapon, the cost of the Manhattan Project was estimated at $23 billion with inflation during 2007.[58]

Several studies funded by the NASA Institute for Advanced Concepts are exploring whether it might be possible to use magnetic scoops to collect the antimatter that occurs naturally in the Van Allen belt of the Earth, and ultimately, the belts of gas giants, like Jupiter, hopefully at a lower cost per gram.[59]

15.6 Uses

15.6.1 Medical

Matter–antimatter reactions have practical applications in medical imaging, such as positron emission tomography (PET). In positive beta decay, a nuclide loses surplus positive charge by emitting a positron (in the same event, a proton becomes a neutron, and a neutrino is also emitted). Nuclides with surplus positive charge are easily made in a cyclotron and are widely generated for medical use. Antiprotons have also been shown within laboratory experiments to have the potential to treat certain cancers, in a similar method currently used for ion (proton) therapy.[60]

15.6.2 Fuel

Isolated and stored anti-matter could be used as a fuel for interplanetary or interstellar travel[61] as part of an antimatter catalyzed nuclear pulse propulsion or other antimatter rocketry, such as the redshift rocket. Since the energy density of antimatter is higher than that of conventional fuels, an antimatter-fueled spacecraft would have a higher thrust-to-weight ratio than a conventional spacecraft.

If matter–antimatter collisions resulted only in photon emission, the entire rest mass of the particles would be converted to kinetic energy. The energy per unit mass (9×10^{16} J/kg) is about 10 orders of magnitude greater than chemical energies,[62] and about 3 orders of magnitude greater than the nuclear potential energy that can be liberated, today, using nuclear fission (about 200 MeV per fission reaction[63] or 8×10^{13} J/kg), and about 2 orders of magnitude greater than the best possible results expected from fusion (about 6.3×10^{14} J/kg for the proton–proton chain). The reaction of 1 kg of antimatter with 1 kg of matter would produce 1.8×10^{17} J (180 petajoules) of energy (by the mass–energy equivalence formula, $E = mc^2$), or the rough equivalent of 43 megatons of TNT – slightly less than the yield of the 27,000 kg Tsar Bomb, the largest thermonuclear weapon ever detonated.

Not all of that energy can be utilized by any realistic propulsion technology because of the nature of the annihilation products. While electron–positron reactions result in gamma ray photons, these are difficult to direct and use for thrust. In reactions between protons and antiprotons, their energy is converted largely into relativistic neutral and charged pions. The neutral pions decay almost immediately (with a half-life of 84 attoseconds) into high-energy photons, but the charged pions decay more slowly (with a half-life of 26 nanoseconds) and can be deflected magnetically to produce thrust.

Note that charged pions ultimately decay into a combination of neutrinos (carrying about 22% of the energy of the charged pions) and unstable charged muons (carrying about 78% of the charged pion energy), with the muons then decaying into a combination of electrons, positrons and neutrinos (cf. muon decay; the neutrinos from this decay carry about 2/3 of the energy of the muons, meaning that from the original charged pions, the total fraction of their energy converted to neutrinos by one route or another would be about $0.22 + (2/3) \cdot 0.78 = 0.74$).[64]

15.6.3 Weapons

Main article: Antimatter weapon

Antimatter has been considered as a trigger mechanism for nuclear weapons.[65] A major obstacle is the difficulty of producing antimatter in large enough quantities, and there is no evidence that it will ever be feasible.[66] However, the U.S. Air Force funded studies of the physics of antimatter in the Cold War, and began considering its possible use in weapons, not just as a trigger, but as the explosive itself.[67]

15.7   See also

- Antimatter comet

- Ambiplasma

- Gravitational interaction of antimatter

15.8   References

[1] Smidgen of Antimatter Surrounds Earth

[2] N. Fornengo, L. Maccione and A. Vittino (2013). "Dark matter searches with cosmic antideuterons: status and perspectives". *Journal of Cosmology and Astroparticle Physics* **2013** (09): 031. doi:10.1088/1475-7516/2013/09/031.

[3] K. Abe et al. (BESS Collaboration) (2012). "Search for Antihelium with the BESS-Polar Spectrometer". *Physical Review Letters* **108** (13): 131301. doi:10.1103/PhysRevLett.108.131301.

[4] David Tenenbaum, David, One step closer: UW-Madison scientists help explain scarcity of anti-matter, University of Wisconsin—Madison News, December 26, 2012

[5] H. Agakishiev et al. (STAR Collaboration) (2011). "Observation of the antimatter helium-4 nucleus". *Nature* **473** (7347): 353–356. doi:10.1038/nature10079.

[6] K. Pearson (1891). "Ether Squirts". *American Journal of Mathematics* **13** (4): 309–72. doi:10.2307/2369570. JSTOR 2369570.

[7] H. Kragh (2002). *Quantum Generations: A History of Physics in the Twentieth Century.* Princeton University Press. pp. 5–6. ISBN 0-691-09552-3.

[8] A. Schuster (1898). "Potential Matter.—A Holiday Dream". *Nature* **58**(1503): 367. Bibcode:1898Natur..58..367S. doi:10.

[9] E. R. Harrison (2000-03-16). *Cosmology: The Science of the Universe* (2nd ed.). Cambridge University Press. pp. 266, 433. ISBN 0-521-66148-X.

[10] P. A. M. Dirac (1928). "The Quantum Theory of the Electron". *Proceedings of the Royal Society A* **117** (778): 610–624. Bibcode:1928RSPSA.117..610D. doi:10.1098/rspa.1928.0023. JSTOR 94981.

[11] M. Kaku, J. T. Thompson; Jennifer Trainer Thompson (1997). *Beyond Einstein: The Cosmic Quest for the Theory of the Universe*. Oxford University Press. pp. 179–180. ISBN 0-19-286196-4.

[12] P. J. Stewart (2010). "Charles Janet: Unrecognized genius of the periodic system". *Foundations of Chemistry* **12** (1): 5–15. doi:10.1007/s10698-008-9062-5.

[13] Canetti, L., Drewes, M., and Shaposhnikov, M. (2012). Matter and antimatter in the universe. New J. Phys. 14 (9), 095012.

[14] E. Sather (1999). "The Mystery of the Matter Asymmetry" (PDF). *Beam Line* **26** (1): 31.

[15] "Integral discovers the galaxy's antimatter cloud is lopsided". European Space Agency. 9 January 2008. Archived from the original on 18 June 2008. Retrieved 24 May 2008.

[16] G. Weidenspointner et al. (2008). "An asymmetric distribution of positrons in the Galactic disk revealed by γ-rays". *Nature* **451** (7175): 159–162. Bibcode:2008Natur.451..159W. doi:10.1038/nature06490. PMID 18185581.

[17] Close, F. E. (2009-01-22). *Antimatter*. Oxford University Press US. p. 114. ISBN 0-19-955016-6.

[18] "Searching for Primordial Antimatter". NASA. 30 October 2008. Retrieved 18 June 2010.

[19] "Antimatter caught streaming from thunderstorms on Earth". BBC. 11 January 2011. Archived from the original on 12 January 2011. Retrieved 11 January 2011.

[20] ScientificAmerican.com. "Rogue Antimatter Found in Thunderclouds". Retrieved 2015-05-14.

[21] Adriani, O.; Barbarino, G. C.; Bazilevskaya, G. A.; Bellotti, R.; Boezio, M.; Bogomolov, E. A.; Bongi, M.; Bonvicini, V.; Borisov, S.; Bottai, S.; Bruno, A.; Cafagna, F.; Campana, D.; Carbone, R.; Carlson, P.; Casolino, M.; Castellini, G.; Consiglio, L.; De Pascale, M. P.; De Santis, C.; De Simone, N.; Di Felice, V.; Galper, A. M.; Gillard, W.; Grishantseva, L.; Jerse, G.; Karelin, A. V.; Kheymits, M. D.; Koldashov, S. V.; Krutkov, S. Y. (2011). "The Discovery of Geomagnetically Trapped Cosmic-Ray Antiprotons". *The Astrophysical Journal Letters* **737** (2): L29. arXiv:1107.4882v1. Bibcode:2011ApJ...737L..29A. doi:10.1088/2041-8205/737/2/L29.

[22] Than, Ker (10 August 2011). "Antimatter Found Orbiting Earth—A First". National Geographic Society. Retrieved 12 August 2011.

[23] "What's the Matter with Antimatter?". NASA. 29 May 2000. Archived from the original on 4 June 2008. Retrieved 24 May 2008.

[24] Serpico, Pasquale D. "Astrophysical models for the origin of the positron "excess"." Astroparticle Physics 39 (2012): 2-11.

[25] L. Accardo (AMS Collaboration) et al. (18 September 2014). "High Statistics Measurement of the Positron Fraction in Primary Cosmic Rays of 0.5–500 GeV with the Alpha Magnetic Spectrometer on the International Space Station" (PDF). *Physical Review Letters* **113**: 121101. Bibcode:2014PhRvL.113l1101A. doi:10.1103/PhysRevLett.113.121101.

[26] Schirber, Michael. "Synopsis: More Dark Matter Hints from Cosmic Rays?". American Physical Society. Retrieved 21 September 2014.

[27] "New results from the Alpha Magnetic$Spectrometer on the International Space Station" (PDF). *AMS-02 at NASA*. Retrieved 21 September 2014.

[28] Aguilar, M.; Alberti, G.; Alpat, B.; Alvino, A.; Ambrosi, G.; Andeen, K.; Anderhub, H.; Arruda, L.; Azzarello, P.; Bachlechner, A.; Barao, F.; Baret, B.; Barrau, A.; Barrin, L.; Bartoloni, A.; Basara, L.; Basili, A.; Batalha, L.; Bates, J.; Battiston, R.; Bazo, J.; Becker, R.; Becker, U.; Behlmann, M.; Beischer, B.; Berdugo, J.; Berges, P.; Bertucci, B.; Bigongiari, G. et al. (2013). "First Result from the Alpha Magnetic Spectrometer on the International Space Station: Precision Measurement of the Positron Fraction in Primary Cosmic Rays of 0.5–350 GeV". *Physical Review Letters* **110** (14): 141102. Bibcode:2013PhRvL.110n1102A. doi:10.1103/PhysRevLett.110.141102.

[29] Moskalenko, I. V.; Strong, A. W.; Ormes, J. F; Potgieter, M. S. (January 2002). "Secondary antiprotons and propagation of cosmic rays in the Galaxy and heliosphere". *The Astrophysical Journal* **565** (1): 280–296. arXiv:astro-ph/0106567v2. Bibcode:2002ApJ...565..280M. doi:10.1086/324402.

[30] AMS Collaboration; Aguilar, M.; Alcaraz, J.; Allaby, J.; Alpat, B.; Ambrosi, G.; Anderhub, H.; Ao, L. et al. (August 2002). "The Alpha Magnetic Spectrometer (AMS) on the International Space Station: Part I – results from the test flight on the space shuttle". *Physics Reports* **366** (6): 331–405. Bibcode:2002PhR...366..331A. doi:10.1016/S0370-1573(02)00013-3.

[31] "Billions of particles of anti-matter created in laboratory" (Press release). Lawrence Livermore National Laboratory. 3 November 2008. Retrieved 19 November 2008.

[32] "Laser creates billions of antimatter particles". Cosmos Magazine. 19 November 2008. Archived from the original on 22 May 2009. Retrieved 1 July 2009.

[33] "All Nobel Prizes in Physics".

[34] "Breaking Through: A Century of Physics at Berkeley, 1868–1968". Regents of the University of California. 2006. Archived from the original on 18 November 2010. Retrieved 18 November 2010.

[35] Massam, T; Muller, Th.; Righini, B.; Schneegans, M.; Zichichi, A. (1965). "Experimental observation of antideuteron production". *Il Nuovo Cimento* **39**: 10–14. Bibcode:1965NCiS..39...10M . doi:10.1007/BF02814251.

[36] Dorfan, D. E; Eades, J.; Lederman, L. M.; Lee, W.; Ting, C. C. (June 1965). "Observation of Antideuterons". *Phys. Rev. Lett.* **14** (24): 1003–1006. Bibcode:1965PhRvL..14.1003D. doi:10.1103/PhysRevLett.14.1003.

[37] Gabrielse, Gerald, and Hartmut Kalinowsky. The production and study of cold antihydrogen. No. SPSLC-I-211. 1996.

[38] M. Amoretti et al. (2002). "Production and detection of cold antihydrogen atoms".*Nature***419**(6906): 456–9.Bibcode:2002. doi:10.1038/nature01096. PMID 12368849.

[39] G. Gabrielse et al. (2002). "Background-free observation of cold antihydrogen with field ionization analysis of its states". *Physical Review Letters* **89** (21): 213401. Bibcode:2002PhRvL..89u3401G. doi:10.1103/PhysRevLett.89.213401. PMID 12443407.

[40] J. H. Malmberg, J. S. deGrassie (1975). "Properties of a nonneutral plasma".*Physical Review Letters***35**(9): 577.Bibcode:197. doi:10.1103/PhysRevLett.35.577.

[41] G. Gabrielse et al. (1989). "Cooling and slowing of trapped antiprotons below 100 meV". *Physical Review Letters* **63** (13): 1360. Bibcode:1989PhRvL..63.1360G. doi:10.1103/PhysRevLett.63.1360.

[42] C. M. Surko, R. G. Greaves (2004). "Emerging science and technology of antimatter plasmas and trap-based beams". *Physics of Plasmas* **11** (5): 2333. Bibcode:2004PhPl...11.2333S. doi:10.1063/1.1651487.

[43] D. E. Pritchard; Heinz, T.; Shen, Y. (1983). "Cooling neutral atoms in a rgnetic trap for precision spectroscopy". *Physical Review Letters* **51** (21): 1983. Bibcode:1983PhRvL..51.1983T. doi:10.1103/PhysRevLett.51.1983.

[44] Andresen; Ashkezari, M. D.; Baquero-Ruiz, M.; Bertsche, W.; Bowe, P. D.; Butler, E.; Cesar, C. L.; Chapman, S. et al. (2010). "Trapped antihydrogen". *Nature* **468** (7324): 673–676. Bibcode:2010Natur.468..673A. doi:10.1038/nature09610. PMID 21085118.

[45] "Antimatter atoms produced and trapped at CERN". CERN. 17 November 2010. Retrieved 20 January 2011.

[46] ALPHA Collaboration (2011). "Confinement of antihydrogen for 1000 seconds".*Nature Physics***7**(7): 558–564.arXiv:110. Bibcode:2011NatPh...7..558A. doi:10.1038/nphys2025.

[47] ALPHA Collaboration (2011). "Confinement of antihydrogen for 1,000 seconds".*Nature Physics***7**(7).Bibcode:2011NatPh... doi:10.1038/nphys2025.

[48] Amole, C.; Ashkezari, M. D.; Baquero-Ruiz, M.; Bertsche, W.; Bowe, P. D.; Butler, E.; Capra, A.; Cesar, C. L.; Charlton, M.; Deller, A.; Donnan, P. H.; Eriksson, S.; Fajans, J.; Friesen, T.; Fujiwara, M. C.; Gill, D. R.; Gutierrez, A.; Hangst, J. S.; Hardy, W. N.; Hayden, M. E.; Humphries, A. J.; Isaac, C. A.; Jonsell, S.; Kurchaninov, L.; Little, A.; Madsen, N.; McKenna, J. T. K.; Menary, S.; Napoli, S. C.; Nolan, P. (2012). "Resonant quantum transitions in trapped antihydrogen atoms". *Nature* **483** (7390): 439–443. Bibcode:2012Natur.483..439A. doi:10.1038/nature10942. PMID 22398451.

[49] N. Madsen (2010). "Cold antihydrogen: a new frontier in fundamental physics". *Philosophical Transactions of the Royal Society A* **368** (1924): 3671–82. Bibcode:2010RSPTA.368.3671M. doi:10.1098/rsta.2010.0026. PMID 20603376.

[50] Y.M. Antipov et al. (1974). "Observation of antihelium3 (in Russian)". *Yad. Fiz.* **12**: 311.

[51] R. Arsenescu et al. (2003). "Antihelium-3 production in lead–lead collisions at 158 *A* GeV/*c*". *New Journal of Physics* **5**: 1. Bibcode:2003NJPh....5....1A. doi:10.1088/1367-2630/5/1/301.

[52] H. Agakishiev et al. (2011). "Observation of the antimatter helium-4 nucleus". arXiv:1103.3312.

[53] Blaum, Klaus, Mark G. Raizen, and Wolfgang Quint. "An experimental test of the weak equivalence principle for antihydrogen at the future FLAIR facility." International Journal of Modern Physics: Conference Series. Vol. 30. The Authors, 2014.

[54] http://www.economist.com/node/18802932 *The Economist.* Antimatter of Fact. 9 June 2011

[55] "Reaching for the stars: Scientists examine using antimatter and fusion to propel future spacecraft". NASA. 12 April 1999. Retrieved 11 June 2010. Antimatter is the most expensive substance on Earth

[56] B. Steigerwald (14 March 2006). "New and Improved Antimatter Spaceship for Mars Missions". NASA. Retrieved 11 June 2010. "A rough estimate to produce the 10 milligrams of positrons needed for a human Mars mission is about 250 million dollars using technology that is currently under development," said Smith.

[57] "Antimatter Questions & Answers". CERN. 2001. Archived from the original on 21 April 2008. Retrieved 24 May 2008.

[58] http://www.ctbto.org/nuclear-testing/history-of-nuclear-testing/manhattan-project/

[59] J. Bickford. "Extraction of Antiparticles Concentrated in Planetary Magnetic Fields" (PDF). NASA. Retrieved 24 May 2008.

[60] "Antiproton portable traps and medical applications" (PDF).

[61] G. R. Schmidt (1999). "Antimatter Production for Near-Term Propulsion Applications". *Nuclear Physics and High-Energy Physics*. Marshall Space Flight Center, NASA. Retrieved 14 December 2012.

[62] (compared to the formation of water at 1.56×10^7 J/kg, for example)

[63] M. G. Sowerby. "§4.7 Nuclear fission and fusion, and neutron interactions". *Kaye & Laby: Table of Physical & Chemical Constants*. National Physical Laboratory. Retrieved 18 June 2010.

[64] S. K. Borowski (1987). "Comparison of Fusion/Antiproton Propulsion systems" (PDF). *NASA Technical Memorandum 107030*. NASA. pp. 5–6 (pp. 6–7 of pdf). AIAA–87–1814. Archived (PDF) from the original on 28 May 2008. Retrieved 24 May 2008.

[65] Page discussing the possibility of using antimatter as a trigger for a thermonuclear explosion

[66] Paper discussing the number of antiprotons required to ignite a thermonuclear weapon.

[67] "Air Force pursuing antimatter weapons: Program was touted publicly, then came official gag order"

15.9   Further reading

- G. Fraser (2000-05-18). *Antimatter, The Ultimate Mirror*. Cambridge University Press. ISBN 978-0-521-65252-0.

- Schmidt, G.R.; Gerrish, H.P.; Martin, J.J.; Smith, G.A.; Meyer, K.J. "Antimatter Production for Near-term Propulsion Applications" (PDF).

15.10   External links

- Antimatter (physics) at *Encyclopædia Britannica*

-

- Antimatter on *In Our Time* at the BBC. (listen now)

- Freeview Video 'Antimatter' by the Vega Science Trust and the BBC/OU

- CERN Webcasts (RealPlayer required)

- What is Antimatter? (from the Frequently Asked Questions at the Center for Antimatter–Matter Studies)

- FAQ from CERN with lots of information about antimatter aimed at the general reader, posted in response to antimatter's fictional portrayal in Angels & Demons

- What is direct CP-violation?

- Animated illustration of antihydrogen production at CERN from the Exploratorium.

Chapter 16

Dark matter

Not to be confused with antimatter, dark energy, dark fluid, or dark flow. For other uses, see Dark Matter (disambiguation)

Dark matter is a hypothetical kind of matter that cannot be seen with telescopes but would account for most of the matter in the universe. The existence and properties of dark matter are inferred from its gravitational effects on visible matter, on radiation, and on the large-scale structure of the universe. Dark matter has not been detected directly, making it one of the greatest mysteries in modern astrophysics.

Dark matter neither emits nor absorbs light or any other electromagnetic radiation at any significant level. According to the Planck mission team, and based on the standard model of cosmology, the total mass–energy of the known universe contains 4.9% ordinary matter, 26.8% dark matter and 68.3% dark energy.[2][3] Thus, dark matter is estimated to constitute 84.5% [note 1] of the total matter in the universe, while dark energy plus dark matter constitute 95.1% of the total mass–energy content of the universe.[4][5][6]

Astrophysicists hypothesized the existence of dark matter to account for discrepancies between the mass of large astronomical objects determined from their gravitational effects, and their mass as calculated from the observable matter (stars, gas, and dust) that they can be seen to contain. Their gravitational effects suggest that their masses are much greater than the observable matter survey suggests.

Dark matter was postulated by Jan Oort in 1932, albeit based upon insufficient evidence, to account for the orbital velocities of stars in the Milky Way. In 1933, Fritz Zwicky was the first to use the virial theorem to infer the existence of unseen matter, which he referred to as *dunkle Materie* 'dark matter'.[7] More robust evidence from galaxy rotation curves was discovered by Horace W. Babcock in 1939, but was not attributed to dark matter. The first hypothesis to postulate "dark matter" based upon robust evidence was formulated by Vera Rubin and Kent Ford in the 1960s–1970s, using galaxy rotation curves.[8][9] Subsequently, many other observations have indicated the presence of dark matter in the universe, including gravitational lensing of background objects by galaxy clusters such as the Bullet Cluster, the temperature distribution of hot gas in galaxies and clusters of galaxies and, more recently, the pattern of anisotropies in the cosmic microwave background. According to consensus among cosmologists, dark matter is composed primarily of a not yet characterized type of subatomic particle.[10][11] The search for this particle, by a variety of means, is one of the major efforts in particle physics today.[12]

Although the existence of dark matter is generally accepted by the mainstream scientific community, some alternative theories of gravity have been proposed, such as MOND and TeVeS, which try to account for the anomalous observations without requiring additional matter. However, these theories cannot account for the properties of galaxy clusters.[13]

16.1   Overview

Dark matter's existence is inferred from gravitational effects on visible matter and gravitational lensing of background radiation, and was originally hypothesized to account for discrepancies between calculations of the mass of galaxies, clusters of galaxies and the entire universe made through dynamical and general relativistic means, and calculations based on the mass of the visible "luminous" matter these objects contain: stars and the gas and dust of the interstellar and

Dark matter *is invisible. Based on the effect of gravitational lensing, a ring of* dark matter *has been inferred in this image of a galaxy cluster (CL0024+17) and has been represented in blue.*[1]

intergalactic medium.[14]

The most widely accepted explanation for these phenomena is that dark matter exists and that it is most probably[10] composed of weakly interacting massive particles (WIMPs) that interact only through gravity and the weak force. Alternative explanations have been proposed, and there is not yet sufficient experimental evidence to determine whether any of them are correct. Many experiments to detect proposed dark matter particles through non-gravitational means are under way.[12]

One other theory suggests the existence of a "Hidden Valley", a parallel world made of dark matter having very little in common with matter we know,[15] and that could only interact with our visible universe through gravity.[16][17]

According to observations of structures larger than star systems, as well as Big Bang cosmology interpreted under the Friedmann equations and the Friedmann–Lemaître–Robertson–Walker metric, dark matter accounts for 26.8% of the mass-energy content of the observable universe. In comparison, ordinary (baryonic) matter accounts for only 4.9% of the mass-energy content of the observable universe, with the remainder being attributable to dark energy.[3] From these figures, matter accounts for 31.7% of the mass-energy content of the universe, and 84.5% of the matter is dark matter.

Dark matter plays a central role in state-of-the-art modeling of cosmic structure formation and galaxy formation and evolution and has measurable effects on the anisotropies observed in the cosmic microwave background (CMB). All these lines of evidence suggest that galaxies, clusters of galaxies, and the universe as a whole contain far more matter than that which is easily visible with electromagnetic radiation.[16]

Important as dark matter is thought to be in the cosmos, direct evidence of its existence and a concrete understanding of its nature have remained elusive. Though the theory of dark matter remains the most widely accepted theory to explain the anomalies in observed galactic rotation, some alternative theoretical approaches have been developed which broadly fall into the categories of modified gravitational laws and quantum gravitational laws.[18]

16.2 Baryonic and nonbaryonic dark matter

There are three separate lines of evidence that suggest the majority of dark matter is not made of baryons (ordinary matter including protons and neutrons):

- The theory of Big Bang nucleosynthesis, which predicts the observed abundance of the chemical elements,[19] predicts that baryonic matter accounts for around 4–5 percent of the critical density of the universe. In contrast, evidence from large-scale structure and other observations indicates that the total matter density is about 30% of the critical density.

- Large astronomical searches for gravitational microlensing, including the MACHO, EROS and OGLE projects, have shown that only a small fraction of the dark matter in the Milky Way can be hiding in dark compact objects; the excluded range covers objects above half the Earth's mass up to 30 solar masses, excluding nearly all the plausible candidates.

- Detailed analysis of the small irregularities (anisotropies) in the cosmic microwave background observed by WMAP and Planck shows that around five-sixths of the total matter is in a form which does not interact significantly with ordinary matter or photons except through gravitational effects.

A small proportion of dark matter may be baryonic dark matter: astronomical bodies, such as massive compact halo objects, which are composed of ordinary matter but emit little or no electromagnetic radiation. The study of nucleosynthesis in the Big Bang gives an upper bound on the amount of baryonic matter in the universe,[20] which indicates that the vast majority of dark matter in the universe cannot be baryons, and thus does not form atoms. It also cannot interact with ordinary matter via electromagnetic forces; in particular, dark matter particles do not carry any electric charge.

Candidates for nonbaryonic dark matter are hypothetical particles such as axions, or supersymmetric particles; neutrinos can only form a small fraction of the dark matter, due to limits from large-scale structure and high-redshift galaxies. Unlike baryonic dark matter, nonbaryonic dark matter does not contribute to the formation of the elements in the early universe ("Big Bang nucleosynthesis")[10] and so its presence is revealed only via its gravitational attraction. In addition, if the particles of which it is composed are supersymmetric, they can undergo annihilation interactions with themselves, possibly resulting in observable by-products such as gamma rays and neutrinos ("indirect detection").[21]

Nonbaryonic dark matter is classified in terms of the mass of the particle(s) that is assumed to make it up, and/or the typical velocity dispersion of those particles (since more massive particles move more slowly). There are three prominent hypotheses on nonbaryonic dark matter, called cold dark matter (CDM), warm dark matter (WDM), and hot dark matter (HDM); some combination of these is also possible. The most widely discussed models for nonbaryonic dark matter are based on the cold dark matter hypothesis, and the corresponding particle is most commonly assumed to be a weakly interacting massive particle (WIMP). Hot dark matter may include (massive) neutrinos, but observations imply that only a small fraction of dark matter can be hot. Cold dark matter leads to a "bottom-up" formation of structure in the universe while hot dark matter would result in a "top-down" formation scenario; since the late 1990s, the latter has been ruled out by observations of high-redshift galaxies such as the Hubble Ultra-Deep Field.[12]

16.3   Observational evidence

The first person to interpret evidence and infer the presence of dark matter was Dutch astronomer Jan Oort, a pioneer in radio astronomy, in 1932.[23] Oort was studying stellar motions in the local galactic neighbourhood and found that the mass in the galactic plane must be greater than what was observed, but this measurement was later determined to be essentially erroneous.[24]

In 1933, the Swiss astrophysicist Fritz Zwicky, who studied clusters of galaxies while working at the California Institute of Technology, made a similar inference.[25][26] Zwicky applied the virial theorem to the Coma cluster of galaxies and obtained evidence of unseen mass. Zwicky estimated the cluster's total mass based on the motions of galaxies near its edge and compared that estimate to one based on the number of galaxies and total brightness of the cluster. He estimated that there was about 400 times more mass than was visually observable. The gravity effect of the visible galaxies in the cluster would be far too small for such fast orbits, unless there was mass hidden from visual observation. This is known as the "missing mass problem". Based on these conclusions, Zwicky inferred that there must be some non-visible form of matter which would provide enough mass and gravitation attraction to hold the cluster together. This was the first formal inference about the existence of dark matter.[27]

Zwicky's estimates were not accurate and were off by more than an order of magnitude.[28] Notwithstanding, although the same calculation today shows a smaller factor, based on greater values for the mass of luminous material, it is still clear that the great majority of matter in Zwicky's calculations was correctly inferred to be dark.[29]

Much of the evidence for dark matter comes from the study of the motions of galaxies.[31] Many of these appear to be fairly uniform, so by the virial theorem, the total kinetic energy should be half the total gravitational binding energy of the galaxies. Observationally, however, the total kinetic energy is found to be much greater. In particular, assuming the gravitational mass is due to only the visible matter of the galaxy, stars far from the center of galaxies have much higher velocities than are predicted by the virial theorem. Galactic rotation curves, which illustrate the velocity of rotation versus the distance from the galactic center, show the well known phenomenology that cannot be explained by the presence of the visible matter only. Assuming that the visible material makes up only a small part of the cluster's mass is the most straightforward way of accounting for this discrepancy. The distribution of dark matter in galaxies required to explain the motion of the observed baryonic matter suggests the presence of a roughly spherically symmetric, centrally concentrated halo of dark matter with the visible matter concentrated in a disc at the center. Low surface brightness dwarf galaxies are important sources of information for studying dark matter, as they have an uncommonly low ratio of visible matter to dark matter, and have few bright stars at the center which would otherwise impair observations of the rotation curve of outlying stars.

Gravitational lensing observations of galaxy clusters allow direct estimates of the gravitational mass based on its effect on light coming from background galaxies, since large collections of matter (dark or otherwise) will gravitationally deflect light. In clusters such as Abell 1689, lensing observations confirm the presence of considerably more mass than is indicated by the clusters' light alone. In the Bullet Cluster, lensing observations show that much of the lensing mass is separated from the X-ray-emitting baryonic mass. In July 2012, lensing observations were used to identify a "filament" of dark matter between two clusters of galaxies, as cosmological simulations have predicted.[32]

16.3.1   Galaxy rotation curves

Main article: Galaxy rotation curve

The first robust indications that the mass to light ratio was anything other than unity came from measurements of galaxy rotation curves. In 1939, Horace W. Babcock reported in his PhD thesis measurements of the rotation curve for the Andromeda nebula which suggested that the mass-to-luminosity ratio increases radially.[33] He, however, attributed it to either absorption of light within the galaxy or modified dynamics in the outer portions of the spiral and not to any form of missing matter.

In the late 1960s and early 1970s, Vera Rubin was the first to both make robust measurements indicating the existence of dark matter and attribute them to dark matter. Rubin worked with a new sensitive spectrograph that could measure the velocity curve of edge-on spiral galaxies to a greater degree of accuracy than previously.[9] Together with fellow staff-member Kent Ford, Rubin announced at a 1975 meeting of the American Astronomical Society the discovery that most stars in spiral galaxies orbit at roughly the same speed, which implied that the mass densities of the galaxies were uniform well beyond the regions containing most of the stars (the galactic bulge), a result independently found in 1978.[34] An influential paper presented Rubin's results in 1980.[35] Rubin's observations and calculations showed that most galaxies must contain about six times as much "dark" mass as could be accounted for by the visible stars. Eventually other astronomers began to corroborate her work. It soon became well-established that most galaxies were dominated by "dark matter":

- Low-surface-brightness (LSB) galaxies.[36] LSB galaxies are probably everywhere dark matter-dominated, with the observed stellar populations making only a small contribution to their total mass. Such a property is extremely important as it allows one to avoid the difficulties associated with the deprojection and disentanglement of the dark and visible matter contributions to the rotation curves.[12]

- Spiral galaxies.[37] Rotation curves of both low and high surface luminosity galaxies suggest a universal rotation curve, which can be expressed as the sum of an exponential distribution of visible matter that is maximum at the center and tapering to zero at great distances, and a spherical dark matter halo with a flat core of radius r_0 and density $\rho_0 = 4.5 \times 10^{-2} (r_0/\text{kpc})^{-2/3} \, M\odot\text{pc}^{-3}$.

- Elliptical galaxies. Some elliptical galaxies show evidence for dark matter via strong gravitational lensing,[38] X-ray evidence reveals the presence of extended atmospheres of hot gas that fill the dark haloes of isolated elliptical galaxies and whose hydrostatic support provides evidence for the existence of dark matter. Other ellipticals have low velocities in their outskirts (tracked for example by the motion of planetary nebulae embedded within) and were interpreted as not having dark matter haloes.[12] However, simulations of disk-galaxy mergers suggest that stars may have been torn by tidal forces from their original galaxies during the first close passage and put on outgoing trajectories, explaining the low velocities of the remaining stars even with the presence of a dark matter halo.[39] More research is needed to clarify this situation.

Simulated dark matter haloes have significantly steeper density profiles (having central cusps) than are inferred from observations, which is a problem for cosmological models with dark matter at the smallest scale of galaxies (as of 2008).[12] This may only be a problem of resolution: star-forming regions which might alter the dark matter distribution via outflows of gas have been too small to resolve and model simultaneously with larger dark matter clumps. A recent simulation[40] of a dwarf galaxy, that included these star-forming regions, reported that strong outflows from supernovae remove low-angular-momentum gas, which inhibits the formation of a galactic bulge and decreases the dark matter density to less than half of what it would have been in the central kiloparsec. These simulation predictions—bulgeless and with shallow central dark matter density profiles—correspond closely to observations of actual dwarf galaxies. There are no such discrepancies at the larger scales of clusters of galaxies and greater, or in the outer regions of haloes of galaxies.

The exceptions to this general picture of dark matter haloes for galaxies appear to be galaxies with mass-to-light ratios that are close to that of the stars they contain. Otherwise, numerous observations have been made that do indicate the presence of dark matter in various parts of the cosmos, such as observations of the cosmic microwave background, of supernovas used as distance measures, of gravitational lensing at various scales, and many types of sky survey. Starting with Rubin's findings for spiral galaxies, such robust observational evidence for dark matter has collected over the decades to the point that by the 1980s most astrophysicists have accepted its existence.[41] As a unifying concept, dark matter is one of the dominant features considered in the analysis of structures on the order of galactic scale and larger.

16.3.2   Velocity dispersions of galaxies

Rubin's pioneering work has stood the test of time. Measurements of velocity curves in spiral galaxies were soon followed up with velocity dispersions of elliptical galaxies.[42] While some elliptical galaxies display lower mass-to-light ratios, measurements of ellipticals generally indicate a relatively high dark matter content. Likewise, measurements of the diffuse interstellar gas found at the edge of galaxies indicate not only dark matter distributions that extend beyond the visible limit of the galaxies, but also that the galaxies are virialized (i.e., gravitationally bound and orbiting each other with velocities which appear to disproportionately correspond to predicted orbital velocities of general relativity) up to ten times their visible radii.[43] This has the effect of pushing up the dark matter as a fraction of the total matter from 50% as measured by Rubin to the now accepted value of nearly 95%.

There are places where dark matter seems to be a small component or totally absent. Globular clusters show little evidence that they contain dark matter,[44] though their orbital interactions with galaxies do show evidence for galactic dark matter. For some time, measurements of the velocity profile of stars seemed to indicate concentration of dark matter in the disk of the Milky Way. It now appears, however, that the high concentration of baryonic matter in the disk of the galaxy (especially in the interstellar medium) can account for this motion. Galaxy mass profiles are thought to look very different from the light profiles. The typical model for dark matter galaxies is a smooth, spherical distribution in virialized halos. Such would have to be the case to avoid small-scale (stellar) dynamical effects. Recent research reported in January 2006 from the University of Massachusetts Amherst would explain the previously mysterious warp in the disk of the Milky Way by the interaction of the Large and Small Magellanic Clouds and the predicted 20 fold increase in mass of the Milky Way taking into account dark matter.[45]

In 2005, astronomers from Cardiff University claimed to have discovered a galaxy made almost entirely of dark matter, 50 million light years away in the Virgo Cluster, which was named VIRGOHI21.[46] Unusually, VIRGOHI21 does not appear to contain any visible stars: it was seen with radio frequency observations of hydrogen. Based on rotation profiles, the scientists estimate that this object contains approximately 1000 times more dark matter than hydrogen and has a total mass of about 1/10 that of the Milky Way. For comparison, the Milky Way is estimated to have roughly 10 times as much dark matter as ordinary matter. Models of the Big Bang and structure formation have suggested that such dark galaxies should be very common in the universe, but none had previously been detected.

There are some galaxies, such as NGC 3379, whose velocity profile indicates an absence of dark matter.[47]

16.3.3   Galaxy clusters and gravitational lensing

Galaxy clusters are especially important for dark matter studies since their masses can be estimated in three independent ways:

- From the scatter in radial velocities of the galaxies within the clusters (as in Zwicky's early observations, but with more accurate measurements and much larger samples).

- From X-rays emitted by very hot gas within the clusters. The temperature and density of the gas can be estimated from the energy and flux of the X-rays, and hence the gas pressure derived; assuming pressure and gravity balance, this enables the mass profile of the cluster to be derived. Many of the experiments of the Chandra X-ray Observatory use this technique to independently determine the mass of clusters. These observations generally indicate that baryonic mass is approximately 12–15 percent, in reasonable agreement with the Planck spacecraft cosmic average of 15.5–16 percent.[48]

- From their gravitational lensing effects on background objects (usually more distant galaxies). This is observed as "strong lensing" (multiple images) near the cluster core, and weak lensing (shape distortions) in the outer parts. Several large Hubble projects have used this method to measure cluster masses.

Generally these three methods are in reasonable agreement, that clusters contain much more matter than suggested by the visible components of galaxies and gas.

A gravitational lens is formed when the light from a more distant source (such as a quasar) is "bent" around a massive object (such as a cluster of galaxies) lying inline between the source object and the observer. The process is known as gravitational lensing.

The galaxy cluster Abell 2029 is composed of thousands of galaxies enveloped in a cloud of hot gas, and an amount of dark matter equivalent to more than 10^{14} $M\odot$. At the center of this cluster is an enormous, elliptically shaped galaxy that is thought to have been formed from the mergers of many smaller galaxies.[49] The measured orbital velocities of galaxies within galactic clusters have been found to be consistent with dark matter observations.

Another important tool for future dark matter observations is gravitational lensing. Lensing relies on the bending of light, as described by general relativity, to predict masses without relying on observations of the distant galaxies dynamics, and so is a completely independent means of measuring the dark matter. Strong lensing, the observed distortion of background galaxies into arcs when their light passes through such a gravitational lens, has been observed around a few distant clusters including Abell 1689 (pictured).[50] By measuring the distortion geometry, the mass of the intervening cluster causing the phenomena can be obtained. In the dozens of cases where this has been done, the mass-to-light ratios obtained correspond to the dynamical dark matter measurements of clusters.[51]

Weak gravitational lensing investigates minute distortions of galaxies, using statistical analyses of vast galaxy surveys, caused by foreground objects. By examining the apparent shear deformation of the adjacent background galaxies, astrophysicists can characterize the mean distribution of dark matter and have found mass-to-light ratios that correspond to dark matter densities predicted by other large-scale structure measurements.[52] The correspondence of the two gravitational lens techniques to other dark matter measurements has convinced almost all astrophysicists that dark matter actually exists as a major component of the universe's composition.

The most direct observational evidence to date for dark matter comes from a system known as the Bullet Cluster. In most regions of the universe, dark matter and visible matter are found together,[53] as expected due to their mutual gravitational attraction. In the Bullet Cluster however, a collision between two galaxy clusters appears to have caused a separation of dark matter and baryonic matter. X-ray observations show that much of the baryonic matter (in the form of 10^7–10^8 Kelvin[54] gas or plasma) in the system is concentrated in the center of the system. Electromagnetic interactions between passing gas particles caused them to slow and settle near the point of impact of those galaxies. However, weak gravitational lensing observations of the same system show that much of the mass resides outside of the central region of baryonic gas. Because dark matter does not interact by electromagnetic forces, it would not have been slowed as the X-ray visible gas, so the dark matter components of the two clusters passed through each other without slowing substantially, throwing the dark matter further out than that of the baryonic gas. This accounts for the separation. Unlike the galactic rotation curves, this evidence for dark matter is independent of the details of Newtonian gravity, so it is claimed to be direct evidence of the existence of dark matter.[54]

Another galaxy cluster, known as the Train Wreck Cluster/Abell 520, initially appeared to have an unusually massive and dark matter core containing few of the cluster's galaxies, which presented problems for standard dark matter models.[55] However, more precise observations since that time have shown that the earlier observations were misleading, and that the distribution of dark matter and its ratio to normal matter are very similar to those in galaxies in general, making novel explanations unnecessary.[54]

The observed behavior of dark matter in clusters constrains whether and how much dark matter scatters off other dark matter particles, quantified as its self-interaction cross section. More simply, the question is whether the dark matter has pressure, and thus can be described as a perfect fluid that has no damping.[56] The distribution of mass (and thus dark matter) in galaxy clusters has been used to argue both for[57] and against[58] the existence of significant self-interaction in dark matter. Specifically, the distribution of dark matter in merging clusters such as the Bullet Cluster shows that dark matter scatters off other dark matter particles only very weakly if at all.[59]

A currently ongoing survey using the Subaru telescope is using weak lensing to analyze background light, bent by dark matter, to determine how dark matter is distributed in the foreground. The analysis of dark matter and its effects could determine how dark matter assembled over time, which can be related to the history of the expansion of the universe, and could reveal some physical properties of dark energy, its strength and how it has changed over time. The survey is observing galaxies more than a billion light-years away, across an area greater than a thousand square degrees (about one fortieth of the entire sky).[60][61]

16.3.4 Cosmic microwave background

Main article: Cosmic microwave background
See also: Wilkinson Microwave Anisotropy Probe
 Angular fluctuations in the cosmic microwave background (CMB) spectrum provide evidence for dark matter. Since the 1964 discovery and confirmation of the CMB radiation,[62] many measurements of the CMB have supported and constrained this theory. The NASA Cosmic Background Explorer (COBE) found that the CMB spectrum to be a blackbody spectrum with a temperature of 2.726 K. In 1992, COBE detected fluctuations (anisotropies) in the CMB spectrum, at a level of about one part in 10^5.[63] In the following decade, CMB anisotropies were further investigated by a large number of ground-based and balloon experiments. The primary goal of those was to measure the angular scale of the first acoustic peak of the power spectrum of the anisotropies, for which COBE did not have sufficient resolution. In 2000–2001, several experiments, most notably BOOMERanG[64] found the universe to be almost spatially flat by measuring the typical angular size of the anisotropies. During the 1990s, the first peak was measured with increasing sensitivity and

by 2000 the BOOMERanG experiment reported that the highest power fluctuations occur at scales of approximately one degree. These measurements were able to rule out cosmic strings as the leading theory of cosmic structure formation, and suggested cosmic inflation was the correct theory.

A number of ground-based interferometers provided measurements of the fluctuations with higher accuracy over the next three years, including the Very Small Array, the Degree Angular Scale Interferometer (DASI) and the Cosmic Background Imager (CBI). DASI made the first detection of the polarization of the CMB,[65][66] and the CBI provided the first E-mode polarization spectrum with compelling evidence that it is out of phase with the T-mode spectrum.[67] COBE's successor, the Wilkinson Microwave Anisotropy Probe (WMAP) has provided the most detailed measurements of (large-scale) anisotropies in the CMB as of 2009 with ESA's Planck spacecraft returning more detailed results in 2012-2014.[68] WMAP's measurements played the key role in establishing the current Standard Model of Cosmology, namely the Lambda-CDM model, a flat universe dominated by dark energy, supplemented by dark matter and atoms with density fluctuations seeded by a Gaussian, adiabatic, nearly scale invariant process. The basic properties of this universe are determined by five numbers: the density of matter, the density of atoms, the age of the universe (or equivalently, the Hubble constant today), the amplitude of the initial fluctuations, and their scale dependence.

A successful Big Bang cosmology theory must fit with all available astronomical observations, including the CMB. In cosmology, the CMB is explained as relic radiation from shortly after the big bang. The anisotropies in the CMB are explained as being the result of acoustic oscillations in the photon-baryon plasma (prior to the emission of the CMB after the photons decouple from the baryons 379,000 years after the Big Bang) whose restoring force is gravity.[69] Ordinary (baryonic) matter interacts strongly by way of radiation whereas dark matter particles, such as WIMPs for example, do not; both affect the oscillations by way of their gravity, so the two forms of matter will have different effects. The typical angular scales of the oscillations in the CMB, measured as the power spectrum of the CMB anisotropies, thus reveal the different effects of baryonic matter and dark matter. The CMB power spectrum shows a large first peak and smaller successive peaks, with three peaks resolved as of 2009.[68] The first peak tells mostly about the density of baryonic matter and the third peak mostly about the density of dark matter, measuring the density of matter and the density of atoms in the universe.

16.3.5   Sky surveys and baryon acoustic oscillations

Main article: Baryon acoustic oscillations

The acoustic oscillations in the early universe (see the previous section) have left their imprint on visible matter by way of Baryon Acoustic Oscillation (BAO) clustering, in a way that can be measured with sky surveys such as the Sloan Digital Sky Survey and the 2dF Galaxy Redshift Survey.[70] These measurements are consistent with those of the CMB derived from the WMAP spacecraft and further constrain the Lambda CDM model and dark matter. Note that the CMB data and the BAO data measure the acoustic oscillations at very different distance scales.[69]

16.3.6   Type Ia supernovae distance measurements

Main article: Type Ia supernova

Type Ia supernovae can be used as "standard candles" to measure extragalactic distances, and extensive data sets of these supernovae can be used to constrain cosmological models.[71] They constrain the dark energy density $\Omega\Lambda = \sim 0.713$ for a flat, Lambda CDM universe and the parameter w for a quintessence model. Once again, the values obtained are roughly consistent with those derived from the WMAP observations and further constrain the Lambda CDM model and (indirectly) dark matter.[69]

16.3.7   Lyman-alpha forest

Main article: Lyman-alpha forest

In astronomical spectroscopy, the Lyman-alpha forest is the sum of the absorption lines arising from the Lyman-alpha transition of the neutral hydrogen in the spectra of distant galaxies and quasars. Observations of the Lyman-alpha forest can also be used to constrain cosmological models.[72] These constraints are again in agreement with those obtained from WMAP data.

16.3.8 Structure formation

Main article: Structure formation

Dark matter is crucial to the Big Bang model of cosmology as a component which corresponds directly to measurements of the parameters associated with Friedmann cosmology solutions to general relativity. In particular, measurements of the cosmic microwave background anisotropies correspond to a cosmology where much of the matter interacts with photons more weakly than the known forces that couple light interactions to baryonic matter. Likewise, a significant amount of non-baryonic, cold matter is necessary to explain the large-scale structure of the universe.

Observations suggest that structure formation in the universe proceeds hierarchically, with the smallest structures collapsing first and followed by galaxies and then clusters of galaxies. As the structures collapse in the evolving universe, they begin to "light up" as the baryonic matter heats up through gravitational contraction and approaches hydrostatic pressure balance. Originally, baryonic matter had too high a temperature, and pressure left over from the Big Bang to allow collapse and form smaller structures, such as stars, via the Jeans instability. Dark matter acts as a compactor allowing the creation of structure where there would not have been any. This model not only corresponds with statistical surveying of the visible structure in the universe but also corresponds precisely to the dark matter predictions of the cosmic microwave background.

This *bottom up* model of structure formation requires something like cold dark matter to succeed. Large computer simulations of billions of dark matter particles have been used[74] to confirm that the cold dark matter model of structure formation is consistent with the structures observed in the universe through galaxy surveys, such as the Sloan Digital Sky Survey and 2dF Galaxy Redshift Survey, as well as observations of the Lyman-alpha forest. These studies have been crucial in constructing the Lambda-CDM model which measures the cosmological parameters, including the fraction of the universe made up of baryons and dark matter.

There are, however, several points of tension between observation and simulations of structure formation driven by dark matter. There is evidence that there exist 10 to 100 times fewer small galaxies than permitted by what the dark matter theory of galaxy formation predicts.[75][76] This is known as the dwarf galaxy problem. In addition, the simulations predict dark matter distributions with a very dense cusp near the centers of galaxies, but the observed halos are smoother than predicted.

16.4 History of the search for its composition

Although dark matter had historically been inferred from many astronomical observations, its composition long remained speculative. Early theories of dark matter concentrated on hidden heavy normal objects (such as black holes, neutron stars, faint old white dwarfs, and brown dwarfs) as the possible candidates for dark matter, collectively known as massive compact halo objects or MACHOs. Astronomical surveys for gravitational microlensing, including the MACHO, EROS and OGLE projects, along with Hubble telescope searches for ultra-faint stars, have not found enough of these hidden MACHOs.[77][78][79] Some hard-to-detect baryonic matter, such as MACHOs and some forms of gas, were additionally speculated to make a contribution to the overall dark matter content, but evidence indicated such would constitute only a small portion.[80][81][82]

Furthermore, data from a number of lines of other evidence, including galaxy rotation curves, gravitational lensing, structure formation, and the fraction of baryons in clusters and the cluster abundance combined with independent evidence for the baryon density, indicated that 85–90% of the mass in the universe does not interact with the electromagnetic force. This "nonbaryonic dark matter" is evident through its gravitational effect. Consequently, the most commonly held view was that dark matter is primarily non-baryonic, made of one or more elementary particles other than the usual electrons, protons, neutrons, and known neutrinos. The most commonly proposed particles then became WIMPs (Weakly Interacting Massive Particles, including neutralinos), axions, or sterile neutrinos, though many other possible candidates have been proposed.

Dark matter candidates can be approximately divided into three classes, called *cold*, *warm* and *hot* dark matter.[83] These categories do not correspond to an actual temperature, but instead refer to how fast the particles were moving, thus how far they moved due to random motions in the early universe, before they slowed due to the expansion of the universe – this is an important distance called the "free streaming length". Primordial density fluctuations smaller than this free-streaming length get washed out as particles move from overdense to underdense regions, while fluctuations larger than the free-streaming length are unaffected; therefore this free-streaming length sets a minimum scale for structure formation.

- Cold dark matter – objects with a free-streaming length much smaller than a protogalaxy.[84]

- Warm dark matter – particles with a free-streaming length similar to a protogalaxy.

- Hot dark matter – particles with a free-streaming length much larger than a protogalaxy.[85]

Though a fourth category had been considered early on, called mixed dark matter, it was quickly eliminated (from the 1990s) since the discovery of dark energy.

As an example, Davis *et al.* wrote in 1985:

> Candidate particles can be grouped into three categories on the basis of their effect on the fluctuation spectrum (Bond *et al.* 1983). If the dark matter is composed of abundant light particles which remain relativistic until shortly before recombination, then it may be termed "hot". The best candidate for hot dark matter is a neutrino ... A second possibility is for the dark matter particles to interact more weakly than neutrinos, to be less abundant, and to have a mass of order 1 keV. Such particles are termed "warm dark matter", because they have lower thermal velocities than massive neutrinos ... there are at present few candidate particles which fit this description. Gravitinos and photinos have been suggested (Pagels and Primack 1982; Bond, Szalay and Turner 1982) ... Any particles which became nonrelativistic very early, and so were able to diffuse a negligible distance, are termed "cold" dark matter (CDM). There are many candidates for CDM including supersymmetric particles.[86]

The full calculations are quite technical, but an approximate dividing line is that "warm" dark matter particles became non-relativistic when the universe was approximately 1 year old and 1 millionth of its present size; standard hot big bang theory implies the universe was then in the radiation-dominated era (photons and neutrinos), with a photon temperature 2.7 million K. Standard physical cosmology gives the particle horizon size as 2ct in the radiation-dominated era, thus 2 light-years, and a region of this size would expand to 2 million light years today (if there were no structure formation). The actual free-streaming length is roughly 5 times larger than the above length, since the free-streaming length continues to grow slowly as particle velocities decrease inversely with the scale factor after they become non-relativistic; therefore, in this example the free-streaming length would correspond to 10 million light-years or 3 Mpc today, which is around the size containing on average the mass of a large galaxy.

The above temperature of 2.7 million K gives a typical photon energy of 250 electron-volts, thereby setting a typical mass scale for "warm" dark matter: particles much more massive than this, such as GeV – TeV mass WIMPs, would become non-relativistic much earlier than 1 year after the Big Bang and thus have a free-streaming length much smaller than a proto-galaxy, making them cold dark matter. Conversely, much lighter particles, such as neutrinos with masses of only a few eV, have a free-streaming length much larger than a proto-galaxy, thus making them hot dark matter.

16.4.1 Cold dark matter

Main article: Cold dark matter

Today, cold dark matter is the simplest explanation for most cosmological observations. "Cold" dark matter is dark matter composed of constituents with a free-streaming length much smaller than the ancestor of a galaxy-scale perturbation. This is currently the area of greatest interest for dark matter research, as hot dark matter does not seem to be viable for galaxy and galaxy cluster formation, and most particle candidates become non-relativistic at very early times, hence are classified as cold.

The composition of the constituents of cold dark matter is currently unknown. Possibilities range from large objects like MACHOs (such as black holes[87]) or RAMBOs, to new particles like WIMPs and axions. Possibilities involving normal baryonic matter include brown dwarfs, other stellar remnants such as white dwarfs, or perhaps small, dense chunks of heavy elements.

Studies of big bang nucleosynthesis and gravitational lensing have convinced most scientists[12][88][89][90][91][92] that MACHOs of any type cannot be more than a small fraction of the total dark matter.[10][88] Black holes of nearly any mass are ruled out as a primary dark matter constituent by a variety of searches and constraints.[88][90] According to A. Peter: "...the only *really plausible* dark-matter candidates are new particles."[89]

The DAMA/NaI experiment and its successor DAMA/LIBRA have claimed to directly detect dark matter particles passing through the Earth, but many scientists remain skeptical, as negative results from similar experiments seem incompatible with the DAMA results.

Many supersymmetric models naturally give rise to stable dark matter candidates in the form of the Lightest Supersymmetric Particle (LSP). Separately, heavy sterile neutrinos exist in non-supersymmetric extensions to the standard model that explain the small neutrino mass through the seesaw mechanism.

16.4.2 Warm dark matter

Main article: Warm dark matter

Warm dark matter refers to particles with a free-streaming length comparable to the size of a region which subsequently evolved into a dwarf galaxy. This leads to predictions which are very similar to cold dark matter on large scales, including the CMB, galaxy clustering and large galaxy rotation curves, but with less small-scale density perturbations. This reduces the predicted abundance of dwarf galaxies and may lead to lower density of dark matter in the central parts of large galaxies; some researchers consider this may be a better fit to observations. A challenge for this model is that there are no very well-motivated particle physics candidates with the required mass ~ 300 eV to 3000 eV.

There have been no particles discovered so far that can be categorized as warm dark matter. There is a postulated candidate for the warm dark matter category, which is the sterile neutrino: a heavier, slower form of neutrino which does not even interact through the Weak force unlike regular neutrinos. Interestingly, some modified gravity theories, such as Scalar-tensor-vector gravity, also require that a warm dark matter exist to make their equations work out.

16.4.3 Hot dark matter

Main article: Hot dark matter

Hot dark matter consists of particles that have a free-streaming length much larger than that of a proto-galaxy.

An example of hot dark matter is already known: the neutrino. Neutrinos were discovered quite separately from the search for dark matter, and long before it seriously began: they were first postulated in 1930, and first detected in 1956. Neutrinos have a very small mass: at least 100,000 times less massive than an electron. Other than gravity, neutrinos only interact with normal matter via the weak force making them very difficult to detect (the weak force only works over a small distance, thus a neutrino will only trigger a weak force event if it hits a nucleus directly head-on). This would make them 'weakly interacting light particles' (WILPs), as opposed to cold dark matter's theoretical candidates, the weakly interacting massive particles (WIMPs).

There are three different known flavors of neutrinos (i.e., the *electron*, *muon*, and *tau* neutrinos), and their masses are slightly different. The resolution to the solar neutrino problem demonstrated that these three types of neutrinos actually change and oscillate from one flavor to the others and back as they are in-flight. It is hard to determine an exact upper bound on the collective average mass of the three neutrinos (let alone a mass for any of the three individually). For example, if the average neutrino mass were chosen to be over 50 eV/c^2 (which is still less than 1/10,000th of the mass of an electron), just by the sheer number of them in the universe, the universe would collapse due to their mass. So other observations have served to estimate an upper-bound for the neutrino mass. Using cosmic microwave background data and other methods, the current conclusion is that their average mass probably does not exceed 0.3 eV/c^2 Thus, the normal forms of neutrinos cannot be responsible for the measured dark matter component from cosmology.[93]

Hot dark matter was popular for a time in the early 1980s, but it suffers from a severe problem: because all galaxy-size density fluctuations get washed out by free-streaming, the first objects that can form are huge supercluster-size pancakes, which then were theorised somehow to fragment into galaxies. Deep-field observations clearly show that galaxies formed at early times, with clusters and superclusters forming later as galaxies clump together, so any model dominated by hot dark matter is seriously in conflict with observations.

16.4.4   Mixed dark matter

Main article: Mixed dark matter

Mixed dark matter is a now obsolete model, with a specifically chosen mass ratio of 80% cold dark matter and 20% hot dark matter (neutrinos) content. Though it is presumable that hot dark matter coexists with cold dark matter in any case, there was a very specific reason for choosing this particular ratio of hot to cold dark matter in this model. During the early 1990s it became steadily clear that a universe with critical density of cold dark matter did not fit the COBE and large-scale galaxy clustering observations; either the 80/20 mixed dark matter model, or LambdaCDM, were able to reconcile these. With the discovery of the accelerating universe from supernovae, and more accurate measurements of CMB anisotropy and galaxy clustering, the mixed dark matter model was essentially ruled out while the concordance LambdaCDM model remained a good fit.

16.5   Detection

If the dark matter within the Milky Way is made up of Weakly Interacting Massive Particles (WIMPs), then millions, possibly billions, of WIMPs must pass through every square centimeter of the Earth each second.[94][95] There are many experiments currently running, or planned, aiming to test this hypothesis by searching for WIMPs. Although WIMPs are the historically more popular dark matter candidate for searches,[12] there are experiments searching for other particle candidates; the Axion Dark Matter eXperiment (ADMX) is currently searching for the dark matter axion, a well-motivated and constrained dark matter source. It is also possible that dark matter consists of very heavy hidden sector particles which only interact with ordinary matter via gravity.

These experiments can be divided into two classes: direct detection experiments, which search for the scattering of dark matter particles off atomic nuclei within a detector; and indirect detection, which look for the products of WIMP annihilations.[21]

An alternative approach to the detection of WIMPs in nature is to produce them in the laboratory. Experiments with the Large Hadron Collider (LHC) may be able to detect WIMPs produced in collisions of the LHC proton beams. Because a WIMP has negligible interactions with matter, it may be detected indirectly as (large amounts of) missing energy and momentum which escape the LHC detectors, provided all the other (non-negligible) collision products are detected.[96] These experiments could show that WIMPs can be created, but it would still require a direct detection experiment to show that they exist in sufficient numbers to account for dark matter.

16.5.1   Direct detection experiments

Direct detection experiments usually operate in deep underground laboratories to reduce the background from cosmic rays. These include: the Soudan mine; the SNOLAB underground laboratory at Sudbury, Ontario (Canada); the Gran Sasso National Laboratory (Italy); the Canfranc Underground Laboratory (Spain); the Boulby Underground Laboratory (United Kingdom); the Deep Underground Science and Engineering Laboratory, South Dakota (United States); and the Particle and Astrophysical Xenon Detector (China).

The majority of present experiments use one of two detector technologies: cryogenic detectors, operating at temperatures below 100mK, detect the heat produced when a particle hits an atom in a crystal absorber such as germanium. Noble liquid detectors detect the flash of scintillation light produced by a particle collision in liquid xenon or argon. Cryogenic detector experiments include: CDMS, CRESST, EDELWEISS, EURECA. Noble liquid experiments include ZEPLIN, XENON, DEAP, ArDM, WARP, DarkSide, PandaX, and LUX, the Large Underground Xenon experiment. Both of these detector techniques are capable of distinguishing background particles which scatter off electrons, from dark matter particles which scatter off nuclei. Other experiments include SIMPLE and PICASSO.

The DAMA/NaI, DAMA/LIBRA experiments have detected an annual modulation in the event rate,[97] which they claim is due to dark matter particles. (As the Earth orbits the Sun, the velocity of the detector relative to the dark matter halo will vary by a small amount depending on the time of year). This claim is so far unconfirmed and difficult to reconcile with the negative results of other experiments assuming that the WIMP scenario is correct.[98]

Directional detection of dark matter is a search strategy based on the motion of the Solar System around the Galactic Center.[99][100][101][102]

By using a low pressure TPC, it is possible to access information on recoiling tracks (3D reconstruction if possible) and to constrain the WIMP-nucleus kinematics. WIMPs coming from the direction in which the Sun is travelling (roughly in the direction of the Cygnus constellation) may then be separated from background noise, which should be isotropic. Directional dark matter experiments include DMTPC, DRIFT, Newage and MIMAC.

On 17 December 2009, CDMS researchers reported two possible WIMP candidate events. They estimate that the probability that these events are due to a known background (neutrons or misidentified beta or gamma events) is 23%, and conclude "this analysis cannot be interpreted as significant evidence for WIMP interactions, but we cannot reject either event as signal."[103]

More recently, on 4 September 2011, researchers using the CRESST detectors presented evidence[104] of 67 collisions occurring in detector crystals from subatomic particles, calculating there is a less than 1 in 10,000 chance that all were caused by known sources of interference or contamination. It is quite possible then that many of these collisions were caused by WIMPs, and/or other unknown particles.

16.5.2 Indirect detection experiments

Indirect detection experiments search for the products of WIMP annihilation or decay. If WIMPs are Majorana particles (WIMPs are their own antiparticle) then two WIMPs could annihilate to produce gamma rays or Standard Model particle-antiparticle pairs. Additionally, if the WIMP is unstable, WIMPs could decay into standard model particles. These processes could be detected indirectly through an excess of gamma rays, antiprotons or positrons emanating from regions of high dark matter density. The detection of such a signal is not conclusive evidence for dark matter, as the production of gamma rays from other sources is not fully understood.[12][21]

The EGRET gamma ray telescope observed more gamma rays than expected from the Milky Way, but scientists concluded that this was most likely due to a mis-estimation of the telescope's sensitivity.[106]

The Fermi Gamma-ray Space Telescope, launched 11 June 2008, is searching for gamma rays from dark matter annihilation and decay.[107] In April 2012, an analysis[108] of previously available data from its Large Area Telescope instrument produced strong statistical evidence of a 130 GeV line in the gamma radiation coming from the center of the Milky Way. At the time, WIMP annihilation was the most probable explanation for that line.[109]

At higher energies, ground-based gamma-ray telescopes have set limits on the annihilation of dark matter in dwarf spheroidal galaxies[110] and in clusters of galaxies.[111]

The PAMELA experiment (launched 2006) has detected a larger number of positrons than expected. These extra positrons could be produced by dark matter annihilation, but may also come from pulsars. No excess of anti-protons has been observed.[112] The Alpha Magnetic Spectrometer on the International Space Station is designed to directly measure the fraction of cosmic rays which are positrons. The first results, published in April 2013, indicate an excess of high-energy cosmic rays which could potentially be due to annihilation of dark matter.[113][114][115][116][117][118]

A few of the WIMPs passing through the Sun or Earth may scatter off atoms and lose energy. This way a large population of WIMPs may accumulate at the center of these bodies, increasing the chance that two will collide and annihilate. This could produce a distinctive signal in the form of high-energy neutrinos originating from the center of the Sun or Earth.[119] It is generally considered that the detection of such a signal would be the strongest indirect proof of WIMP dark matter.[12] High-energy neutrino telescopes such as AMANDA, IceCube and ANTARES are searching for this signal.

WIMP annihilation from the Milky Way Galaxy as a whole may also be detected in the form of various annihilation products.[120] The Galactic Center is a particularly good place to look because the density of dark matter may be very high there.[121]

16.6 Alternative theories

16.6.1 Mass in extra dimensions

In some multidimensional theories, the force of gravity is the unique force able to have an effect across all the various extra dimensions,[17] which would explain the relative weakness of the force of gravity compared to the other known forces of nature that would not be able to cross into extra dimensions: electromagnetism, strong interaction, and weak interaction.

In that case, dark matter would be a perfect candidate for matter that would exist in other dimensions and that could only interact with the matter on our dimensions through gravity. That dark matter located on different dimensions could potentially aggregate in the same way as the matter in our visible universe does, forming exotic galaxies.[16]

16.6.2 Topological defects

Dark matter could consist of primordial defects (defects originating with the birth of the universe) in the topology of quantum fields, which would contain energy and therefore gravitate. This possibility may be investigated by the use of an orbital network of atomic clocks, which would register the passage of topological defects by monitoring the synchronization of the clocks. The Global Positioning System may be able to operate as such a network.[122]

16.6.3 Modified gravity

Numerous alternative theories have been proposed to explain these observations without the need for a large amount of undetected matter. Most of these theories modify the laws of gravity established by Newton and Einstein.

The earliest modified gravity model to emerge was Mordehai Milgrom's Modified Newtonian Dynamics (MOND) in 1983, which adjusts Newton's laws to create a stronger gravitational field when gravitational acceleration levels become tiny (such as near the rim of a galaxy). It had some success explaining galactic-scale features, such as rotational velocity curves of elliptical galaxies, and dwarf elliptical galaxies, but did not successfully explain galaxy cluster gravitational lensing. However, MOND was not relativistic, since it was just a straight adjustment of the older Newtonian account of gravitation, not of the newer account in Einstein's general relativity. Soon after 1983, attempts were made to bring MOND into conformity with general relativity; this is an ongoing process, and many competing hypotheses have emerged based around the original MOND model—including TeVeS, MOG or STV gravity, and phenomenological covariant approach,[123] among others.

In 2007, John W. Moffat proposed a modified gravity hypothesis based on the nonsymmetric gravitational theory (NGT) that claims to account for the behavior of colliding galaxies.[124] This model requires the presence of non-relativistic neutrinos, or other candidates for (cold) dark matter, to work.

Another proposal uses a gravitational backreaction in an emerging theoretical field that seeks to explain gravity between objects as an action, a reaction, and then a back-reaction. Simply, an object A affects an object B, and the object B then re-affects object A, and so on: creating a sort of feedback loop that strengthens gravity.[125]

In 2008, another group has proposed a modification of large-scale gravity in a hypothesis named "dark fluid". In this formulation, the attractive gravitational effects attributed to dark matter are instead a side-effect of dark energy. Dark fluid combines dark matter and dark energy in a single energy field that produces different effects at different scales. This treatment is a simplified approach to a previous fluid-like model called the generalized Chaplygin gas model where the whole of spacetime is a compressible gas.[126] Dark fluid can be compared to an atmospheric system. Atmospheric pressure causes air to expand, but part of the air can collapse to form clouds. In the same way, the dark fluid might generally expand, but it also could collect around galaxies to help hold them together.[126]

Another set of proposals is based on the possibility of a double metric tensor for space-time.[127] It has been argued that time-reversed solutions in general relativity require such double metric for consistency, and that both dark matter and dark energy can be understood in terms of time-reversed solutions of general relativity.[128]

16.6.4 Fractality of Spacetime

Applying relativity to fractal non-differentiable spacetime, Laurent Nottale, in his Scale Relativity theory, suggests that potential energy arises due to the fractality of space, and accounts for the missing mass-energy observed at cosmological scales.

16.7 Popular culture

Main article: Dark matter in fiction

Mention of dark matter is made in some video games and other works of fiction. In such cases, it is usually attributed extraordinary physical or magical properties. Such descriptions are often inconsistent with the properties of dark matter proposed in physics and cosmology.

16.8 See also

- Chameleon particle

- Conformal gravity

- General Antiparticle Spectrometer

- Illustris project

- Light dark matter

- Mirror matter

- Multidark (research program)

- Scalar field dark matter

- Self-interacting dark matter

- SIMP

- Unparticle physics

16.9 Notes

[1] $26.8/(4.9 + 26.8) = 0.845$

16.10   References

[1]   "Hubble Finds Dark Matter Ring in Galaxy Cluster".

[2]   Ade, P. A. R.; Aghanim, N.; Armitage-Caplan, C.; (Planck Collaboration) et al. (22 March 2013). "Planck 2013 results. I. Overview of products and scientific results – Table 9". *Astronomy and Astrophysics* **1303**: 5062. arXiv:1303.5062.B .doi:10.1051/0004-6361/201321529.

[3]   Francis, Matthew (22 March 2013). "First Planck results: the Universe is still weird and interesting". *Arstechnica*.

[4]   "Planck captures portrait of the young Universe, revealing earliest light". University of Cambridge. 21 March 2013. Retrieved 21 March 2013.

[5]   Sean Carroll, Ph.D., Cal Tech, 2007, The Teaching Company, *Dark Matter, Dark Energy: The Dark Side of the Universe*, Guidebook Part 2 page 46, Accessed Oct. 7, 2013, "...dark matter: An invisible, essentially collisionless component of matter that makes up about 25 percent of the energy density of the universe... it's a different kind of particle... something not yet observed in the laboratory..."

[6]   Ferris, Timothy. "Dark Matter". Retrieved 2015-06-10.

[7]   Zwicky, F. (1933), "Die Rotverschiebung von extragalaktischen Nebeln",*Helvetica Physica Acta***6**: 110–127,Bibcode:1933AcZ See also Zwicky, F. (1937), "On the Masses of Nebulae and of Clusters of Nebulae", *Astrophysical Journal* **86**: 217, Bibcode: doi:10.1086/143864

[8]   First observational evidence of dark matter. Darkmatterphysics.com. Retrieved on 6 August 2013.

[9]   Rubin, Vera C.; Ford, W. Kent, Jr. (February 1970). "Rotation of the Andromeda Nebula from a Spectroscopic Survey of Emission Regions". *The Astrophysical Journal* **159**: 379–403. Bibcode:1970ApJ...159..379R. doi:10.1086/150317.

[10] Copi, C. J.; Schramm, D. N.; Turner, M. S. (1995). "Big-Bang Nucleosynthesis and the Baryon Density of the Universe". *Science* **267** (5195): 192–199. arXiv:astro-ph/9407006. Bibcode:1995Sci...267..192C. doi:10.1126/science.7809624. PMID 7809624.

[11] Bergstrom, L. (2000). "Non-baryonic dark matter: Observational evidence and detection methods". *Reports on Progress in Physics* **63** (5): 793–841. arXiv:hep-ph/0002126. Bibcode:2000RPPh...63..793B. doi:10.1088/0034-4885/63/5/2r3.

[12] Bertone, G.; Hooper, D.; Silk, J. (2005). "Particle dark matter: Evidence, candidates and constraints". *Physics Reports* **405** (5–6): 279–390. arXiv:hep-ph/0404175. Bibcode:2005PhR...405..279B. doi:10.1016/j.physrep.2004.08.031.

[13] Angus, G. (2013). "Cosmological simulations in MOND: the cluster scale halo mass function with light sterile neutrinos". *Monthly Notices of the Royal Astronomical Society***436**: 202–211.arXiv:1309.6094.Bibcode:2013MNRAS.436..202A.doi:.

[14] Trimble, V. (1987). "Existence and nature of dark matter in the universe". *Annual Review of Astronomy and Astrophysics* **25**: 425–472. Bibcode:1987ARA&A..25..425T. doi:10.1146/annurev.aa.25.090187.002233.

[15] Dark matter. CERN. Retrieved on 17 November 2014.

[16] Siegfried, T. (5 July 1999). "Hidden Space Dimensions May Permit Parallel Universes, Explain Cosmic Mysteries". *The Dallas Morning News*.

[17] Extra dimensions, gravitons, and tiny black holes. CERN. Retrieved on 17 November 2014.

[18] Kroupa, P. et al. (2010). "Local-Group tests of dark-matter Concordance Cosmology: Towards a new paradigm for structure formation". *Astronomy and Astrophysics* **523**: 32–54. arXiv:1006.1647. Bibcode:2010A&A...523A..32K. doi:10.1051/0004-6361/201014892.

[19] Achim Weiss, "Big Bang Nucleosynthesis: Cooking up the first light elements" in: Einstein Online Vol. 2 (2006), 1017

[20] Raine, D.; Thomas, T. (2001). *An Introduction to the Science of Cosmology*. IOP Publishing. p. 30. ISBN 0-7503-0405-7.

[21] Bertone, G.; Merritt, D. (2005). "Dark Matter Dynamics and Indirect Detection". *Modern Physics Letters A* **20** (14): 1021–1036. arXiv:astro-ph/0504422. Bibcode:2005MPLA...20.1021B. doi:10.1142/S0217732305017391.

[22] "Serious Blow to Dark Matter Theories?" (Press release). European Southern Observatory. 18 April 2012.

[23] "The Hidden Lives of Galaxies: Hidden Mass". *Imagine the Universe!*. NASA/GSFC.

[24] Kuijken K. and Gilmore G. (1989). "The Mass Distribution in the Galactic Disc - Part III - the Local Volume Mass Density". *Monthly Notices of the Royal Astronomical Society* **239**: 651. Bibcode:1989MNRAS.239..651K. doi:10.1093/mnras/239.2.651.

[25]Zwicky, F. (1933). "Die Rotverschiebung von extragalaktischen Nebeln".*Helvetica Physica Acta***6**: 110–127.Bibcode:1.

[26] Zwicky, F. (1937). "On the Masses of Nebulae and of Clusters of Nebulae". *The Astrophysical Journal* **86**: 217. Bibcode: doi:10.1086/143864.

[27] Some details of Zwicky's calculation and of more modern values are given in Richmond, M., *Using the virial theorem: the mass of a cluster of galaxies*, retrieved 2007-07-10

[28] Freese, Katerine (2014). *The Cosmic Cocktail: Three Parts Dark Matter*. Princeton, New Jersey: Princeton University Press. ISBN 978-0691153353.

[29] Some details of Zwicky's calculation and of more modern values are given in Richmond, M., *Using the virial theorem: the mass of a cluster of galaxies*, retrieved 2007-07-10.

[30] "First Signs of Self-interacting Dark Matter?". *ESO Press Release*. European Southern Observatory. Retrieved 15 April 2015.

[31] Freeman, K.; McNamara, G. (2006). *In Search of Dark Matter*. Birkhäuser. p. 37. ISBN 0-387-27616-5.

[32] Jörg, D. et al. (2012). "A filament of dark matter between two clusters of galaxies". *Nature* **487** (7406): 202. arXiv:1207.0809. Bibcode:2012Natur.487..202D. doi:10.1038/nature11224.

[33] Babcock, H, 1939, "The rotation of the Andromeda Nebula", Lick Observatory bulletin ; no. 498

[34] Bosma, A. (1978). "The distribution and kinematics of neutral hydrogen in spiral galaxies of various morphological types" (Ph.D. Thesis). Rijksuniversiteit Groningen.

[35] Rubin, V.; Thonnard, W. K. Jr.; Ford, N. (1980). "Rotational Properties of 21 Sc Galaxies with a Large Range of Luminosities and Radii from NGC 4605 ($R = 4$kpc) to UGC 2885 ($R = 122$kpc)". *The Astrophysical Journal* **238**: 471. Bibcode:1980Ap. doi:10.1086/158003.

[36] de Blok, W. J. G.; McGaugh, S. S.; Bosma, A.; Rubin, V. C. (2001). "Mass Density Profiles of Low Surface Brightness Galaxies". *The Astrophysical Journal Letters* **552** (1): L23–L26. arXiv:astro-ph/0103102. Bibcode:2001ApJ...552L..23D. doi:10.1086/320262.

[37] Salucci, P.; Borriello, A. (2003). "The Intriguing Distribution of Dark Matter in Galaxies". *Lecture Notes in Physics*. Lecture Notes in Physics **616**: 66–77. arXiv:astro-ph/0203457. Bibcode:2003LNP...616...66S. doi:10.1007/3-540-36539-7_5. ISBN 978-3-540-00711-1.

[38] Koopmans, L. V. E.; Treu, T. (2003). "The Structure and Dynamics of Luminous and Dark Matter in the Early-Type Lens Galaxy of 0047−281 at $z = 0.485$". *The Astrophysical Journal* **583** (2): 606–615. arXiv:astro-ph/0205281. Bibcode:2003ApJ. doi:10.1086/345423.

[39] Dekel, A. et al. (2005). "Lost and found dark matter in elliptical galaxies". *Nature* **437** (7059): 707–710. arXiv:astro-ph/0501622. Bibcode:2005Natur.437..707D. doi:10.1038/nature03970. PMID 16193046.

[40] Governato, F. et al. (2010). "Bulgeless dwarf galaxies and dark matter cores from supernova-driven outflows". *Nature* **463** (7278): 203–206. arXiv:0911.2237. Bibcode:2010Natur.463..203G. doi:10.1038/nature08640.

[41] Ostriker, J. P.; Steinhardt, P. (2003). "New Light on Dark Matter". *Science* **300** (5627): 1909–1913. doi:10.1126/science.1085976. PMID 12817140.

[42] Faber, S. M.; Jackson, R. E. (1976). "Velocity dispersions and mass-to-light ratios for elliptical galaxies". *The Astrophysical Journal* **204**: 668–683. Bibcode:1976ApJ...204..668F. doi:10.1086/154215.

[43] Collins, G. W. (1978). "The Virial Theorem in Stellar Astrophysics". Pachart Press.

[44] Rejkuba, M.; Dubath, P.; Minniti, D.; Meylan, G. (2008). "Masses and M/L Ratios of Bright Globular Clusters in NGC 5128". *Proceedings of the International Astronomical Union***246**: 418–422.Bibcode:2008IAUS..246..418R.doi:10.1017/S17439.

[45] Weinberg, M. D.; Blitz, L. (2006). "A Magellanic Origin for the Warp of the Galaxy". *The Astrophysical Journal Letters* **641** (1): L33–L36. arXiv:astro-ph/0601694. Bibcode:2006ApJ...641L..33W. doi:10.1086/503607.

[46] Minchin, R. et al. (2005). "A Dark Hydrogen Cloud in the Virgo Cluster". *The Astrophysical Journal Letters* **622**: L21–L24. arXiv:astro-ph/0502312. Bibcode:2005ApJ...622L..21M. doi:10.1086/429538.

[47] Ciardullo, R.; Jacoby, G. H.; Dejonghe, H. B. (1993). "The radial velocities of planetary nebulae in NGC 3379". *The Astrophysical Journal* **414**: 454–462. Bibcode:1993ApJ...414..454C. doi:10.1086/173092.

[48] Vikhlinin, A. et al. (2006). "Chandra Sample of Nearby Relaxed Galaxy Clusters: Mass, Gas Fraction, and Mass–Temperature Relation". *The Astrophysical Journal* **640** (2): 691–709. arXiv:astro-ph/0507092. Bibcode:2006ApJ...640..691V.doi:1.

[49] "Abell 2029: Hot News for Cold Dark Matter". Chandra X-ray Observatory. 11 June 2003.

[50] Taylor, A. N. et al. (1998). "Gravitational Lens Magnification and the Mass of Abell 1689". *The Astrophysical Journal* **501** (2): 539. arXiv:astro-ph/9801158. Bibcode:1998ApJ...501..539T. doi:10.1086/305827.

[51] Wu, X.; Chiueh, T.; Fang, L.; Xue, Y. (1998). "A comparison of different cluster mass estimates: consistency or discrepancy?". *Monthly Notices of the Royal Astronomical Society* **301** (3): 861–871. arXiv:astro-ph/9808179.Bibcode:1998MNRAS.30. doi:10.1046/j.1365-8711.1998.02055.x.

[52] Refregier, A. (2003). "Weak gravitational lensing by large-scale structure". *Annual Review of Astronomy and Astrophysics* **41** (1): 645–668. arXiv:astro-ph/0307212. Bibcode:2003ARA&A..41..645R. doi:10.1146/annurev.astro.41.111302.102207.

[53] Massey, R. et al. (2007). "Dark matter maps reveal cosmic scaffolding". *Nature* **445** (7125): 286–290. arXiv:astro-ph/0701594. Bibcode:2007Natur.445..286M. doi:10.1038/nature05497. PMID 17206154.

[54] Clowe, D. et al. (2006). "A direct empirical proof of the existence of dark matter". *The Astrophysical Journal* **648** (2): 109–113. arXiv:astro-ph/0608407. Bibcode:2006ApJ...648L.109C. doi:10.1086/508162.

[55] "Dark Matter Mystery Deepens in Cosmic "Train Wreck"". Chandra X-Ray Observatory. 16 August 2007.

[56] Tiberiu, H.; Lobo, F. S. N. (2011). "Two-fluid dark matter models". *Physical Review D* **83** (12): 124051. arXiv:1106.2642. Bibcode:2011PhRvD..83l4051H. doi:10.1103/PhysRevD.83.124051.

[57] Spergel, D. N.; Steinhardt, P. J. (2000). "Observational evidence for self-interacting cold dark matter". *Physical Review Letters* **84** (17): 3760–3763. arXiv:astro-ph/9909386. Bibcode:2000PhRvL..84.3760S. doi:10.1103/PhysRevLett.84.3760.

[58] Allen, S. W.; Evrard, A. E.; Mantz, A. B. (2011). "Cosmological Parameters from Observations of Galaxy Clusters". *Annual Review of Astronomy & Astrophysics* **49**: 409–470. arXiv:1103.4829. Bibcode:2011ARA&A..49..409A. doi:10.1146/annurev-astro-081710-102514.

[59] Markevitch, M. et al. (2004). "Direct Constraints on the Dark Matter Self-Interaction Cross Section from the Merging Galaxy Cluster 1E 0657-56". *The Astrophysical Journal* **606** (2): 819–824. arXiv:astro-ph/0309303. Bibcode:2004ApJ...606..819M. doi:10.1086/383178.

[60] "Press Release - Dark Matter Map Begins to Reveal the Universe's Early History - Subaru Telescope". *www.subarutelescope.org*. Retrieved 2015-07-03.

[61] Miyazaki, Satoshi; Oguri, Masamune; Hamana, Takashi; Tanaka, Masayuki; Miller, Lance; Utsumi, Yousuke; Komiyama, Yutaka; Furusawa, Hisanori; Sakurai, Junya (2015-07-01). "Properties of Weak Lensing Clusters Detected on Hyper Suprime-Cam's 2.3 deg2 field". *The Astrophysical Journal* **807** (1): 22. arXiv:1504.06974. Bibcode:2015ApJ...807...22M.doi:10.1088-637X/807/1/22. ISSN 0004-637X.

[62] Penzias, A.A.; Wilson, R. W. (1965). "A Measurement of Excess Antenna Temperature at 4080 Mc/s". *The Astrophysical Journal* **142**: 419. Bibcode:1965ApJ...142..419P. doi:10.1086/148307.

[63] Boggess, N. W. et al. (1992). "The COBE Mission: Its Design and Performance Two Years after the launch". *The Astrophysical Journal* **397**: 420. Bibcode:1992ApJ...397..420B. doi:10.1086/171797.

[64] Melchiorri, A. et al. (2000). "A Measurement of Ω from the North American Test Flight of Boomerang". *The Astrophysical Journal Letters* **536** (2): L63–L66. arXiv:astro-ph/9911445. Bibcode:2000ApJ...536L..63M. doi:10.1086/312744.

[65] Leitch, E. M. et al. (2002). "Measurement of polarization with the Degree Angular Scale Interferometer". *Nature* **420** (6917): 763–771. arXiv:astro-ph/0209476. Bibcode:2002Natur.420..763L. doi:10.1038/nature01271. PMID 12490940.

[66] Leitch, E. M. et al. (2005). "Degree Angular Scale Interferometer 3 Year Cosmic Microwave Background Polarization Results". *The Astrophysical Journal* **624** (1): 10–20. arXiv:astro-ph/0409357. Bibcode:2005ApJ...624...10L. doi:10.1086/428825.

[67] Readhead, A. C. S. et al. (2004). "Polarization Observations with the Cosmic Background Imager". *Science* **306** (5697): 836–844. arXiv:astro-ph/0409569. Bibcode:2004Sci...306..836R. doi:10.1126/science.1105598. PMID 15472038.

[68] Hinshaw, G. et al. (2009). "Five-Year Wilkinson Microwave Anisotropy Probe Observations: Data Processing, Sky Maps, and Basic Results". *The Astrophysical Journal Supplement* **180** (2): 225–245. arXiv:0803.0732. Bibcode:2009ApJS..180..225H. doi:10.1088/0067-0049/180/2/225.

[69] Komatsu, E. et al. (2009). "Five-Year Wilkinson Microwave Anisotropy Probe Observations: Cosmological Interpretation". *The Astrophysical Journal Supplement* **180** (2): 330–376. arXiv:0803.0547. Bibcode:2009ApJS..180..330K. doi:10.1088/0067-0049/180/2/330.

[70] Percival, W. J. et al. (2007). "Measuring the Baryon Acoustic Oscillation scale using the Sloan Digital Sky Survey and 2dF Galaxy Redshift Survey". *Monthly Notices of the Royal Astronomical Society* **381** (3): 1053–1066. arXiv:0705.3323. Bibcode:2007MNRAS.381.1053P. doi:10.1111/j.1365-2966.2007.12268.x.

[71] Kowalski, M. et al. (2008). "Improved Cosmological Constraints from New, Old, and Combined Supernova Data Sets". *The Astrophysical Journal* **686** (2): 749–778. arXiv:0804.4142. Bibcode:2008ApJ...686..749K. doi:10.1086/589937.

[72] Viel, M.; Bolton, J. S.; Haehnelt, M. G. (2009). "Cosmological and astrophysical constraints from the Lyman α forest flux probability distribution function". *Monthly Notices of the Royal Astronomical Society* **399** (1): L39–L43. arXiv:0907.2927. Bibcode:2009MNRAS.399L..39V. doi:10.1111/j.1745-3933.2009.00720.x.

[73] "Hubble Maps the Cosmic Web of "Clumpy" Dark Matter in 3-D" (Press release). NASA. 7 January 2007.

[74] Springel, V. et al. (2005). "Simulations of the formation, evolution and clustering of galaxies and quasars". *Nature* **435** (7042): 629–636. arXiv:astro-ph/0504097. Bibcode:2005Natur.435..629S. doi:10.1038/nature03597. PMID 15931216.

[75] Mateo, M. L. (1998). "Dwarf Galaxies of the Local Group". *Annual Review of Astronomy and Astrophysics* **36** (1): 435–506. arXiv:astro-ph/9810070. Bibcode:1998ARA&A..36..435M. doi:10.1146/annurev.astro.36.1.435.

[76] Moore, B. et al. (1999). "Dark Matter Substructure within Galactic Halos". *The Astrophysical Journal Letters* **524** (1): L19–L22. arXiv:astro-ph/9907411. Bibcode:1999ApJ...524L..19M. doi:10.1086/312287.

[77] Tisserand, P.; Le Guillou, L.; Afonso, C.; Albert, J. N.; Andersen, J.; Ansari, R.; Aubourg, É.; Bareyre, P.; Beaulieu, J. P.; Charlot, X.; Coutures, C.; Ferlet, R.; Fouqué, P.; Glicenstein, J. F.; Goldman, B.; Gould, A.; Graff, D.; Gros, M.; Haissinski, J.; Hamadache, C.; De Kat, J.; Lasserre, T.; Lesquoy, É.; Loup, C.; Magneville, C.; Marquette, J. B.; Maurice, É.; Maury, A.; Milsztajn, A.; Moniez, M. (2007). "Limits on the Macho content of the Galactic Halo from the EROS-2 Survey of the Magellanic Clouds". *Astronomy and Astrophysics* **469** (2): 387. arXiv:astro-ph/0607207. Bibcode:2007A&A...469..387T. doi:10.1051/0004-6361:20066017.

[78] Graff, D. S.; Freese, K. (1996). "Analysis of a *Hubble Space Telescope* Search for Red Dwarfs: Limits on Baryonic Matter in the Galactic Halo". *The Astrophysical Journal* **456**. arXiv:astro-ph/9507097. Bibcode:1996ApJ...456L..49G. doi:10.1086/309850.

[79] Najita, J. R.; Tiede, G. P.; Carr, J. S. (2000). "From Stars to Superplanets: The Low-Mass Initial Mass Function in the Young Cluster IC 348". *The Astrophysical Journal* **541** (2): 977. doi:10.1086/309477.

[80] Wyrzykowski, Lukasz et al. (2011) The OGLE view of microlensing towards the Magellanic Clouds – IV. OGLE-III SMC data and final conclusions on MACHOs, MNRAS, 416, 2949

[81] Freese, Katherine; Fields, Brian; Graff, David (2000). "Death of Stellar Baryonic Dark Matter Candidates". arXiv:astro-ph/0007444 [astro-ph].

[82] Freese, Katherine; Fields, Brian; Graff, David (2000). "Death of Stellar Baryonic Dark Matter". *The First Stars*. ESO Astrophysics Symposia. p. 18. arXiv:astro-ph/0002058. Bibcode:2000fist.conf...18F. doi:10.1007/10719504_3. ISBN 3-540-67222-2.

[83] Silk, Joseph (1980). *The Big Bang* (1989 ed.). San Francisco: Freeman. chapter ix, page 182. ISBN 0-7167-1085-4.

[84] Vittorio, N.; J. Silk (1984). "Fine-scale anisotropy of the cosmic microwave background in a universe dominated by cold dark matter". *Astrophysical Journal, Part 2 – Letters to the Editor* **285**: L39–L43. Bibcode:1984ApJ...285L..39V. doi:10.1086/184361.

[85] Umemura, Masayuki; Satoru Ikeuchi (1985). "Formation of Subgalactic Objects within Two-Component Dark Matter". *Astrophysical Journal* **299**: 583–592. Bibcode:1985ApJ...299..583U. doi:10.1086/163726.

[86] Davis, M.; Efstathiou, G., Frenk, C. S., & White, S. D. M. (May 15, 1985). "The evolution of large-scale structure in a universe dominated by cold dark matter". *Astrophysical Journal* **292**: 371–394. Bibcode:1985ApJ...292..371D. doi:10.1086/163168.

[87] Hawkins, M. R. S. (2011). "The case for primordial black holes as dark matter". *Monthly Notices of the Royal Astronomical Society* **415** (3): 2744–2757. arXiv:1106.3875. Bibcode:2011MNRAS.415.2744H. doi:10.1111/j.1365-2966.2011.18890.x.

[88] Carr, B. J. et al. (May 2010). "New cosmological constraints on primordial black holes". *Physical Review D* **81** (10): 104019. arXiv:0912.5297. Bibcode:2010PhRvD..81j4019C. doi:10.1103/PhysRevD.81.104019.

[89] Peter, A. H. G. (2012). "Dark Matter: A Brief Review". arXiv:1201.3942 [astro-ph.CO].

[90] Garrett, Katherine; Dūda, Gintaras (2011). "Dark Matter: A Primer". *Advances in Astronomy* **2011**: 1. arXiv:1006.2483. Bibcode:2011AdAst2011E...8G. doi:10.1155/2011/968283. MACHOs can only account for a very small percentage of the nonluminous mass in our galaxy, revealing that most dark matter cannot be strongly concentrated or exist in the form of baryonic astrophysical objects. Although microlensing surveys rule out baryonic objects like brown dwarfs, black holes, and neutron stars in our galactic halo, can other forms of baryonic matter make up the bulk of dark matter? The answer, surprisingly, is no...

[91] Bertone, G. (2010). "The moment of truth for WIMP dark matter". *Nature* **468** (7322): 389–393. doi:10.1038/nature09509. PMID 21085174.

[92] Olive, Keith A. (2003). "TASI Lectures on Dark Matter". p. 21

[93] "Neutrinos as Dark Matter". Astro.ucla.edu. 21 September 1998. Retrieved 6 January 2011.

[94] Gaitskell, Richard J. (2004). "Direct Detection of Dark Matter". *"Annual Review of Nuclear and Particle Systems"* **54**: 315–359. Bibcode:2004ARNPS..54..315G. doi:10.1146/annurev.nucl.54.070103.181244.

[95] "NEUTRALINO DARK MATTER". Retrieved 26 December 2011. Griest, Kim. "WIMPs and MACHOs" (PDF). Retrieved 26 December 2011.

[96] Kane, G. and Watson, S. (2008). "Dark Matter and LHC:. what is the Connection?". *Modern Physics Letters A* **23** (26): 2103–2123. arXiv:0807.2244. Bibcode:2008MPLA...23.2103K. doi:10.1142/S0217732308028314.

[97] Drukier, A.; Freese, K. and Spergel, D. (1986). "Detecting Cold Dark Matter Candidates". *Physical Review D* **33** (12): 3495–3508. Bibcode:1986PhRvD..33.3495D. doi:10.1103/PhysRevD.33.3495.

[98] Bernabei, R.; Belli, P.; Cappella, F.; Cerulli, R.; Dai, C. J.; d'Angelo, A.; He, H. L.; Incicchitti, A.; Kuang, H. H.; Ma, J. M.; Montecchia, F.; Nozzoli, F.; Prosperi, D.; Sheng, X. D.; Ye, Z. P. (2008). "First results from DAMA/LIBRA and the combined results with DAMA/NaI". *Eur. Phys. J. C* **56** (3): 333–355. arXiv:0804.2741. doi:10.1140/epjc/s10052-008-0662-y.

[99] Stonebraker, Alan (2014-01-03). "Synopsis: Dark-Matter Wind Sways through the Seasons". *Physics - Synopses* (American Physical Society). Retrieved 6 January 2014.

[100] Lee, Samuel K.; Mariangela Lisanti, Annika H. G. Peter, and Benjamin R. Safdi (2014-01-03). "Effect of Gravitational Focusing on Annual Modulation in Dark-Matter Direct-Detection Experiments". *Phys. Rev. Lett.* (American Physical Society) **112** (1): 011301 (2014) [5 pages]. arXiv:1308.1953. Bibcode:2014PhRvL.112a1301L. doi:10.1103/PhysRevLett.112.011301.

[101] The Dark Matter Group. "An Introduction to Dark Matter". *Dark Matter Research* (Sheffield, UK: University of Sheffield). Retrieved 7 January 2014.

[102] "Blowing in the Wind". *Kavli News* (Sheffield, UK: Kavli Foundation). Retrieved 7 January 2014. Scientists at Kavli MIT are working on...a tool to track the movement of dark matter.

[103] The CDMS II Collaboration; Ahmed, Z.; Akerib, D. S.; Arrenberg, S.; Bailey, C. N.; Balakishiyeva, D.; Baudis, L.; Bauer, D. A.; Brink, P. L.; Bruch, T.; Bunker, R.; Cabrera, B.; Caldwell, D. O.; Cooley, J.; Cushman, P.; Daal, M.; Dejongh, F.; Dragowsky, M. R.; Duong, L.; Fallows, S.; Figueroa-Feliciano, E.; Filippini, J.; Fritts, M.; Golwala, S. R.; Grant, D. R.; Hall, J.; Hennings-Yeomans, R.; Hertel, S. A.; Holmgren, D.; Hsu, L. (2010). "Dark Matter Search Results from the CDMS II Experiment". *Science* **327** (5973): 1619–1621. doi:10.1126/science.1186112. PMID 20150446.

[104] Angloher, G.; Bauer; Bavykina; Bento; Bucci; Ciemniak; Deuter; von Feilitzsch; Hauff et al. (2011). "Results from 730kg days of the CRESST-II Dark Matter Search". arXiv:1109.0702v1 [astro-ph.CO].

[105] "Dark matter even darker than once thought". Retrieved 16 June 2015.

[106] Stecker, F.W.; Hunter, S; Kniffen, D (2008). "The likely cause of the EGRET GeV anomaly and its implications". *Astroparticle Physics* **29** (1): 25–29. arXiv:0705.4311. Bibcode:2008APh....29...25S. doi:10.1016/j.astropartphys.2007.11.002.

[107] Atwood, W.B.; Abdo, A. A.; Ackermann, M.; Althouse, W.; Anderson, B.; Axelsson, M.; Baldini, L.; Ballet, J. et al. (2009). "The large area telescope on the Fermi Gamma-ray Space Telescope Mission". *Astrophysical Journal* **697** (2): 1071–1102. arXiv:0902.1089. Bibcode:2009ApJ...697.1071A. doi:10.1088/0004-637X/697/2/1071.

[108] Weniger, Christoph (2012). "A Tentative Gamma-Ray Line from Dark Matter Annihilation at the Fermi Large Area Telescope". *Journal of Cosmology and Astroparticle Physics* **2012** (8): 7. arXiv:1204.2797v2. Bibcode:2012JCAP...08..007W. doi:10.1088/1475-7516/2012/08/007.

[109] Cartlidge, Edwin (24 April 2012). "Gamma rays hint at dark matter". Institute Of Physics. Retrieved 23 April 2013.

[110] Albert, J.; Aliu, E.; Anderhub, H.; Antoranz, P.; Backes, M.; Baixeras, C.; Barrio, J. A.; Bartko, H.; Bastieri, D.; Becker, J. K.; Bednarek, W.; Berger, K.; Bigongiari, C.; Biland, A.; Bock, R. K.; Bordas, P.; Bosch-Ramon, V.; Bretz, T.; Britvitch, I.; Camara, M.; Carmona, E.; Chilingarian, A.; Commichau, S.; Contreras, J. L.; Cortina, J.; Costado, M. T.; Curtef, V.; Danielyan, V.; Dazzi, F.; De Angelis, A. (2008). "Upper Limit for γ-Ray Emission above 140 GeV from the Dwarf Spheroidal Galaxy Draco". *The Astrophysical Journal* **679**: 428. doi:10.1086/529135.

[111] Aleksić, J.; Antonelli, L. A.; Antoranz, P.; Backes, M.; Baixeras, C.; Balestra, S.; Barrio, J. A.; Bastieri, D.; González, J. B.; Bednarek, W.; Berdyugin, A.; Berger, K.; Bernardini, E.; Biland, A.; Bock, R. K.; Bonnoli, G.; Bordas, P.; Tridon, D. B.; Bosch-Ramon, V.; Bose, D.; Braun, I.; Bretz, T.; Britzger, D.; Camara, M.; Carmona, E.; Carosi, A.; Colin, P.; Commichau, S.; Contreras, J. L.; Cortina, J. (2010). "Magic Gamma-Ray Telescope Observation of the Perseus Cluster of Galaxies: Implications for Cosmic Rays, Dark Matter, and Ngc 1275". *The Astrophysical Journal* **710**: 634. doi:10.1088/0004-637X/710/1/634.

[112] Adriani, O.; Barbarino, G. C.; Bazilevskaya, G. A.; Bellotti, R.; Boezio, M.; Bogomolov, E. A.; Bonechi, L.; Bongi, M.; Bonvicini, V.; Bottai, S.; Bruno, A.; Cafagna, F.; Campana, D.; Carlson, P.; Casolino, M.; Castellini, G.; De Pascale, M. P.; De Rosa, G.; De Simone, N.; Di Felice, V.; Galper, A. M.; Grishantseva, L.; Hofverberg, P.; Koldashov, S. V.; Krutkov, S. Y.; Kvashnin, A. N.; Leonov, A.; Malvezzi, V.; Marcelli, L.; Menn, W. (2009). "An anomalous positron abundance in cosmic rays with energies 1.5–100 GeV". *Nature* **458** (7238): 607–609. doi:10.1038/nature07942. PMID 19340076.

[113] Aguilar, M. (AMS Collaboration) et al. (3 April 2013). "First Result from the Alpha Magnetic Spectrometer on the International Space Station: Precision Measurement of the Positron Fraction in Primary Cosmic Rays of 0.5–350 GeV". *Physical Review Letters*. Bibcode:2013PhRvL.110n1102A. doi:10.1103/PhysRevLett.110.141102. Retrieved 3 April 2013.

[114] "First Result from the Alpha Magnetic Spectrometer Experiment". *AMS Collaboration*. 3 April 2013. Retrieved 3 April 2013.

[115] Heilprin, John; Borenstein, Seth (3 April 2013). "Scientists find hint of dark matter from cosmos". Associated Press. Retrieved 3 April 2013.

[116] Amos, Jonathan (3 April 2013). "Alpha Magnetic Spectrometer zeroes in on dark matter". *BBC*. Retrieved 3 April 2013.

[117] Perrotto, Trent J.; Byerly, Josh (2 April 2013). "NASA TV Briefing Discusses Alpha Magnetic Spectrometer Results". *NASA*. Retrieved 3 April 2013.

[118] Overbye, Dennis (3 April 2013). "New Clues to the Mystery of Dark Matter". *New York Times*. Retrieved 3 April 2013.

[119] Freese, K. (1986). "Can Scalar Neutrinos or Massive Dirac Neutrinos be the Missing Mass?". *Physics Letters B* **167** (3): 295–300. Bibcode:1986PhLB..167..295F. doi:10.1016/0370-2693(86)90349-7.

[120] Ellis, J.; Flores, R. A.; Freese, K.; Ritz, S.; Seckel, D.; Silk, J. (1988). "Cosmic ray constraints on the annihilations of relic particles in the galactic halo". *Physics Letters B* **214** (3): 403. Bibcode:1988PhLB..214..403E. doi:10.1016/0370-2693(88)91385-8.

[121] Merritt, David (1 January 2010). "Dark Matter at the Centers of Galaxies". In Bertone, Gianfranco. *Particle Dark Matter : Observations, Models and Searches*. Cambridge University Press. pp. 83–104. arXiv:1001.3706. ISBN 978-0-521-76368-4.

[122] Rzetelny, Xaq (19 November 2014). "Looking for a different sort of dark matter with GPS satellites". *Ars Technica*. Retrieved 24 November 2014.

[123] Exirifard, Q. (2010). "Phenomenological covariant approach to gravity". *General Relativity and Gravitation* **43** (1): 93–106. arXiv:0808.1962. Bibcode:2011GReGr..43...93E. doi:10.1007/s10714-010-1073-6.

[124] Brownstein, J.R.; Moffat, J. W. (2007). "The Bullet Cluster 1E0657-558 evidence shows modified gravity in the absence of dark matter". *Monthly Notices of the Royal Astronomical Society* **382** (1): 29–47. arXiv:astro-ph/0702146.Bibcode:2007. doi:10.1111/j.1365-2966.2007.12275.x.

[125] Anastopoulos, C. (2009). "Gravitational backreaction in cosmological spacetimes". *Physical Review D* **79** (8): 084029. arXiv:0902.0159. Bibcode:2009PhRvD..79h4029A. doi:10.1103/PhysRevD.79.084029.

[126] "New Cosmic Theory Unites Dark Forces". SPACE.com. 11 February 2008. Retrieved 6 January 2011.

[127] Hossenfelder, S. (2008). "A Bi-Metric Theory with Exchange Symmetry". *Physical Review D* **78** (4): 044015. arXiv:gr-qc/0603005. Bibcode:2008PhRvD..78d4015H. doi:10.1103/PhysRevD.78.044015.

[128] Ripalda, Jose M. (1999). "Time reversal and negative energies in general relativity". arXiv:gr-qc/9906012 [gr-qc].

16.11   External links

- Dark matter at DMOZ

- What is dark matter? at cosmosmagazine.com

- The Dark Matter Crisis

- The European astroparticle physics network

- Helmholtz Alliance for Astroparticle Physics

- "NASA Finds Direct Proof of Dark Matter" (Press release). NASA. 21 August 2006.

- Tuttle, Kelen (22 August 2006). "Dark Matter Observed". SLAC (Stanford Linear Accelerator Center) Today.

- "Astronomers claim first 'dark galaxy' find". New Scientist. 23 February 2005.

- Sample, Ian (17 December 2009). "Dark Matter Detected". London: Guardian. Retrieved 1 May 2010.

- Video lecture on dark matter by Scott Tremaine, IAS professor

- Science Daily story "Astronomers' Doubts About the Dark Side ..."

- Gray, Meghan; Merrifield, Mike; Copeland, Ed (2010). "Dark Matter". *Sixty Symbols*. Brady Haran for the University of Nottingham.

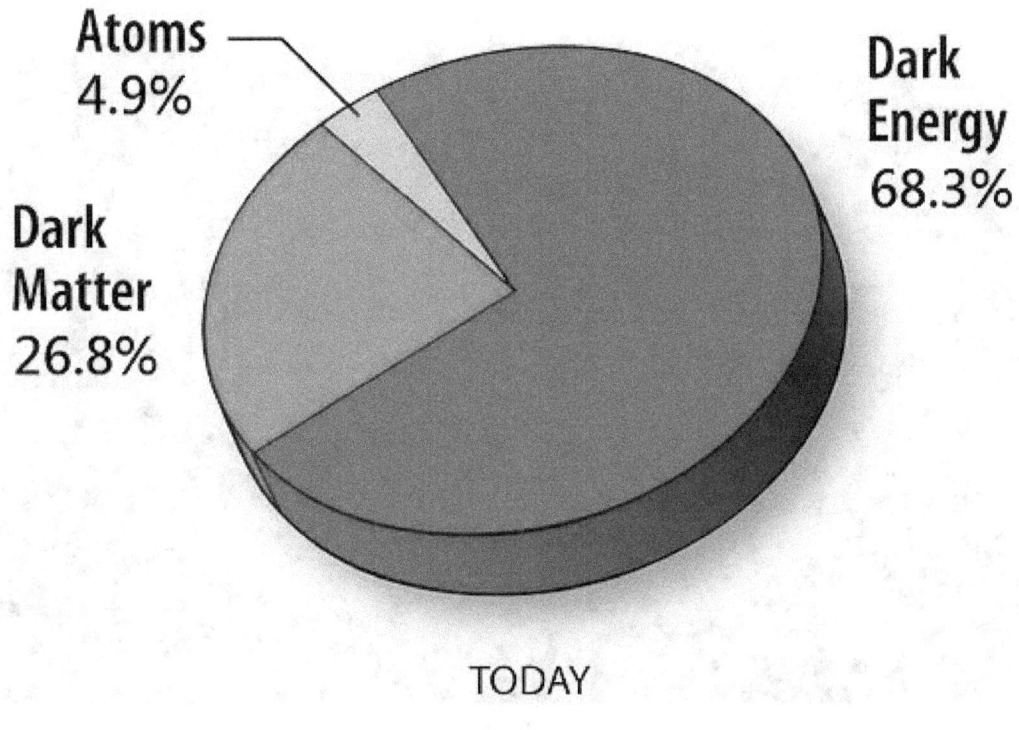

TODAY

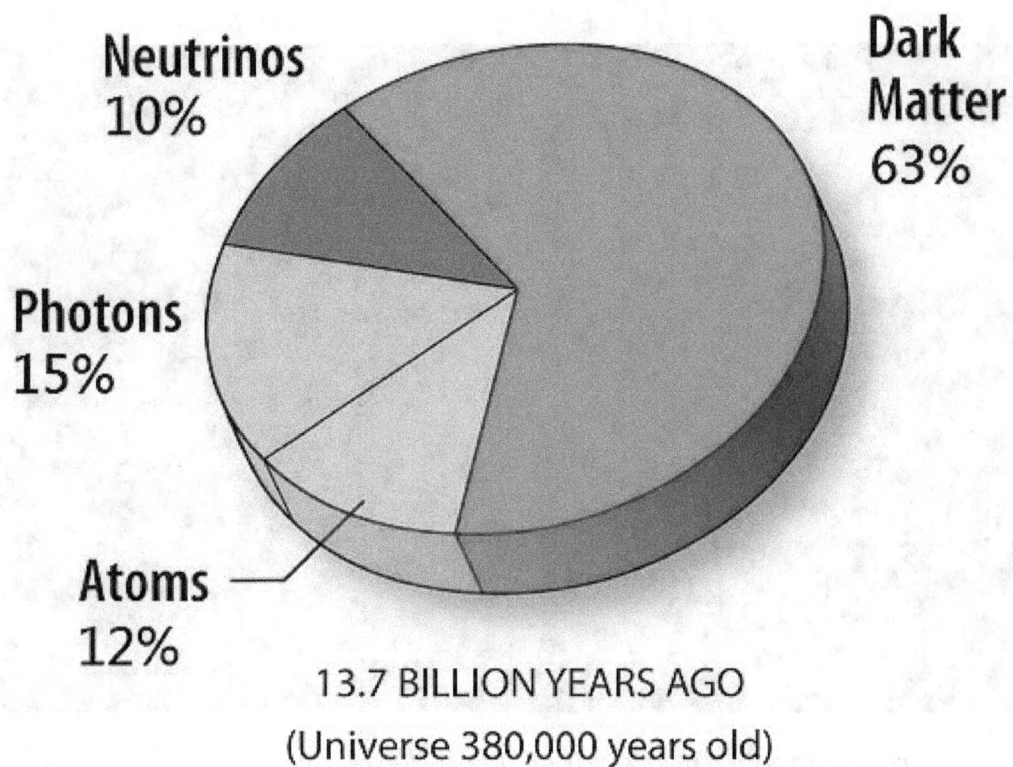

13.7 BILLION YEARS AGO
(Universe 380,000 years old)

Estimated distribution of matter and energy in the universe, today (top) and when the CMB was released (bottom)

Fermi-LAT observations of dwarf galaxies provide new insights on dark matter.

This artist's impression shows the expected distribution of dark matter in the Milky Way galaxy as a blue halo of material surrounding the galaxy.[22]

Observations have provided hints that the dark matter around one of the central four merging galaxies is not moving with the galaxy itself.[30]

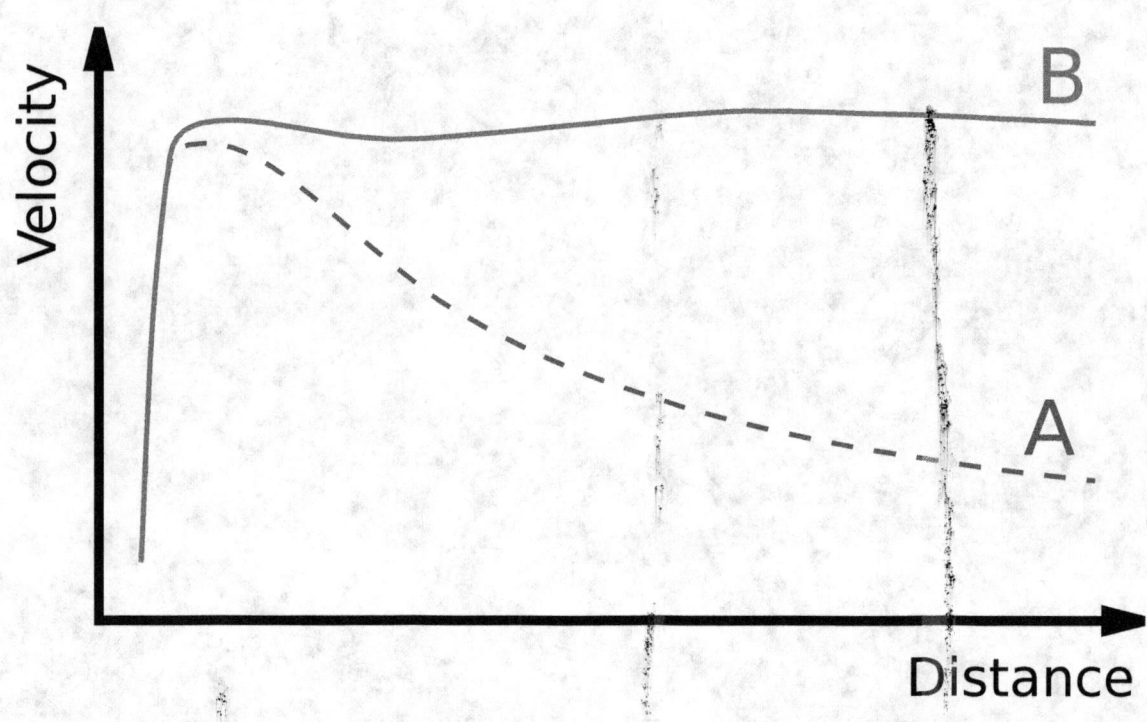

*Rotation curve of a typical spiral galaxy: predicted (**A**) and observed (**B**). Dark matter can explain the 'flat' appearance of the velocity curve out to a large radius*

Strong gravitational lensing as observed by the Hubble Space Telescope in Abell 1689 indicates the presence of dark matter—enlarge the image to see the lensing arcs.

The Bullet Cluster: HST image with overlays. The total projected mass distribution reconstructed from strong and weak gravitational lensing is shown in blue, while the X-ray emitting hot gas observed with Chandra is shown in red.

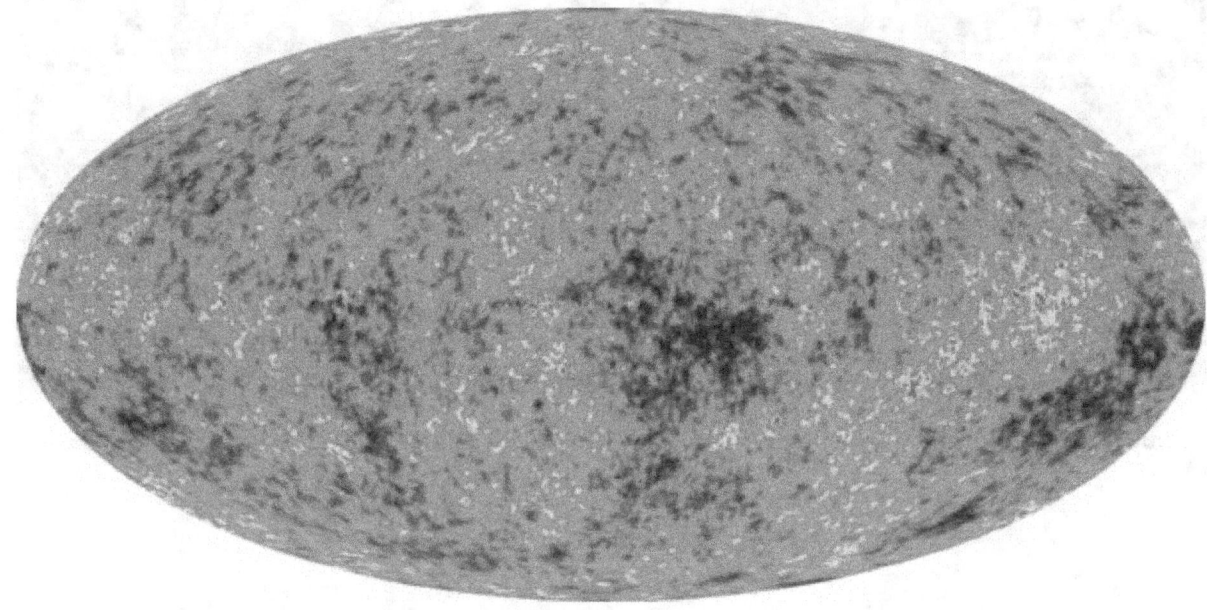

The cosmic microwave background by WMAP

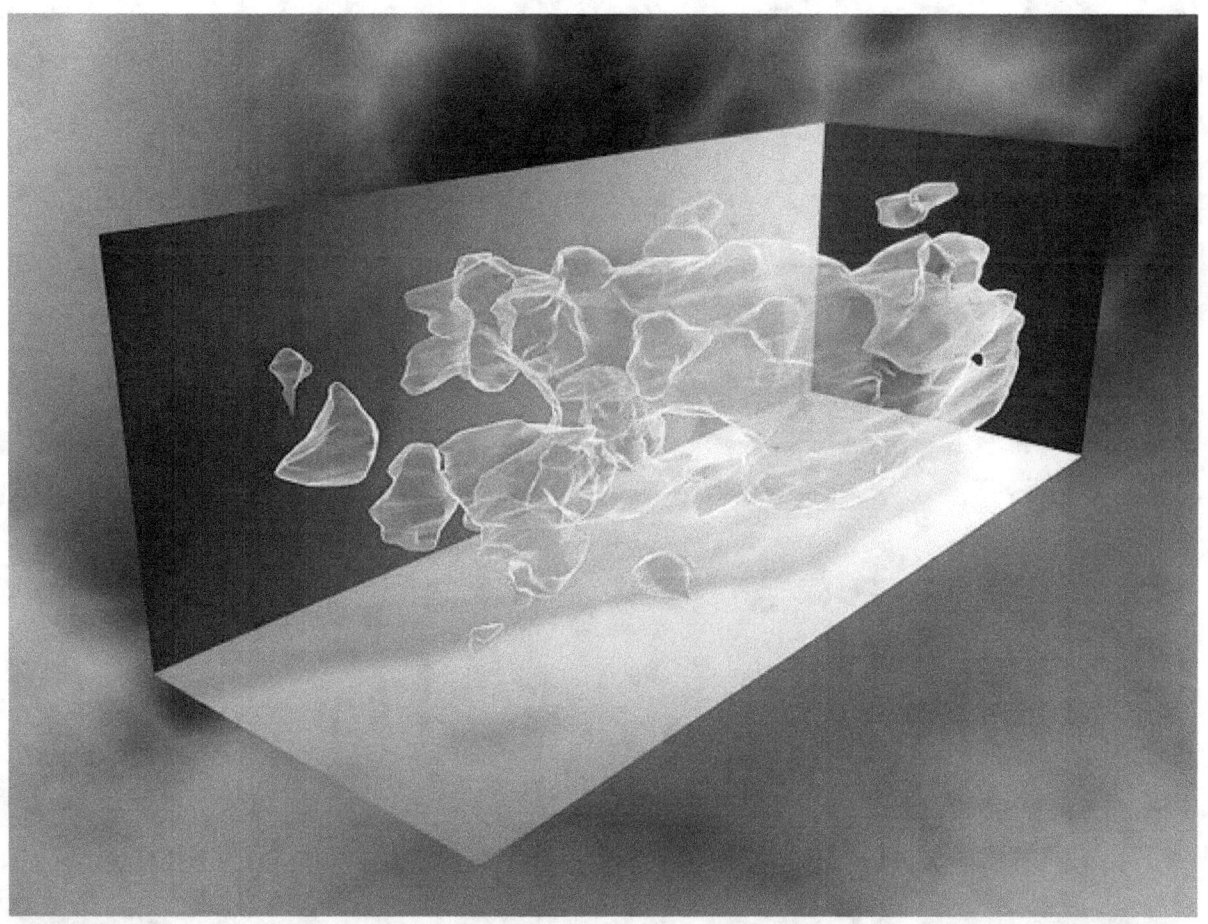

3D map of the large-scale distribution of dark matter, reconstructed from measurements of weak gravitational lensing with the Hubble Space Telescope.[73]

Collage of six cluster collisions with dark matter maps. The clusters were observed in a study of how dark matter in clusters of galaxies behaves when the clusters collide.[105]

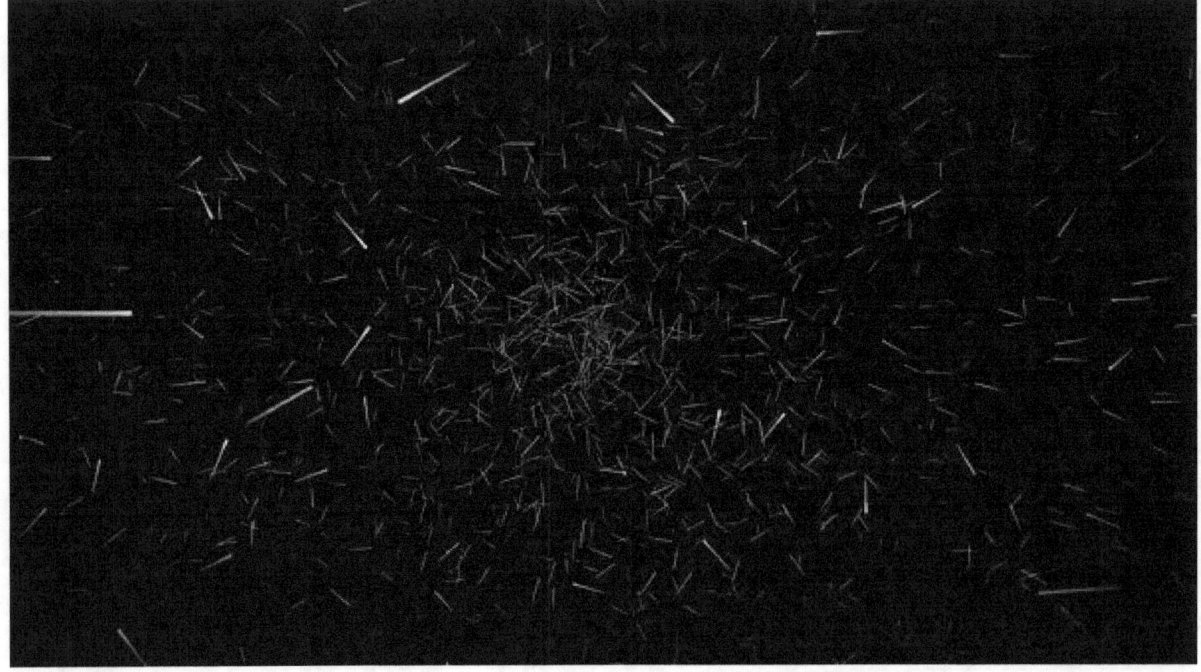

Video about the potential gamma-ray detection of dark matter annihilation around supermassive black holes. (Duration 3:13, also see file description.)

Chapter 17

Exotic matter

In physics, **exotic matter** is matter that somehow deviates from normal matter and has "exotic" properties. A more broad definition of exotic matter is any kind of non-baryonic matter—that is not made of baryons, the subatomic particles, such as protons and neutrons, of which the ordinary matter is composed.[1] Exotic mass has been considered a colloquial term for matters such as dark matter, negative mass, or imaginary mass.

17.1 Types of exotic matter

There are several types of exotic matter:

- Hypothetical particles that have "exotic" physical properties that would violate known laws of physics, such as a particle having a negative mass.

- Hypothetical particles that have not yet been encountered, such as exotic baryons, but whose properties would be within the realm of mainstream physics if found to exist.

- States of matter that are not commonly encountered, such as Bose–Einstein condensates and quark–gluon plasma, but whose properties are perfectly within the realm of mainstream physics.

- States of matter that are poorly understood, such as dark matter.

- Ordinary matter placed under high pressure.

17.2 Negative mass

Main article: Negative mass

Negative mass would possess some strange properties, such as accelerating in the direction opposite of applied force. For example, an object with negative inertial mass and positive electric charge would accelerate away from objects with negative charge, and towards objects with positive charge, the opposite of the normal rule that like charges repel and opposite charges attract. This behaviour can produce bizarre results: for instance, a gas containing a mixture of positive and negative matter particles will have the positive matter portion increase in temperature without bound. However, the negative matter portion gains negative temperature at the same rate, again balancing out.

Despite being inconsistent with the expected behavior of "normal" matter, negative mass is mathematically consistent and introduces no violation of conservation of momentum or energy. It is used in certain speculative theories, such as on the construction of wormholes and the Alcubierre drive. The closest known real representative of such exotic matter is the region of pseudo-negative-pressure density produced by the Casimir effect.

17.3   Imaginary mass

Main article: Tachyon § Mass

A hypothetical particle with imaginary rest mass would always travel faster than the speed of light. Such particles are called tachyons. There is no confirmed existence of tachyons.

$$E = \frac{m \cdot c^2}{\sqrt{1 - \frac{|\mathbf{v}|^2}{c^2}}}$$

If the rest mass m is imaginary this implies that the denominator is imaginary because the total energy is an observable and thus must be real. Therefore the quantity under the square root must be negative, which can only happen if v is greater than c. As noted by Gregory Benford *et al.,* special relativity implies that tachyons, if they existed, could be used to communicate backwards in time[2] (see tachyonic antitelephone). Because time travel is considered to be non-physical, tachyons are believed by physicists either to not exist, or else to be incapable of interacting with normal matter.

In quantum field theory, imaginary mass would induce tachyon condensation.

17.4   Materials at high pressure

At high pressure, materials such as NaCl in the presence of an excess of either chlorine or sodium were transformed into compounds "forbidden" by classical chemistry, such as Na
3Cl and NaCl
3. Quantum mechanical calculations predict the possibility of other compounds, such as NaCl
7, Na
3Cl
2, Na
2Cl, and Na
3Cl. The materials are thermodynamically stable at high pressures. Such compounds may exist in natural environments that exist at high pressure, such as the deep ocean or inside planetary cores. The materials have potentially useful properties. For instance, Na
3Cl is a two-dimensional metal, made of layers of pure sodium and salt that can conduct electricity. The salt layers act as insulators while the sodium layers act as conductors.[3][4]

17.5   See also

- Antimatter

- Dark energy

- Dark matter

- Gravitational interaction of antimatter

- Mirror matter

- Negative energy

- Negative mass

- Strange matter

- QCD matter

17.6   References

[1]   "Exotic matter". daviddarling.info. Retrieved 2015-06-24.

[2]   G. A. Benford, D. L. Book, and W. A. Newcomb (1970). "The Tachyonic Antitelephone". *Physical Review D* **2**: 263. Bibcode:1970PhRvD...2..263B. doi:10.1103/PhysRevD.2.263.

[3]   "Scientists turn table salt into forbidden compounds that violate textbook rules". Gizmag.com. Retrieved 2014-01-21.

[4]   Zhang, W.; Oganov, A. R.; Goncharov, A. F.; Zhu, Q.; Boulfelfel, S. E.; Lyakhov, A. O.; Stavrou, E.; Somayazulu, M.; Prakapenka, V. B.; Konôpková, Z. (2013). "Unexpected Stable Stoichiometries of Sodium Chlorides". *Science* **342** (6165): 1502–1505. doi:10.1126/science.1244989. PMID 24357316.

17.7   External links

- Exotic Matter and Negative Energy on YouTube
- The Riddle of AntiMatter on YouTube

Chapter 18

Periodic table

This article is about the table used in chemistry. For other uses, see Periodic table (disambiguation).

Group→	1	2	3	4	5	6	7	8	9	10	11	12	13	14	15	16	17	18
↓Period																		
1	1 H																	2 He
2	3 Li	4 Be											5 B	6 C	7 N	8 O	9 F	10 Ne
3	11 Na	12 Mg											13 Al	14 Si	15 P	16 S	17 Cl	18 Ar
4	19 K	20 Ca	21 Sc	22 Ti	23 V	24 Cr	25 Mn	26 Fe	27 Co	28 Ni	29 Cu	30 Zn	31 Ga	32 Ge	33 As	34 Se	35 Br	36 Kr
5	37 Rb	38 Sr	39 Y	40 Zr	41 Nb	42 Mo	43 Tc	44 Ru	45 Rh	46 Pd	47 Ag	48 Cd	49 In	50 Sn	51 Sb	52 Te	53 I	54 Xe
6	55 Cs	56 Ba	*	72 Hf	73 Ta	74 W	75 Re	76 Os	77 Ir	78 Pt	79 Au	80 Hg	81 Tl	82 Pb	83 Bi	84 Po	85 At	86 Rn
7	87 Fr	88 Ra	**	104 Rf	105 Db	106 Sg	107 Bh	108 Hs	109 Mt	110 Ds	111 Rg	112 Cn	113 Uut	114 Fl	115 Uup	116 Lv	117 Uus	118 Uuo

	57 La	58 Ce	59 Pr	60 Nd	61 Pm	62 Sm	63 Eu	64 Gd	65 Tb	66 Dy	67 Ho	68 Er	69 Tm	70 Yb	71 Lu
*															
**	89 Ac	90 Th	91 Pa	92 U	93 Np	94 Pu	95 Am	96 Cm	97 Bk	98 Cf	99 Es	100 Fm	101 Md	102 No	103 Lr

Standard form of the periodic table (color legend below)

The **periodic table** is a tabular arrangement of the chemical elements, ordered by their atomic number (number of protons in the nucleus), electron configurations, and recurring chemical properties. The table also shows four rectangular blocks: s-, p- d- and f-block. In general, within one row (period) the elements are metals on the lefthand side, and non-metals on the righthand side.

The rows of the table are called periods; the columns are called groups. Six groups (columns) have names as well as numbers: for example, group 17 elements are the halogens; and group 18, the noble gases. The periodic table can be used to derive relationships between the properties of the elements, and predict the properties of new elements yet to be discovered or synthesized. The periodic table provides a useful framework for analyzing chemical behavior, and is widely used in chemistry and other sciences.

Although precursors exist, Dmitri Mendeleev is generally credited with the publication, in 1869, of the first widely recognized periodic table. He developed his table to illustrate periodic trends in the properties of the then-known elements.

Mendeleev also predicted some properties of then-unknown elements that would be expected to fill gaps in this table. Most of his predictions were proved correct when the elements in question were subsequently discovered. Mendeleev's periodic table has since been expanded and refined with the discovery or synthesis of further new elements and the development of new theoretical models to explain chemical behavior.

All elements from atomic numbers 1 (hydrogen) to 118 (ununoctium) have been discovered or reportedly synthesized, with elements 113, 115, 117, and 118 having yet to be confirmed. The first 98 elements exist naturally, although some are found only in trace amounts and were synthesized in laboratories before being found in nature.[n 1] Elements with atomic numbers from 99 to 118 have only been synthesized in laboratories. It has been shown that einsteinium and fermium once occurred in nature but currently do not.[1] Synthesis of elements having higher atomic numbers is being pursued. Numerous synthetic radionuclides of naturally occurring elements have also been produced in laboratories.

18.1 Overview

For large cell versions, see Periodic table (large cells).

Some presentations include an element zero (i.e. a substance composed purely of neutrons), although this is uncommon. See, for example. Philip Stewart's Chemical Galaxy. Each chemical element has a unique atomic number representing the number of protons in its nucleus. Most elements have differing numbers of neutrons among different atoms, with these variants being referred to as isotopes. For example, carbon has three naturally occurring isotopes: all of its atoms have six protons and most have six neutrons as well, but about one per cent have seven neutrons, and a very small fraction have eight neutrons. Isotopes are never separated in the periodic table; they are always grouped together under a single element. Elements with no stable isotopes have the atomic masses of their most stable isotopes, where such masses are shown, listed in parentheses.[2]

In the standard periodic table, the elements are listed in order of increasing atomic number (the number of protons in the nucleus of an atom). A new row (*period*) is started when a new electron shell has its first electron. Columns (*groups*) are determined by the electron configuration of the atom; elements with the same number of electrons in a particular subshell fall into the same columns (e.g. oxygen and selenium are in the same column because they both have four electrons in the outermost p-subshell). Elements with similar chemical properties generally fall into the same group in the periodic table, although in the f-block, and to some respect in the d-block, the elements in the same period tend to have similar properties, as well. Thus, it is relatively easy to predict the chemical properties of an element if one knows the properties of the elements around it.[3]

As of 2014, the periodic table has 114 confirmed elements, comprising elements 1 (hydrogen) to 112 (copernicium), 114 (flerovium) and 116 (livermorium). Elements 113, 115, 117 and 118 have reportedly been synthesised in laboratories, but none of these claims have been officially confirmed by the International Union of Pure and Applied Chemistry (IUPAC), nor are they named. As such these elements are currently identified by their atomic number (e.g., "element 113"), or by their provisional systematic name ("ununtrium", symbol "Uut").[4]

A total of 98 elements occur naturally; the remaining 16 elements, from einsteinium to copernicium, and flerovium and livermorium, occur only when synthesised in laboratories. Of the 98 elements that occur naturally, 84 are primordial. The other 14 naturally occurring elements occur only in decay chains of primordial elements.[1] No element heavier than einsteinium (element 99) has ever been observed in macroscopic quantities in its pure form.[5]

18.1.1 Layout variants

In the most common graphic presentation of the periodic table, the main table has 18 columns and the lanthanides and the actinides are shown as two additional rows below the main body of the table,[6] with two placeholders shown in the main table, between barium and hafnium, and radium and rutherfordium, respectively. These placeholders can be asterisk-like markers, or a contracted range description of elements ("57–71"). This convention is entirely a matter of formatting practicality. The same table structure can be shown in a 32-column format, with the lanthanides and actinides in the main table's row 6 and 7.

However, based on the chemical and physical properties of elements, many alternative table *structures* have been constructed.

18.2   Grouping methods

18.2.1   Groups

Main article: Group (periodic table)

A *group* or *family* is a vertical column in the periodic table. Groups usually have more significant periodic trends than periods and blocks, explained below. Modern quantum mechanical theories of atomic structure explain group trends by proposing that elements within the same group generally have the same electron configurations in their valence shell.[7] Consequently, elements in the same group tend to have a shared chemistry and exhibit a clear trend in properties with increasing atomic number.[8] However, in some parts of the periodic table, such as the d-block and the f-block, horizontal similarities can be as important as, or more pronounced than, vertical similarities.[9][10][11]

Under an international naming convention, the groups are numbered numerically from 1 to 18 from the leftmost column (the alkali metals) to the rightmost column (the noble gases).[12] Previously, they were known by roman numerals. In America, the roman numerals were followed by either an "A" if the group was in the s- or p-block, or a "B" if the group was in the d-block. The roman numerals used correspond to the last digit of today's naming convention (e.g. the group 4 elements were group IVB, and the group 14 elements was group IVA). In Europe, the lettering was similar, except that "A" was used if the group was before group 10, and "B" was used for groups including and after group 10. In addition, groups 8, 9 and 10 used to be treated as one triple-sized group, known collectively in both notations as group VIII. In 1988, the new IUPAC naming system was put into use, and the old group names were deprecated.[13]

Some of these groups have been given trivial (unsystematic) names, as seen in the table below, although some are rarely used. Groups 3–10 have no trivial names and are referred to simply by their group numbers or by the name of the first member of their group (such as 'the scandium group' for Group 3), since they display fewer similarities and/or vertical trends.[12]

Elements in the same group tend to show patterns in atomic radius, ionization energy, and electronegativity. From top to bottom in a group, the atomic radii of the elements increase. Since there are more filled energy levels, valence electrons are found farther from the nucleus. From the top, each successive element has a lower ionization energy because it is easier to remove an electron since the atoms are less tightly bound. Similarly, a group has a top to bottom decrease in electronegativity due to an increasing distance between valence electrons and the nucleus.[14] There are exceptions to these trends, however, an example of which occurs in group 11 where electronegativity increases farther down the group.[15]

18.2.2   Periods

Main article: Period (periodic table)

A *period* is a horizontal row in the periodic table. Although groups generally have more significant periodic trends, there are regions where horizontal trends are more significant than vertical group trends, such as the f-block, where the lanthanides and actinides form two substantial horizontal series of elements.[16]

Elements in the same period show trends in atomic radius, ionization energy, electron affinity, and electronegativity. Moving left to right across a period, atomic radius usually decreases. This occurs because each successive element has an added proton and electron, which causes the electron to be drawn closer to the nucleus.[17] This decrease in atomic radius also causes the ionization energy to increase when moving from left to right across a period. The more tightly bound an element is, the more energy is required to remove an electron. Electronegativity increases in the same manner as ionization energy because of the pull exerted on the electrons by the nucleus.[14] Electron affinity also shows a slight trend across a period. Metals (left side of a period) generally have a lower electron affinity than nonmetals (right side of a period), with the exception of the noble gases.[18]

18.2.3 Blocks

Main article: Block (periodic table)
 Specific regions of the periodic table can be referred to as *blocks* in recognition of the sequence in which the electron

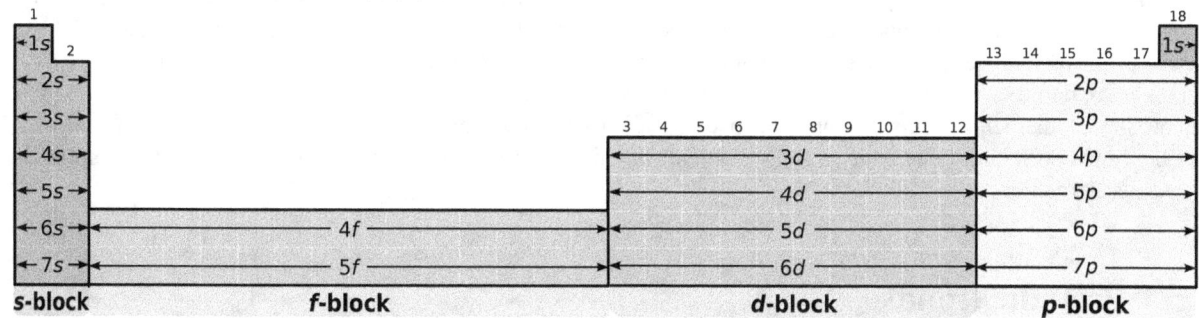

Left to right: s-, f-, d-, p-block in the periodic table

shells of the elements are filled. Each block is named according to the subshell in which the "last" electron notionally resides.[19][n 2] The s-block comprises the first two groups (alkali metals and alkaline earth metals) as well as hydrogen and helium. The p-block comprises the last six groups, which are groups 13 to 18 in IUPAC (3A to 8A in American) and contains, among other elements, all of the metalloids. The d-block comprises groups 3 to 12 (or 3B to 2B in American group numbering) and contains all of the transition metals. The f-block, often offset below the rest of the periodic table, has no group numbers and comprises lanthanides and actinides.[20]

18.2.4 Metals, metalloids and nonmetals

Metals, metalloids, nonmetals, and elements with unknown chemical properties in the periodic table. Sources disagree on the classification of some of these elements.

According to their shared physical and chemical properties, the elements can be classified into the major categories of metals, metalloids and nonmetals. Metals are generally shiny, highly conducting solids that form alloys with one another and salt-like ionic compounds with nonmetals (other than the noble gases). The majority of nonmetals are coloured or colourless insulating gases; nonmetals that form compounds with other nonmetals feature covalent bonding. In between metals and nonmetals are metalloids, which have intermediate or mixed properties.[21]

Metal and nonmetals can be further classified into subcategories that show a gradation from metallic to non-metallic properties, when going left to right in the rows. The metals are subdivided into the highly reactive alkali metals, through the less reactive alkaline earth metals, lanthanides and actinides, via the archetypal transition metals, and ending in the physically and chemically weak post-transition metals. The nonmetals are simply subdivided into the polyatomic nonmetals, which, being nearest to the metalloids, show some incipient metallic character; the diatomic nonmetals, which are essentially nonmetallic; and the monatomic noble gases, which are nonmetallic and almost completely inert. Specialized groupings such as the refractory metals and the noble metals, which are subsets (in this example) of the transition metals, are also known[22] and occasionally denoted.[23]

Placing the elements into categories and subcategories based on shared properties is imperfect. There is a spectrum of properties within each category and it is not hard to find overlaps at the boundaries, as is the case with most classification schemes.[24] Beryllium, for example, is classified as an alkaline earth metal although its amphoteric chemistry and tendency to mostly form covalent compounds are both attributes of a chemically weak or post transition metal. Radon is classified as a nonmetal and a noble gas yet has some cationic chemistry that is more characteristic of a metal. Other classification schemes are possible such as the division of the elements into mineralogical occurrence categories, or crystalline structures. Categorising the elements in this fashion dates back to at least 1869 when Hinrichs[25] wrote that simple boundary lines could be drawn on the periodic table to show elements having like properties, such as the metals and the nonmetals, or the gaseous elements.

18.3   Periodic trends

Main article: Periodic trends

18.3.1   Electron configuration

Main article: Electronic configuration
 The electron configuration or organisation of electrons orbiting neutral atoms shows a recurring pattern or periodicity. The electrons occupy a series of electron shells (numbered shell 1, shell 2, and so on). Each shell consists of one or more subshells (named s, p, d, f and g). As atomic number increases, electrons progressively fill these shells and subshells more or less according to the Madelung rule or energy ordering rule, as shown in the diagram. The electron configuration for neon, for example, is $1s^2 2s^2 2p^6$. With an atomic number of ten, neon has two electrons in the first shell, and eight electrons in the second shell—two in the s subshell and six in the p subshell. In periodic table terms, the first time an electron occupies a new shell corresponds to the start of each new period, these positions being occupied by hydrogen and the alkali metals.[26][27]

Since the properties of an element are mostly determined by its electron configuration, the properties of the elements likewise show recurring patterns or periodic behaviour, some examples of which are shown in the diagrams below for atomic radii, ionization energy and electron affinity. It is this periodicity of properties, manifestations of which were noticed well before the underlying theory was developed, that led to the establishment of the periodic law (the properties of the elements recur at varying intervals) and the formulation of the first periodic tables.[26][27]

18.3.2   Atomic radii

Main article: Atomic radius
 Atomic radii vary in a predictable and explainable manner across the periodic table. For instance, the radii generally decrease along each period of the table, from the alkali metals to the noble gases; and increase down each group. The radius increases sharply between the noble gas at the end of each period and the alkali metal at the beginning of the next period. These trends of the atomic radii (and of various other chemical and physical properties of the elements) can be explained by the electron shell theory of the atom; they provided important evidence for the development and confirmation of quantum theory.[28]

The electrons in the 4f-subshell, which is progressively filled from cerium (element 58) to ytterbium (element 70), are not particularly effective at shielding the increasing nuclear charge from the sub-shells further out. The elements immediately following the lanthanides have atomic radii that are smaller than would be expected and that are almost identical to the atomic radii of the elements immediately above them.[29] Hence hafnium has virtually the same atomic radius (and chemistry) as zirconium, and tantalum has an atomic radius similar to niobium, and so forth. This is known as the lanthanide contraction. The effect of the lanthanide contraction is noticeable up to platinum (element 78), after which

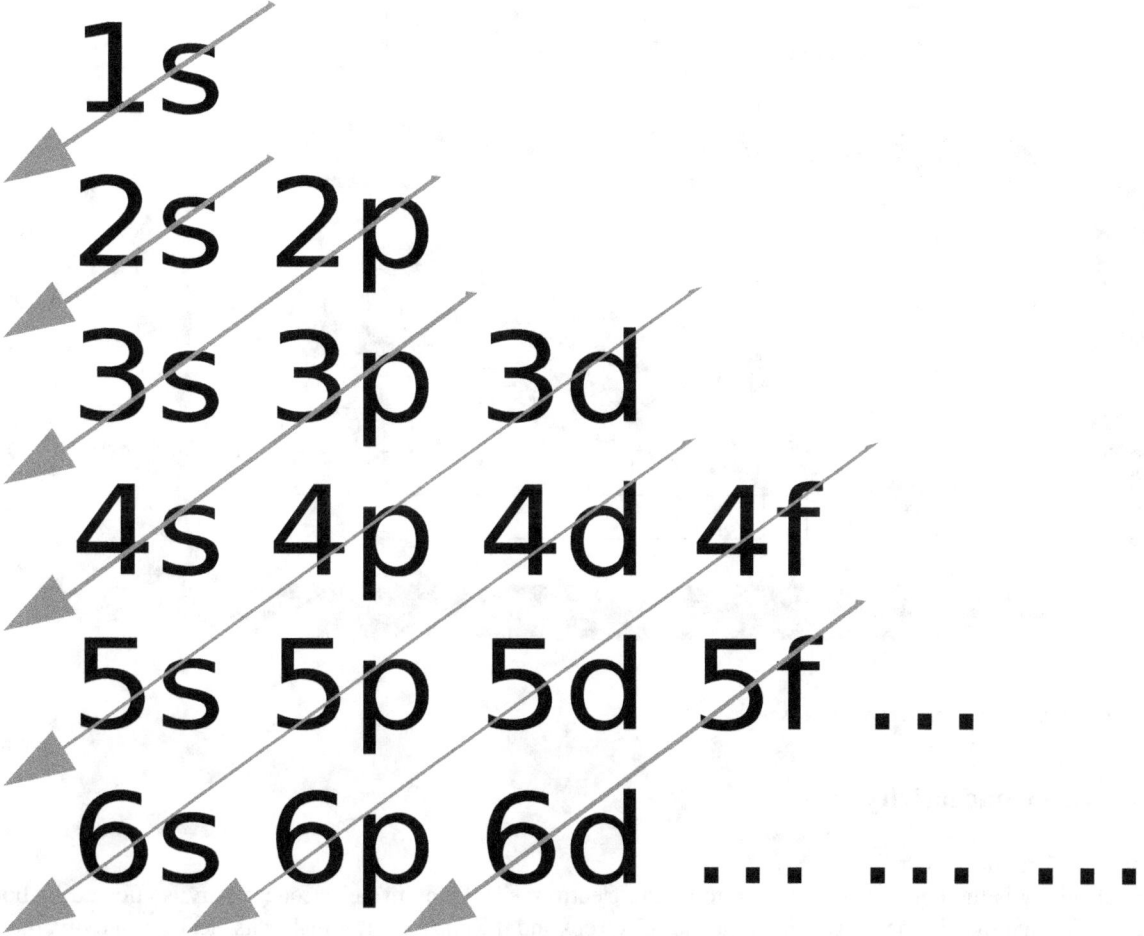

Approximate order in which shells and subshells are arranged by increasing energy according to the Madelung rule

it is masked by a relativistic effect known as the inert pair effect.[30] The d-block contraction, which is a similar effect between the d-block and p-block, is less pronounced than the lanthanide contraction but arises from a similar cause.[29]

18.3.3 Ionization energy

Main article: Ionization energy

The first ionization energy is the energy it takes to remove one electron from an atom, the second ionization energy is the energy it takes to remove a second electron from the atom, and so on. For a given atom, successive ionization energies increase with the degree of ionization. For magnesium as an example, the first ionization energy is 738 kJ/mol and the second is 1450 kJ/mol. Electrons in the closer orbitals experience greater forces of electrostatic attraction; thus, their removal requires increasingly more energy. Ionization energy becomes greater up and to the right of the periodic table.[30]

Large jumps in the successive molar ionization energies occur when removing an electron from a noble gas (complete electron shell) configuration. For magnesium again, the first two molar ionization energies of magnesium given above correspond to removing the two 3s electrons, and the third ionization energy is a much larger 7730 kJ/mol, for the removal of a 2p electron from the very stable neon-like configuration of Mg^{2+}. Similar jumps occur in the ionization energies of other third-row atoms.[30]

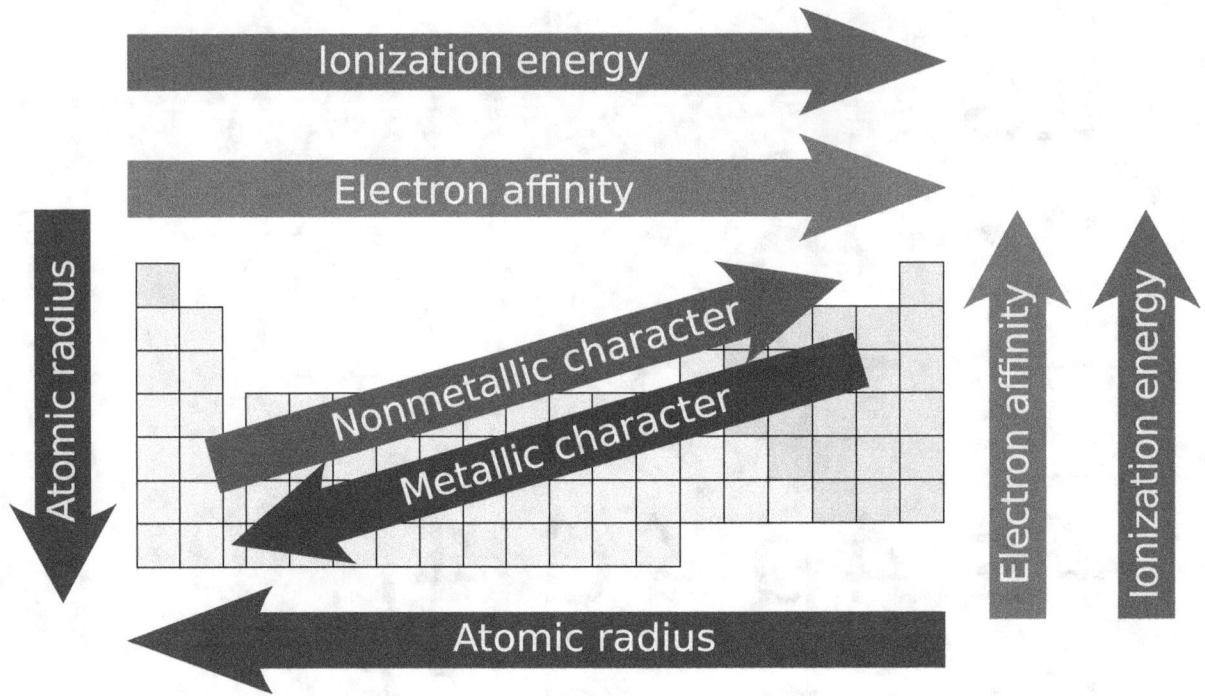

Periodic table trends (arrows direct an increase)

18.3.4 Electronegativity

Main article: Electronegativity

 Electronegativity is the tendency of an atom to attract electrons.[31] An atom's electronegativity is affected by both its atomic number and the distance between the valence electrons and the nucleus. The higher its electronegativity, the more an element attracts electrons. It was first proposed by Linus Pauling in 1932.[32] In general, electronegativity increases on passing from left to right along a period, and decreases on descending a group. Hence, fluorine is the most electronegative of the elements,[n 4] while caesium is the least, at least of those elements for which substantial data is available.[15]

There are some exceptions to this general rule. Gallium and germanium have higher electronegativities than aluminium and silicon respectively because of the d-block contraction. Elements of the fourth period immediately after the first row of the transition metals have unusually small atomic radii because the 3d-electrons are not effective at shielding the increased nuclear charge, and smaller atomic size correlates with higher electronegativity.[15] The anomalously high electronegativity of lead, particularly when compared to thallium and bismuth, appears to be an artifact of data selection (and data availability)—methods of calculation other than the Pauling method show the normal periodic trends for these elements.[33]

18.3.5 Electron affinity

Main article: Electron affinity

 The electron affinity of an atom is the amount of energy released when an electron is added to a neutral atom to form a negative ion. Although electron affinity varies greatly, some patterns emerge. Generally, nonmetals have more positive electron affinity values than metals. Chlorine most strongly attracts an extra electron. The electron affinities of the noble gases have not been measured conclusively, so they may or may not have slightly negative values.[36]

Electron affinity generally increases across a period. This is caused by the filling of the valence shell of the atom; a group 17 atom releases more energy than a group 1 atom on gaining an electron because it obtains a filled valence shell and is therefore more stable.[36]

A trend of decreasing electron affinity going down groups would be expected. The additional electron will be entering an

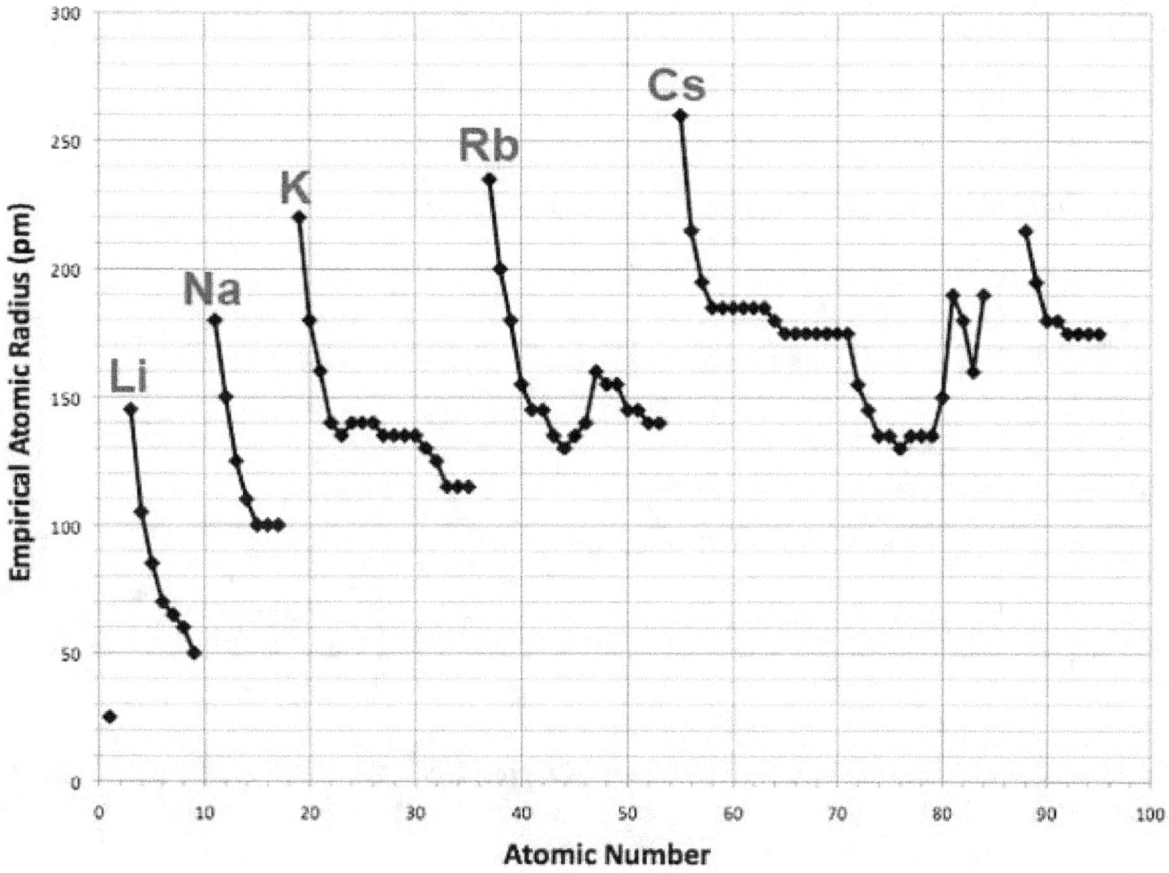

Atomic number plotted against atomic radius[n 3]

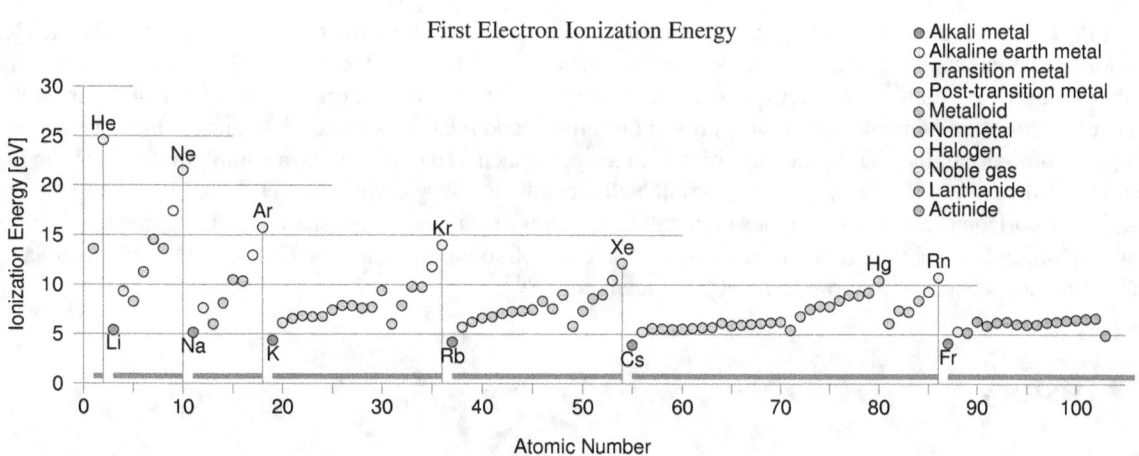

Ionization energy: each period begins at a minimum for the alkali metals, and ends at a maximum for the noble gases

orbital farther away from the nucleus. As such this electron would be less attracted to the nucleus and would release less energy when added. However, in going down a group, around one-third of elements are anomalous, with heavier elements having higher electron affinities than their next lighter congenors. Largely, this is due to the poor shielding by d and f electrons. A uniform decrease in electron affinity only applies to group 1 atoms.[37]

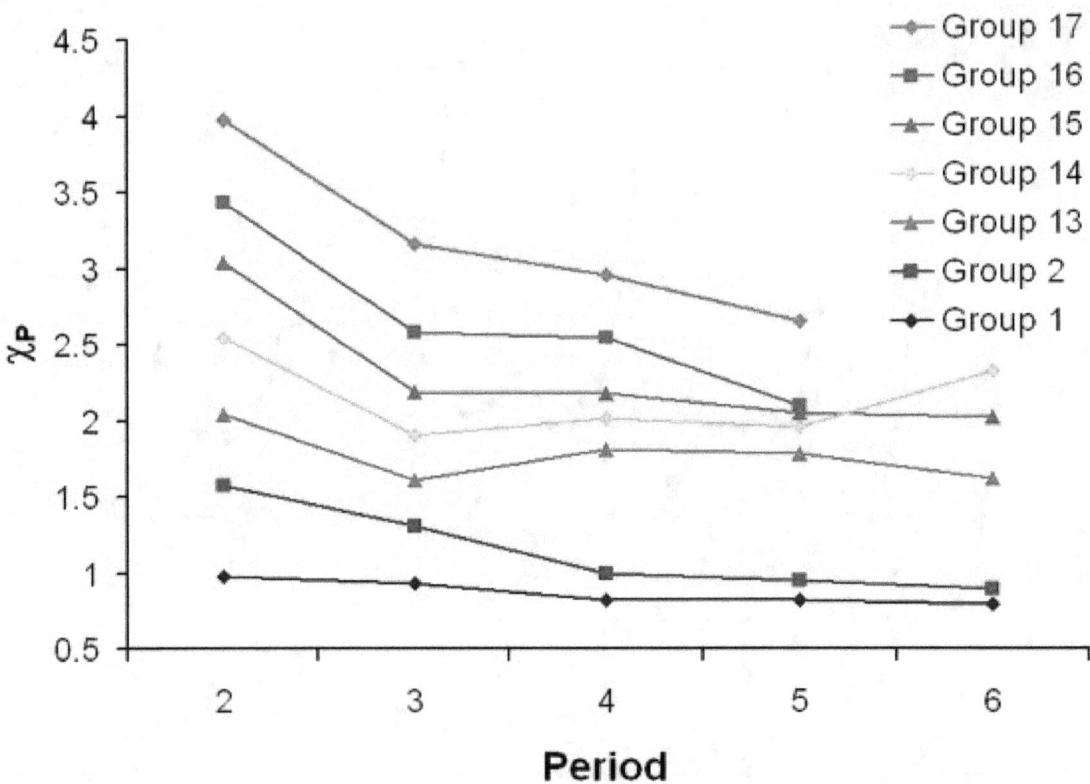

Graph showing increasing electronegativity with growing number of selected groups

18.3.6   Metallic character

The lower the values of ionization energy, electronegativity and electron affinity, the more metallic character the element has. Conversely, nonmetallic character increases with higher values of these properties.[38] Given the periodic trends of these three properties, metallic character tends to decrease going across a period (or row) and, with some irregularities (mostly) due to poor screening of the nucleus by d and f electrons, and relativistic effects,[39] tends to increase going down a group (or column or family). Thus, the most metallic elements (such as caesium and francium) are found at the bottom left of traditional periodic tables and the most nonmetallic elements (oxygen, fluorine, chlorine) at the top right. The combination of horizontal and vertical trends in metallic character explains the stair-shaped dividing line between metals and nonmetals found on some periodic tables, and the practice of sometimes categorizing several elements adjacent to that line, or elements adjacent to those elements, as metalloids.[40][41]

18.4   History

Main article: History of the periodic table

18.4.1   First systemization attempts

In 1789, Antoine Lavoisier published a list of 33 chemical elements, grouping them into gases, metals, nonmetals, and earths.[42] Chemists spent the following century searching for a more precise classification scheme. In 1829, Johann Wolfgang Döbereiner observed that many of the elements could be grouped into triads based on their chemical properties.

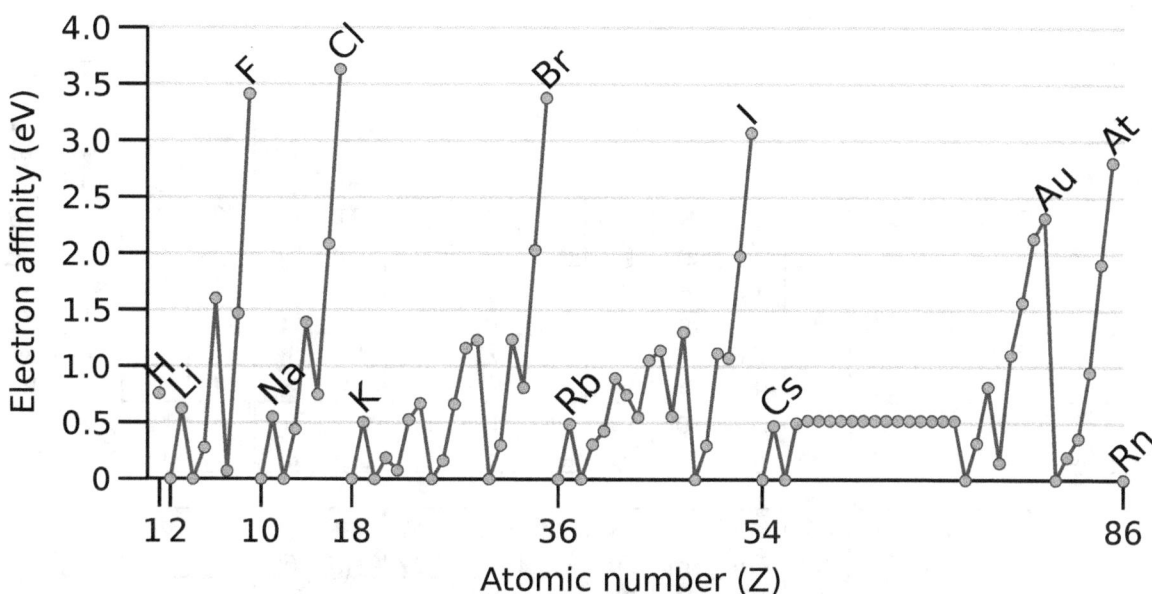

Dependence of electron affinity on atomic number.[34] Values generally increase across each period, culminating with the halogens before decreasing precipitously with the noble gases. Examples of localized peaks seen in hydrogen, the alkali metals and the group 11 elements are caused by a tendency to complete the s-shell (with the 6s shell of gold being further stabilized by relativistic effects and the presence of a filled 4f sub shell). Examples of localized troughs seen in the alkaline earth metals, and nitrogen, phosphorus, manganese and rhenium are caused by filled s-shells, or half-filled p- or d-shells.[35]

Lithium, sodium, and potassium, for example, were grouped together in a triad as soft, reactive metals. Döbereiner also observed that, when arranged by atomic weight, the second member of each triad was roughly the average of the first and the third;[43] this became known as the Law of Triads.[44] German chemist Leopold Gmelin worked with this system, and by 1843 he had identified ten triads, three groups of four, and one group of five. Jean-Baptiste Dumas published work in 1857 describing relationships between various groups of metals. Although various chemists were able to identify relationships between small groups of elements, they had yet to build one scheme that encompassed them all.[43]

In 1858, German chemist August Kekulé observed that carbon often has four other atoms bonded to it. Methane, for example, has one carbon atom and four hydrogen atoms. This concept eventually became known as valency; different elements bond with different numbers of atoms.[45]

In 1862, Alexandre-Emile Béguyer de Chancourtois, a French geologist, published an early form of periodic table, which he called the telluric helix or screw. He was the first person to notice the periodicity of the elements. With the elements arranged in a spiral on a cylinder by order of increasing atomic weight, de Chancourtois showed that elements with similar properties seemed to occur at regular intervals. His chart included some ions and compounds in addition to elements. His paper also used geological rather than chemical terms and did not include a diagram; as a result, it received little attention until the work of Dmitri Mendeleev.[46]

In 1864, Julius Lothar Meyer, a German chemist, published a table with 44 elements arranged by valency. The table showed that elements with similar properties often shared the same valency.[47] Concurrently, William Odling (an English chemist) published an arrangement of 57 elements, ordered on the basis of their atomic weights. With some irregularities and gaps, he noticed what appeared to be a periodicity of atomic weights among the elements and that this accorded with 'their usually received groupings.'[48] Odling alluded to the idea of a periodic law but did not pursue it.[49] He subsequently proposed (in 1870) a valence-based classification of the elements.[50]

English chemist John Newlands produced a series of papers from 1863 to 1866 noting that when the elements were listed in order of increasing atomic weight, similar physical and chemical properties recurred at intervals of eight; he likened such periodicity to the octaves of music.[51][52] This so termed Law of Octaves, however, was ridiculed by Newlands' contemporaries, and the Chemical Society refused to publish his work.[53] Newlands was nonetheless able to draft a table

1 H																	2 He
3 Li	4 Be											5 B	6 C	7 N	8 O	9 F	10 Ne
11 Na	12 Mg											13 Al	14 Si	15 P	16 S	17 Cl	18 Ar
19 K	20 Ca	21 Sc	22 Ti	23 V	24 Cr	25 Mn	26 Fe	27 Co	28 Ni	29 Cu	30 Zn	31 Ga	32 Ge	33 As	34 Se	35 Br	36 Kr
37 Rb	38 Sr	39 Y	40 Zr	41 Nb	42 Mo	43 Tc	44 Ru	45 Rh	46 Pd	47 Ag	48 Cd	49 In	50 Sn	51 Sb	52 Te	53 I	54 Xe
55 Cs	56 Ba	57 -71	72 Hf	73 Ta	74 W	75 Re	76 Os	77 Ir	78 Pt	79 Au	80 Hg	81 Tl	82 Pb	83 Bi	84 Po	85 At	86 Rn
87 Fr	88 Ra	89 -103	104 Rf	105 Db	106 Sg	107 Bh	108 Hs	109 Mt	110 Ds	111 Rg	112 Cn	113 Uut	114 Fl	115 Uup	116 Lv	117 Uus	118 Uuo

57 La	58 Ce	59 Pr	60 Nd	61 Pm	62 Sm	63 Eu	64 Gd	65 Tb	66 Dy	67 Ho	68 Er	69 Tm	70 Yb	71 Lu
89 Ac	90 Th	91 Pa	92 U	93 Np	94 Pu	95 Am	96 Cm	97 Bk	98 Cf	99 Es	100 Fm	101 Md	102 No	103 Lr

Known in antiquity

also known when (akw) Levoisier published his list of elements (1789)

akw Mendeleev published his periodic table (1869)

akw Deming published his periodic table (1923)

akw Seaborg published his periodic table (1945)

also known (ak) up to 2000

ak to 2012

The discovery of the elements mapped to significant periodic table development dates (pre-, per- and post-)

No.	No.	No.	No.	No.	No.	No.	No.
H 1	F 8	Cl 15	Co & Ni 22	Br 29	Pd 36	I 42	Pt & Ir 50
Li 2	Na 9	K 16	Cu 23	Rb 30	Ag 37	Cs 44	Os 51
G 3	Mg 10	Ca 17	Zn 24	Sr 31	Cd 38	Ba & V 45	Hg 52
Bo 4	Al 11	Cr 19	Y 25	Ce & La 33	U 40	Ta 46	Tl 53
C 5	Si 12	Ti 18	In 26	Zr 32	Sn 39	W 47	Pb 54
N 6	P 13	Mn 20	As 27	Di & Mo 34	Nb 41	Bi 48	Bi 55
O 7	S 14	Fe 21	Se 28	Ro & Ru 35	Tc 43	Au 49	Th 56

Newlands's periodic table, as presented to the Chemical Society in 1866, and based on the law of octaves

of the elements and used it to predict the existence of missing elements, such as germanium.[54] The Chemical Society only acknowledged the significance of his discoveries five years after they credited Mendeleev.[55]

In 1867, Gustavus Hinrichs, a Danish born academic chemist based in America, published a spiral periodic system based on atomic spectra and weights, and chemical similarities. His work was regarded as idiosyncratic, ostentatious and labyrinthine and this may have militated against its recognition and acceptance.[56][57]

18.4.2 Mendeleev's table

Russian chemistry professor Dmitri Mendeleev and German chemist Julius Lothar Meyer independently published their periodic tables in 1869 and 1870, respectively.[58] Mendeleev's table was his first published version; that of Meyer was an

Dmitri Mendeleev

expanded version of his (Meyer's) table of 1864.[59] They both constructed their tables by listing the elements in rows or columns in order of atomic weight and starting a new row or column when the characteristics of the elements began to repeat.[60]

The recognition and acceptance afforded to Mendeleev's table came from two decisions he made. The first was to leave

ОПЫТЪ СИСТЕМЫ ЭЛЕМЕНТОВЪ,

ОСНОВАННОЙ НА ИХЪ АТОМНОМЪ ВѢСѢ И ХИМИЧЕСКОМЪ СХОДСТВѢ.

			Ti=50	Zr=90	?=180.
			V=51	Nb=94	Ta=182.
			Cr=52	Mo=96	W=186.
			Mn=55	Rh=104,4	Pt=197,1.
			Fe=56	Ru=104,4	Ir=198.
			Ni=Co=59	Pd=106,6	Os=199.
H=1			Cu=63,4	Ag=108	Hg=200.
	Be= 9,4	Mg=24	Zn=65,2	Cd=112	
	B=11	Al=27,3	?=68	Ur=116	Au=197?
	C=12	Si=28	?=70	Sn=118	
	N=14	P=31	As=75	Sb=122	Bi=210?
	O=16	S=32	Se=79,4	Te=128?	
	F=19	Cl=35,5	Br=80	I=127	
Li=7	Na=23	K=39	Rb=85,4	Cs=133	Tl=204.
		Ca=40	Sr=87,6	Ba=137	Pb=207.
		?=45	Ce=92		
		?Er=56	La=94		
		?Yt=60	Di=95		
		?In=75,6	Th=118?		

Д. Менделѣевъ

A version of Mendeleev's 1869 periodic table: An experiment on a system of elements. Based on their atomic weights and chemical similarities. *This early arrangement presents the periods vertically, and the groups horizontally.*

gaps in the table when it seemed that the corresponding element had not yet been discovered.[61] Mendeleev was not the first chemist to do so, but he was the first to be recognized as using the trends in his periodic table to predict the properties of those missing elements, such as gallium and germanium.[62] The second decision was to occasionally ignore the order suggested by the atomic weights and switch adjacent elements, such as tellurium and iodine, to better classify them into chemical families. Later in 1913, Henry Moseley determined experimental values of the nuclear charge or atomic number of each element, and showed that Mendeleev's ordering actually corresponds to the order of increasing atomic number.[63]

The significance of atomic numbers to the organization of the periodic table was not appreciated until the existence and properties of protons and neutrons became understood. Mendeleev's periodic tables used atomic weight instead of atomic number to organize the elements, information determinable to fair precision in his time. Atomic weight worked well enough in most cases to (as noted) give a presentation that was able to predict the properties of missing elements more accurately than any other method then known. Substitution of atomic numbers, once understood, gave a definitive, integer-based sequence for the elements, and Moseley predicted that the only missing elements (in 1913) between aluminum (Z=13) and gold (Z=79) (in 1913) were Z = 43, 61, 72 and 75, which were all later discovered. The sequence of atomic numbers is still used today even as new synthetic elements are being produced and studied.[64]

18.4.3 Second version and further development

Reihen	Gruppe I. — R²O	Gruppe II. — RO	Gruppe III. — R²O³	Gruppe IV. RH⁴ RO²	Gruppe V. RH³ R²O⁵	Gruppe VI. RH² RO³	Gruppe VII. RH R²O⁷	Gruppe VIII. — RO⁴
1	H=1							
2	Li=7	Be=9.4	B=11	C=12	N=14	O=16	F=19	
3	Na=23	Mg=24	Al=27.3	Si=28	P=31	S=32	Cl=35.5	
4	K=39	Ca=40	—=44	Ti=48	V=51	Cr=52	Mn=55	Fe=56, Co=59, Ni=59, Cu=63.
5	(Cu=63)	Zn=65	—=68	—=72	As=75	Se=78	Br=80	
6	Rb=85	Sr=87	?Yt=88	Zr=90	Nb=94	Mo=96	—=100	Ru=104, Rh=104, Pd=106, Ag=108.
7	(Ag=108)	Cd=112	In=113	Sn=118	Sb=122	Te=125	J=127	
8	Cs=133	Ba=137	?Di=138	?Ce=140	—	—	—	— — — —
9	(—)	—	—	—	—	—	—	
10	—	—	?Er=178	?La=180	Ta=182	W=184	—	Os=195, Ir=197, Pt=198, Au=199.
11	(Au=199)	Hg=200	Tl=204	Pb=207	Bi=208	—	—	
12	—	—	—	Th=231	—	U=240	—	— — — —

Mendeleev's 1871 periodic table with eight groups of elements. Dashes represented elements unknown in 1871.

In 1871, Mendeleev published his periodic table in a new form, with groups of similar elements arranged in columns rather than in rows, and those columns numbered I to VIII corresponding with the element's oxidation state. He also gave detailed predictions for the properties of elements he had earlier noted were missing, but should exist.[65] These gaps were subsequently filled as chemists discovered additional naturally occurring elements.[66] It is often stated that the last naturally occurring element to be discovered was francium (referred to by Mendeleev as *eka-caesium*) in 1939.[67] However, plutonium, produced synthetically in 1940, was identified in trace quantities as a naturally occurring primordial element in 1971,[68] and by 2011 it was known that all the elements up to californium can occur naturally as trace amounts in uranium ores by neutron capture and beta decay.[1]

The popular[69] periodic table layout, also known as the common or standard form (as shown at various other points in this article), is attributable to Horace Groves Deming. In 1923, Deming, an American chemist, published short (Mendeleev style) and medium (18-column) form periodic tables.[70][n 5] Merck and Company prepared a handout form of Deming's 18-column medium table, in 1928, which was widely circulated in American schools. By the 1930s Deming's table was appearing in handbooks and encyclopaedias of chemistry. It was also distributed for many years by the Sargent-Welch Scientific Company.[71][72][73]

With the development of modern quantum mechanical theories of electron configurations within atoms, it became apparent that each period (row) in the table corresponded to the filling of a quantum shell of electrons. Larger atoms have more electron sub-shells, so later tables have required progressively longer periods.[74]

In 1945, Glenn Seaborg, an American scientist, made the suggestion that the actinide elements, like the lanthanides, were filling an f sub-level. Before this time the actinides were thought to be forming a fourth d-block row. Seaborg's colleagues advised him not to publish such a radical suggestion as it would most likely ruin his career. As Seaborg considered he did

Period	Series	Ia	Ib	IIa	IIb	IIIa	IIIb	IVa	IVb	Va	Vb	VIa	VIb	VIIa	VIIb	VIII (a)	VIIIb
1	I	1 H															2 He
2	II	3 Li		4 Be		5 B		6 C		7 N		8 O		9 F			10 Ne
3	III	11 Na		12 Mg		13 Al		14 Si		15 P		16 S		17 Cl			18 Ar
4	IV	19 K		20 Ca			21 Sc		22 Ti		23 V		24 Cr		25 Mn	26 Fe, 27 Co, 28 Ni	
	V		29 Cu		30 Zn	31 Ga		32 Ge		33 As		34 Se		35 Br			36 Kr
5	VI	37 Rb		38 Sr			39 Y		40 Zr		41 Nb		42 Mo		43 Tc	44 Ru, 45 Rh, 46 Pd	
	VII		47 Ag		48 Cd	49 In		50 Sn		51 Sb		52 Te		53 I			54 Xe
6	VIII	55 Cs		56 Ba			57–71		72 Hf		73 Ta		74 W		75 Re	76 Os, 77 Ir, 78 Pt	
	IX		79 Au		80 Hg	81 Tl		82 Pb		83 Bi		84 Po		85 At			86 Rn
7	X	87 Fr		88 Ra			89–103		104 Rf		105 Db		106 Sg		107 Bh	108 Hs, 109 Mt, 110 Ds	
	XI		111 Rg		112 Cn	113 Uut		114 Fl		115 Uup		116 Lv		117 Uus	118 Uuo	

	I	II	III	IV	V	VI	VII	VIII
Higher oxides	R_2O	RO	R_2O_3	RO_2	R_2O_5	RO_3	R_2O_7	RO_4
Volatile hydrogen compounds			$[(RH_3)_x]$	RH_4	RH_3	RH_2	RH	

57 La	58 Ce	59 Pr	60 Nd	61 Pm	62 Sm	63 Eu	64 Gd	65 Tb	66 Dy	67 Ho	68 Er	69 Tm	70 Yb	71 Lu
89 Ac	90 Th	91 Pa	92 U	93 Np	94 Pu	95 Am	96 Cm	97 Bk	98 Cf	99 Es	100 Fm	101 Md	102 No	103 Lr

Eight-column form of periodic table, updated with all elements discovered to 2015

not then have a career to bring into disrepute, he published anyway. Seaborg's suggestion was found to be correct and he subsequently went on to win the 1951 Nobel Prize in chemistry for his work in synthesizing actinide elements.[75][76][n 6]

Although minute quantities of some transuranic elements occur naturally,[1] they were all first discovered in laboratories. Their production has expanded the periodic table significantly, the first of these being neptunium, synthesized in 1939.[77] Because many of the transuranic elements are highly unstable and decay quickly, they are challenging to detect and characterize when produced. There have been controversies concerning the acceptance of competing discovery claims for some elements, requiring independent review to determine which party has priority, and hence naming rights. The most recently accepted and named elements are flerovium (element 114) and livermorium (element 116), both named on 31 May 2012.[78] In 2010, a joint Russia–US collaboration at Dubna, Moscow Oblast, Russia, claimed to have synthesized six atoms of ununseptium (element 117), making it the most recently claimed discovery.[79]

18.5   Alternative structures

Main article: Alternative periodic tables
 There are many periodic tables with structures other than that of the standard form. Within 100 years of the appearance of Mendeleev's table in 1869 it has been estimated that around 700 different periodic table versions were published.[80] As well as numerous rectangular variations, other periodic table formats have included, for example,[n 7] circular, cubic, cylindrical, edificial (building-like), helical, lemniscate, octagonal prismatic, pyramidal, separated, spherical, spiral, and triangular forms. Such alternatives are often developed to highlight or emphasize chemical or physical properties of the elements that are not as apparent in traditional periodic tables.[80]

Glenn T. Seaborg who, in 1945, suggested a new periodic table showing the actinides as belonging to a second f-block series

A popular[81] alternative structure is that of Theodor Benfey (1960). The elements are arranged in a continuous spiral, with hydrogen at the center and the transition metals, lanthanides, and actinides occupying peninsulas.[82]

Most periodic tables are two-dimensional;[1] however, three-dimensional tables are known to as far back as at least 1862 (pre-dating Mendeleev's two-dimensional table of 1869). More recent examples include Courtines' Periodic Classifica-

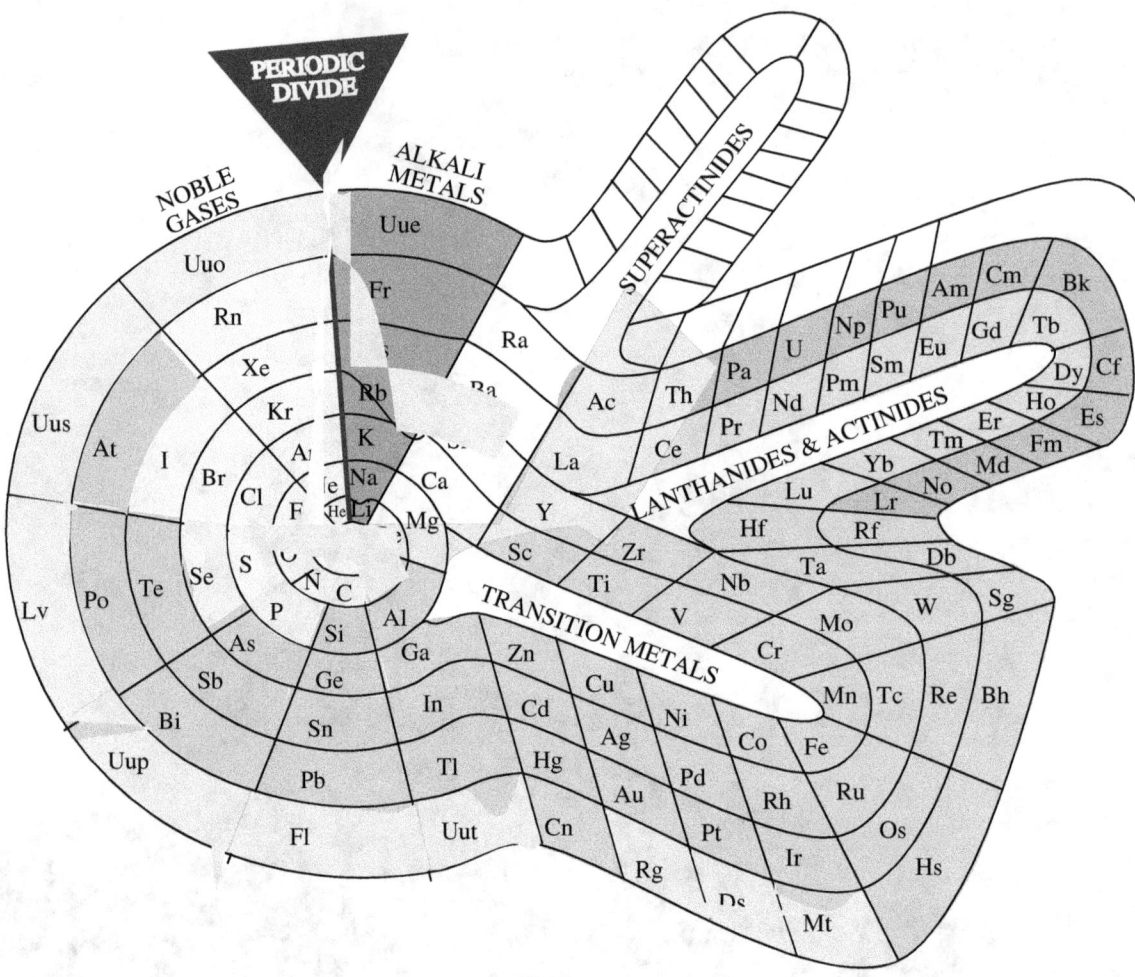

Theodor Benfey's spiral periodic table

tion (1925),[83] Wringley's Lamina System (1949),[84] Giguère's Periodic helix (1965)[85][n 8] and Dufour's Periodic Tree (1996).[86] Going one better, Stowe's Physicist's Periodic Table (1989)[87] has been described as being four-dimensional (having three spatial dimensions and one colour dimension).[88]

The various forms of periodic tables can be thought of as lying on a chemistry–physics continuum.[89] Towards the chemistry end of the continuum can be found, as an example, Rayner-Canham's 'unruly'[90] Inorganic Chemist's Periodic Table (2002),[91] which emphasizes trends and patterns, and unusual chemical relationships and properties. Near the physics end of the continuum is Janet's Left-Step Periodic Table (1928). This has a structure that shows a closer connection to the order of electron-shell filling and, by association, quantum mechanics.[92] Somewhere in the middle of the continuum is the ubiquitous common or standard form of periodic table. This is regarded as better expressing empirical trends in physical state, electrical and thermal conductivity, and oxidation numbers, and other properties easily inferred from traditional techniques of the chemical laboratory.[93]

18.6   Open questions and controversies

18.6.1 Elements with unknown chemical properties

Although all elements up to ununoctium have been discovered, of the elements above hassium (element 108), only copernicium (element 112) and flerovium (element 114) have known chemical properties. The other elements may behave differently from what would be predicted by extrapolation, due to relativistic effects; for example, flerovium has been predicted to possibly exhibit some noble-gas-like properties, even though it is currently placed in the carbon group.[94] More recent experiments have suggested, however, that flerovium behaves chemically like lead, as expected from its periodic table position.[95]

18.6.2 Further periodic table extensions

Main article: Extended periodic table

It is unclear whether new elements will continue the pattern of the current periodic table as period 8, or require further adaptations or adjustments. Seaborg expected the eighth period to follow the previously established pattern exactly, so that it would include a two-element s-block for elements 119 and 120, a new g-block for the next 18 elements, and 30 additional elements continuing the current f-, d-, and p-blocks.[97] More recently, physicists such as Pekka Pyykkö have theorized that these additional elements do not follow the Madelung rule, which predicts how electron shells are filled and thus affects the appearance of the present periodic table.[98]

18.6.3 Element with the highest possible atomic number

The number of possible elements is not known. A very early suggestion made by Elliot Adams in 1911, and based on the arrangement of elements in each horizontal periodic table row, was that elements of atomic weight greater than 256± (which would equate to between elements 99 and 100 in modern-day terms) did not exist.[99] A higher—more recent—estimate is that the periodic table may end soon after the island of stability,[100] which is expected to center around element 126, as the extension of the periodic and nuclides tables is restricted by proton and neutron drip lines.[101] Other predictions of an end to the periodic table include at element 128 by John Emsley,[1] at element 137 by Richard Feynman,[102] and at element 155 by Albert Khazan.[1][n 9]

Bohr model

The Bohr model exhibits difficulty for atoms with atomic number greater than 137, as any element with an atomic number greater than 137 would require 1s electrons to be traveling faster than c, the speed of light.[103] Hence the non-relativistic Bohr model is inaccurate when applied to such an element.

Relativistic Dirac equation

The relativistic Dirac equation has problems for elements with more than 137 protons. For such elements, the wave function of the Dirac ground state is oscillatory rather than bound, and there is no gap between the positive and negative energy spectra, as in the Klein paradox.[104] More accurate calculations taking into account the effects of the finite size of the nucleus indicate that the binding energy first exceeds the limit for elements with more than 173 protons. For heavier elements, if the innermost orbital (1s) is not filled, the electric field of the nucleus will pull an electron out of the vacuum, resulting in the spontaneous emission of a positron;[105] however, this does not happen if the innermost orbital is filled, so that element 173 is not necessarily the end of the periodic table.[106]

18.6.4 Placement of hydrogen and helium

Hydrogen and helium are often placed in different places than their electron configurations would indicate; hydrogen is usually placed above lithium, in accordance with its electron configuration, but is sometimes placed above fluorine,[107] or even carbon,[107] as it also behaves somewhat similarly to them. Hydrogen is also sometimes placed in its own group, as it does not behave similarly enough to any element to be placed in a group with another.[108] Helium is almost always placed above neon, as they are very similar chemically, although it is occasionally placed above beryllium on account of having a comparable electron shell configuration (helium: $1s^2$; beryllium: $[He]\,2s^2$).[19]

18.6.5   Groups included in the transition metals

The definition of a transition metal, as given by IUPAC, is an element whose atom has an incomplete d sub-shell, or which can give rise to cations with an incomplete d sub-shell.[109] By this definition all of the elements in groups 3–11 are transition metals. The IUPAC definition therefore excludes group 12, comprising zinc, cadmium and mercury, from the transition metals category.

Some chemists treat the categories "d-block elements" and "transition metals" interchangeably, thereby including groups 3–12 among the transition metals. In this instance the group 12 elements are treated as a special case of transition metal in which the d electrons are not ordinarily involved in chemical bonding. The recent discovery that mercury can use its d electrons in the formation of mercury(IV) fluoride (HgF_4) has prompted some commentators to suggest that mercury can be regarded as a transition metal.[110] Other commentators, such as Jensen,[111] have argued that the formation of a compound like HgF_4 can occur only under highly abnormal conditions. As such, mercury could not be regarded as a transition metal by any reasonable interpretation of the ordinary meaning of the term.[111]

Still other chemists further exclude the group 3 elements from the definition of a transition metal. They do so on the basis that the group 3 elements do not form any ions having a partially occupied d shell and do not therefore exhibit any properties characteristic of transition metal chemistry.[112] In this case, only groups 4–11 are regarded as transition metals.

18.6.6   Period 6 and 7 elements in group 3

Although scandium and yttrium are always the first two elements in group 3 the identity of the next two elements is not settled. They are either lanthanum and actinium; or lutetium and lawrencium. There are strong chemical and physical arguments supporting the latter arrangement[113][114] but not all authors have been convinced.[115] Most working chemists are not aware there is any controversy.[116]

Lanthanum and actinium are traditionally depicted as the remaining group 3 members.[117][118] It has been suggested that this layout originated in the 1940s, with the appearance of periodic tables relying on the electron configurations of the elements and the notion of the differentiating electron. The configurations of caesium, barium and lanthanum are $[Xe]6s^1$, $[Xe]6s^2$ and $[Xe]5d^16s^2$. Lanthanum thus has a $5d$ differentiating electron and this establishes "it in group 3 as the first member of the d-block for period 6."[111] A consistent set of electron configurations is then seen in group 3: scandium $[Ar]3d^14s^2$, yttrium $[Kr]4d^15s^2$ and lanthanum $[Xe]5d^16s^2$. Still in period 6, ytterbium was assigned an electron configuration of $[Xe]4f^{13}5d^16s^2$ and lutetium $[Xe]4f^{14}5d^16s^2$, "resulting in a $4f$ differentiating electron for lutetium and firmly establishing it as the last member of the f-block for period 6."[111]

In other tables, lutetium and lawrencium are the remaining group 3 members.[119] It has been known since the early 20th century that, "yttrium and (to a lesser degree) scandium are closer in their chemical properties to lutetium and the other heavy rare earths [i.e. lanthanides] than they are to lanthanum."[111] Accordingly, lutetium rather than lanthanum was assigned to group 3 by some chemists in the 1920s and 30s. Later spectroscopic work found that the electron configuration of ytterbium was in fact $[Xe]4f^{14}6s^2$. This meant that ytterbium and lutetium—the latter with $[Xe]4f^{14}5d^16s^2$—both had 14 f electrons, "resulting in a d rather than an f differentiating electron" for lutetium and making it an "equally valid candidate" with $[Xe]5d^16s^2$ lanthanum, for the group 3 periodic table position below yttrium.[111] Several physicists in the 1950s and 60s opted for lutetium, in light of a comparison of several of its physical properties with those of lanthanum.[111] This arrangement, in which lanthanum is the first member of the f-block, is disputed by some authors since lanthanum lacks any f electrons. However, it has been argued that this is not valid concern given other periodic table anomalies— thorium, for example, has no f electrons yet is part of the f-block.[120] As to lawrencium, its electron configuration was confirmed in 2015 as $[Rn]5f^{14}7s^27p^1$. Such a configuration represents another periodic table anomaly, regardless of whether lawrencium is located in the f-block or the d-block, as the only potentially applicable p-block position has been reserved for ununtrium with its predicted electron configuration of $[Rn]5f^{14}6d^{10}7s^27p^1$.[121]

Some tables, including the table on the IUPAC site,[122][n 10] place footnote markers in the two positions below scan-

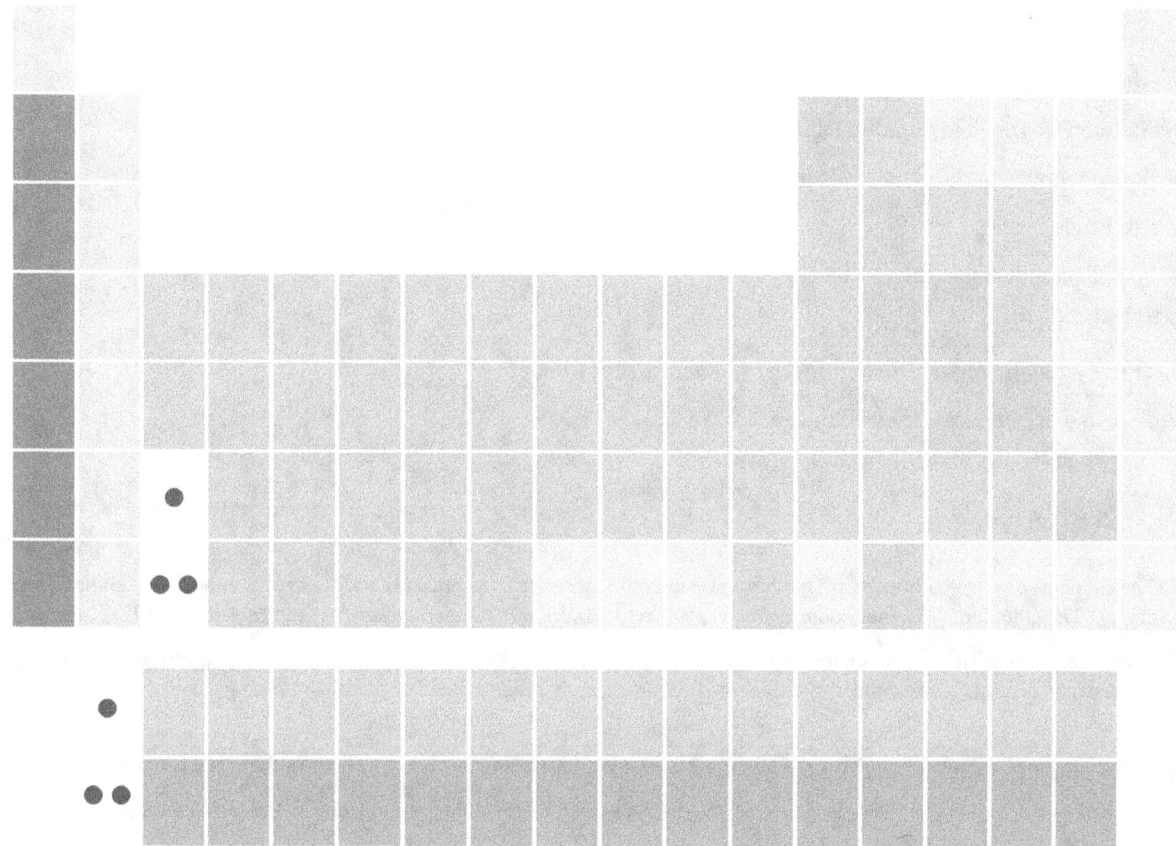

dium and yttrium, and show both lanthanum and lutetium, and actinium and lawrencium as being part of, respectively, the lanthanide series and the actinide series of elements. This arrangement emphasizes similarities in the chemistry of the 15 lanthanide elements (La–Lu) over electron configuration arguments. The actinides are more diverse in their behavior. Most early members show some similarities to transition metals; actinium and the later members are more like lanthanides.[123]

18.6.7   Optimal form

The many different forms of periodic table have prompted the question of whether there is an optimal or definitive form of periodic table. The answer to this question is thought to depend on whether the chemical periodicity seen to occur among the elements has an underlying truth, effectively hard-wired into the universe, or if any such periodicity is instead the product of subjective human interpretation, contingent upon the circumstances, beliefs and predilections of human observers. An objective basis for chemical periodicity would settle the questions about the location of hydrogen and helium, and the composition of group 3. Such an underlying truth, if it exists, is thought to have not yet been discovered. In its absence, the many different forms of periodic table can be regarded as variations on the theme of chemical periodicity, each of which explores and emphasizes different aspects, properties, perspectives and relationships of and among the elements.[n 11] The ubiquity of the standard or medium-long periodic table is thought to be a result of this layout having a good balance of features in terms of ease of construction and size, and its depiction of atomic order and periodic trends.[49][124]

18.7   See also

- Abundance of the chemical elements

- Atomic electron configuration table

- Element collecting

- List of elements

- List of periodic table-related articles

- Table of nuclides

- The Mystery of Matter: Search for the Elements (PBS film)

- Timeline of chemical element discoveries

18.8   Notes

[1] The elements discovered initially by synthesis and later in nature are technetium (Z=43), promethium (61), astatine (85), francium (87), neptunium (93), plutonium (94), americium (95), curium (96), berkelium (97) and californium (98).

[2] There is an inconsistency and some irregularities in this convention. Thus, helium is shown in the p-block but is actually an s-block element, and (for example) the d-subshell in the d-block is actually filled by the time group 11 is reached, rather than group 12.

[3] The noble gases, astatine, francium, and all elements heavier than americium were left out as there is no data for them.

[4] While fluorine is the most electronegative of the elements under the Pauling scale, neon is the most electronegative element under other scales, such as the Allen scale.

[5] An antecedent of Deming's 18-column table may be seen in Adams' 16-column Periodic Table of 1911. Adams omits the rare earths and the 'radioactive elements' (i.e. the actinides) from the main body of his table and instead shows them as being 'careted in only to save space' (rare earths between Ba and eka-Yt; radioactive elements between eka-Te and eka-I). See: Elliot Q. A. (1911). "A modification of the periodic table". *Journal of the American Chemical Society.* **33**(5): 684–688 (687).

[6] A second extra-long periodic table row, to accommodate known and undiscovered elements with an atomic weight greater than bismuth (thorium, protactinium and uranium, for example), had been postulated as far back as 1892. Most investigators, however, considered that these elements were analogues of the third series transition elements, hafnium, tantalum and tungsten. The existence of a second inner transition series, in the form of the actinides, was not accepted until similarities with the electron structures of the lanthanides had been established. See: van Spronsen, J. W. (1969). *The periodic system of chemical elements.* Amsterdam: Elsevier. p. 315–316, ISBN 0-444-40776-6.

[7] See *The Internet database of periodic tables* for depictions of these kinds of variants.

[8] The animated depiction of Giguère's periodic table that is widely available on the internet (including from here) is erroneous, as it does not include hydrogen and helium. Giguère included hydrogen, above lithium, and helium, above beryllium. See: Giguère P.A. (1966). "The "new look" for the periodic system". *Chemistry in Canada* **18** (12): 36–39 (see p. 37).

[9] Karol (2002, p. 63) contends that gravitational effects would become significant when atomic numbers become astronomically large, thereby overcoming other super-massive nuclei instability phenomena, and that neutron stars (with atomic numbers on the order of 10^{21}) can arguably be regarded as representing the heaviest known elements in the universe. See: Karol P. J. (2002). "The Mendeleev–Seaborg periodic table: Through Z = 1138 and beyond". *Journal of Chemical Education* **79** (1): 60–63.

[10] Although this form of the table is sometimes referred to as the "approved" or "official" IUPAC periodic table, "IUPAC has not approved any specific form of the periodic table..." See: Leigh, G. J. (January–February 2009). "Periodic Tables and IUPAC". *Chemistry International* **31** (1).

[11] Scerri, one of the foremost authorities on the history of the periodic table (Sella 2013), favoured the concept of an optimal form of periodic table but has recently changed his mind and now supports the value of a plurality of periodic tables. See: Sella A. (2013). 'An elementary history lesson'. *New Scientist.* 2929, 13 August: 51, accessed 4 September 2013; and Scerri, E. (2013). 'Is there an optimal periodic table and other bigger questions in the philosophy of science.'. 9 August, accessed 4 September 2013.

18.9 References

[1] Emsley, John (2011). *Nature's Building Blocks: An A-Z Guide to the Elements* (New ed.). New York, NY: Oxford University Press. ISBN 978-0-19-960563-7.

[2] Greenwood, pp. 24–27

[3] Gray, p. 6

[4] Koppenol, W. H. (2002). "Naming of New Elements (IUPAC Recommendations 2002)" (PDF). *Pure and Applied Chemistry* **74** (5): 787–791. doi:10.1351/pac200274050787.

[5] Silva, Robert J. (2006). "Fermium, Mendelevium, Nobelium and Lawrencium". In Morss; Edelstein, Norman M.; Fuger, Jean. *The Chemistry of the Actinide and Transactinide Elements* (3rd ed.). Dordrecht, The Netherlands: Springer Science+Business Media. ISBN 1-4020-3555-1.

[6] Gray, p. 11

[7] Scerri 2007, p. 24

[8] Messler, R. W. (2010). *The essence of materials for engineers.* Sudbury, MA: Jones & Bartlett Publishers. p. 32. ISBN 0-7637-7833-8.

[9] Bagnall, K. W. (1967). "Recent advances in actinide and lanthanide chemistry". In Fields, P.R.; Moeller, T. *Advances in chemistry, Lanthanide/Actinide chemistry.* Advances in Chemistry **71**. American Chemical Society. pp. 1–12. doi:10.1021/ba-1967-0071. ISBN 0-8412-0072-6.

[10] Day, M. C., Jr.; Selbin, J. (1969). *Theoretical inorganic chemistry* (2nd ed.). New York: Nostrand-Rienhold Book Corporation. p. 103. ISBN 0-7637-7833-8.

[11] Holman, J.; Hill, G. C. (2000). *Chemistry in context* (5th ed.). Walton-on-Thames: Nelson Thornes. p. 40. ISBN 0-17-448276-0.

[12] Leigh, G. J. (1990). *Nomenclature of Inorganic Chemistry: Recommendations 1990.* Blackwell Science. ISBN 0-632-02494-1.

[13] Fluck, E. (1988). "New Notations in the Periodic Table" (PDF). *Pure Appl. Chem.* (IUPAC) **60** (3): 431–436. doi:10.135. Retrieved 24 March 2012.

[14] Moore, p. 111

[15] Greenwood, p. 30

[16] Stoker, Stephen H. (2007). *General, organic, and biological chemistry.* New York: Houghton Mifflin. p. 68. ISBN 978-0-618-73063-6. OCLC 52445586.

[17] Mascetta, Joseph (2003). *Chemistry The Easy Way* (4th ed.). New York: Hauppauge. p. 50. ISBN 978-0-7641-1978-1. OCLC 52047235.

[18] Kotz, John; Treichel, Paul; Townsend, John (2009). *Chemistry and Chemical Reactivity, Volume 2* (7th ed.). Belmont: Thomson Brooks/Cole. p. 324. ISBN 978-0-495-38712-1. OCLC 220756597.

[19] Gray, p. 12

[20] Jones, Chris (2002). *d- and f-block chemistry.* New York: J. Wiley & Sons. p. 2. ISBN 978-0-471-22476-1. OCLC 300468713.

[21] Silberberg, M. S. (2006). *Chemistry: The molecular nature of matter and change* (4th ed.). New York: McGraw-Hill. p. 536. ISBN 0-07-111658-3.

[22] Manson, S. S.; Halford, G. R. (2006). *Fatigue and durability of structural materials.* Materials Park, Ohio: ASM International. p. 376. ISBN 0-87170-825-6.

[23] Bullinger, Hans-Jörg (2009). *Technology guide: Principles, applications, trends.* Berlin: Springer-Verlag. p. 8. ISBN 978-3-540-88545-0.

[24] Jones, B. W. (2010). *Pluto: Sentinel of the outer solar system*. Cambridge: Cambridge University Press. pp. 169–71. ISBN 978-0-521-19436-5.

[25] Hinrichs, G. D. (1869). "On the classification and the atomic weights of the so-called chemical elements, with particular reference to Stas's determinations". *Proceedings of the American Association for the Advancement of Science* **18** (5): 112–124.

[26] Myers, R. (2003). *The basics of chemistry*. Westport, CT: Greenwood Publishing Group. pp. 61–67. ISBN 0-313-31664-3.

[27] Chang, Raymond (2002). *Chemistry* (7 ed.). New York: McGraw-Hill. pp. 289–310; 340–42. ISBN 0-07-112072-6.

[28] Greenwood, p. 27

[29] Jolly, W. L. (1991). *Modern Inorganic Chemistry* (2nd ed.). McGraw-Hill. p. 22. ISBN 978-0-07-112651-9.

[30] Greenwood, p. 28

[31] IUPAC, *Compendium of Chemical Terminology*, 2nd ed. (the "Gold Book") (1997). Online corrected version: (2006–) "Electronegativity".

[32] Pauling, L. (1932). "The Nature of the Chemical Bond. IV. The Energy of Single Bonds and the Relative Electronegativity of Atoms". *Journal of the American Chemical Society* **54** (9): 3570–3582. doi:10.1021/ja01348a011.

[33] Allred, A. L. (1960). "Electronegativity values from thermochemical data". *Journal of Inorganic and Nuclear Chemistry* (Northwestern University) **17** (3–4): 215–221. doi:10.1016/0022-1902(61)80142-5. Retrieved 11 June 2012.

[34] Huheey, Keiter & Keiter, p. 42

[35] Siekierski, Slawomir; Burgess, John (2002). *Concise chemistry of the elements*. Chichester: Horwood Publishing. pp. 35–36. ISBN 1-898563-71-3.

[36] Chang, pp. 307–309

[37] Huheey, Keiter & Keiter, pp. 42, 880–81

[38] Yoder, C. H.; Suydam, F. H.; Snavely, F. A. (1975). *Chemistry* (2nd ed.). Harcourt Brace Jovanovich. p. 58. ISBN 0-15-506465-7.

[39] Huheey, Keiter & Keiter, pp. 880–85

[40] Sacks, O (2009). *Uncle Tungsten: Memories of a chemical boyhood*. New York: Alfred A. Knopf. pp. 191, 194. ISBN 0-375-70404-3.

[41] Gray, p. 9

[42] Siegfried, Robert (2002). *From elements to atoms a history of chemical composition*. Philadelphia, Pennsylvania: Library of Congress Cataloging-in-Publication Data. p. 92. ISBN 0-87169-924-9.

[43] Ball, p. 100

[44] Horvitz, Leslie (2002). *Eureka!: Scientific Breakthroughs That Changed The World*. New York: John Wiley. p. 43. ISBN 978-0-471-23341-1. OCLC 50766822.

[45] van Spronsen, J. W. (1969). *The periodic system of chemical elements*. Amsterdam: Elsevier. p. 19. ISBN 0-444-40776-6.

[46] "Alexandre-Emile Bélguier de Chancourtois (1820-1886)" (in French). Annales des Mines history page. Retrieved 18 September 2014.

[47] Venable, pp. 85–86; 97

[48] Odling, W. (2002). "On the proportional numbers of the elements". *Quarterly Journal of Science* **1**: 642–648 (643).

[49] Scerri, Eric R. (2011). *The periodic table: A very short introduction*. Oxford: Oxford University Press. ISBN 978-0-19-958249-5.

[50] Kaji, M. (2004). "Discovery of the periodic law: Mendeleev and other researchers on element classification in the 1860s". In Rouvray, D. H.; King, R. Bruce. *The periodic table: Into the 21st Century*. Research Studies Press. pp. 91–122 (95). ISBN 0-86380-292-3.

[51] Newlands, John A. R. (20 August 1864). "On Relations Among the Equivalents". *Chemical News* **10**: 94–95.

[52] Newlands, John A. R. (18 August 1865). "On the Law of Octaves". *Chemical News* **12**: 83.

[53] Bryson, Bill (2004). *A Short History of Nearly Everything*. Black Swan. pp. 141–142. ISBN 978-0-552-15174-0.

[54] Scerri 2007, p. 306

[55] Brock, W. H.; Knight, D. M. (1965). "The Atomic Debates: 'Memorable and Interesting Evenings in the Life of the Chemical Society'". *Isis* (The University of Chicago Press) **56** (1): 5–25. doi:10.1086/349922.

[56] Scerri 2007, pp. 87, 92

[57] Kauffman, George B. (March 1969). "American forerunners of the periodic law". *Journal of Chemical Education* **46** (3): 128–135 (132). Bibcode:1969JChEd..46..128K. doi:10.1021/ed046p128.

[58] Mendelejew, Dimitri (1869). "Über die Beziehungen der Eigenschaften zu den Atomgewichten der Elemente". *Zeitschrift für Chemie* (in German): 405–406.

[59] Venable, pp. 96–97; 100–102

[60] Ball, pp. 100–102

[61] Pullman, Bernard (1998). *The Atom in the History of Human Thought*. Translated by Axel Reisinger. Oxford University Press. p. 227. ISBN 0-19-515040-6.

[62] Ball, p. 105

[63] Atkins, P. W. (1995). *The Periodic Kingdom*. HarperCollins Publishers, Inc. p. 87. ISBN 0-465-07265-8.

[64] Samanta, C.; Chowdhury, P. Roy; Basu, D.N. (2007). "Predictions of alpha decay half lives of heavy and superheavy elements". *Nucl. Phys. A* **789**: 142–154. arXiv:nucl-th/0703086. Bibcode:2007NuPhA.789..142S. doi:10.1016/j.nuclphysa.2007.04.001.

[65] Scerri 2007, p. 112

[66] Kaji, Masanori (2002). "D.I. Mendeleev's Concept of Chemical Elements and the Principle of Chemistry" (PDF). *Bull. Hist. Chem.* (Tokyo Institute of Technology) **27** (1): 4–16. Retrieved 11 June 2012.

[67] Adloff, Jean-Pierre; Kaufman, George B. (25 September 2005). "Francium (Atomic Number 87), the Last Discovered Natural Element". The Chemical Educator. Retrieved 26 March 2007.

[68] Hoffman, D. C.; Lawrence, F. O.; Mewherter, J. L.; Rourke, F. M. (1971). "Detection of Plutonium-244 in Nature". *Nature* **234** (5325): 132–134. Bibcode:1971Natur.234..132H. doi:10.1038/234132a0.

[69] Gray, p. 12

[70] Deming, Horace G (1923). *General chemistry: An elementary survey*. New York: J. Wiley & Sons. pp. 160, 165.

[71] Abraham, M; Coshow, D; Fix, W. *Periodicity:A source book module, version 1.0* (PDF). New York: Chemsource, Inc. p. 3.

[72] Emsley, J (7 March 1985). "Mendeleyev's dream table". *New Scientist*: 32–36(36).

[73] Fluck, E (1988). "New notations in the period table". *Pure & Applied Chemistry* **60** (3): 431–436 (432). doi:10.1351/pa.

[74] Ball, p. 111

[75] Scerri 2007, pp. 270–71

[76] Masterton, William L.; Hurley, Cecile N.; Neth, Edward J. *Chemistry: Principles and reactions* (7th ed.). Belmont, CA: Brooks/Cole Cengage Learning. p. 173. ISBN 1-111-42710-0.

[77] Ball, p. 123

[78] Barber, Robert C.; Karol, Paul J; Nakahara, Hiromichi; Vardaci, Emanuele; Vogt, Erich W. (2011). "Discovery of the elements with atomic numbers greater than or equal to 113 (IUPAC Technical Report)". *Pure Appl. Chem.* **83** (7): 1485. doi:10.1351/PAC-REP-10-05-01.

[79] Эксперимент по синтезу 117-го элемента получает продолжение[Experiment on sythesis of the 117th element is to be continued] (in Russian). JINR. 2012.

[80] Scerri 2007, p. 20

[81] Emsely, J; Sharp, R (21 June 2010). "The periodic table: Top of the charts". *The Independent.*

[82] Seaborg, Glenn (1964). "Plutonium: The Ornery Element". *Chemistry* **37** (6): 14.

[83] Mark R. Leach. "1925 Courtines' Periodic Classification". Retrieved 16 October 2012.

[84] Mark R. Leach. "1949 Wringley's Lamina System". Retrieved 16 October 2012.

[85] Mazurs, E.G. (1974). *Graphical Representations of the Periodic System During One Hundred Years.* Alabama: University of Alabama Press. p. 111. ISBN 978-0-8173-3200-6.

[86] Mark R. Leach. "1996 Dufour's Periodic Tree". Retrieved 16 October 2012.

[87] Mark R. Leach. "1989 Physicist's Periodic Table by Timothy Stowe". Retrieved 16 October 2012.

[88] Bradley, David (20 July 2011). "At last, a definitive periodic table?". *Chem Views Magazine.* doi:10.1002/chemv.201000107.

[89] Scerri 2007, pp. 285–86

[90] Scerri 2007, p. 285

[91] Mark R. Leach. "2002 Inorganic Chemist's Periodic Table". Retrieved 16 October 2012.

[92] Scerri, Eric (2008). "The role of triads in the evolution of the periodic table: Past and present". *Journal of Chemical Education* **85** (4): 585–89 (see p.589). Bibcode:2008JChEd..85..585S. doi:10.1021/ed085p585.

[93] Bent, H. A.; Weinhold, F (2007). "Supporting information: News from the periodic table: An introduction to "Periodicity symbols, tables, and models for higher-order valency and donor–acceptor kinships"". *Journal of Chemical Education* **84** (7): 3–4. doi:10.1021/ed084p1145.

[94] Schändel, Matthias (2003). *The Chemistry of Superheavy Elements.* Dordrecht: Kluwer Academic Publishers. p. 277. ISBN 1-4020-1250-0.

[95] Scerri 2011, pp. 142–143

[96] Fricke, B.; Greiner, W.; Waber, J. T. (1971). "The continuation of the periodic table up to Z = 172. The chemistry of superheavy elements". *Theoretica chimica acta* (Springer-Verlag) **21** (3): 235–260. doi:10.1007/BF01172015. Retrieved 28 November 2012.

[97] Frazier, K. (1978). "Superheavy Elements". *Science News* **113** (15): 236–238. doi:10.2307/3963006. JSTOR 3963006.

[98] Pyykkö, Pekka (2011). "A suggested periodic table up to Z ≤ 172, based on Dirac–Fock calculations on atoms and ions". *Physical Chemistry Chemical Physics* **13** (1): 161–168. Bibcode:2011PCCP...13..161P. doi:10.1039/c0cp01575j. PMID 20967377.

[99] Elliot, Q. A. (1911). "A modification of the periodic table". *Journal of the American Chemical Society* **33** (5): 684–688 (688). doi:10.1021/ja02218a004.

[100] Glenn Seaborg (c. 2006). "transuranium element (chemical element)". Encyclopædia Britannica. Retrieved 16 March 2010.

[101] Cwiok, S.; Heenen, P.-H.; Nazarewicz, W. (2005). "Shape coexistence and triaxiality in the superheavy nuclei". *Nature* **433** (7027): 705–9. Bibcode:2005Natur.433..705C. doi:10.1038/nature03336. PMID 15716943.

[102] Column: The crucible Ball, Philip in Chemistry World, Royal Society of Chemistry, Nov. 2010

[103] Eisberg, R.; Resnick, R. (1985). *Quantum Physics of Atoms, Molecules, Solids, Nuclei and Particles.* Wiley.

[104] Bjorken, J. D.; Drell, S. D. (1964). *Relativistic Quantum Mechanics.* McGraw-Hill.

[105] Greiner, W.; Schramm, S. (2008). "American Journal of Physics" **76**. p. 509., and references therein.

[106] Ball, Philip (November 2010). "Would Element 137 Really Spell the End of the Periodic Table? Philip Ball Examines the Evidence". Royal Society of Chemistry. Retrieved 30 September 2012.

[107] Cronyn, Marshall W. (August 2003). "The Proper Place for Hydrogen in the Periodic Table". *Journal of Chemical Education* **80** (8): 947–951. Bibcode:2003JChEd..80..947C. doi:10.1021/ed080p947.

[108] Gray, p. 14

[109] IUPAC, *Compendium of Chemical Terminology*, 2nd ed. (the "Gold Book") (1997). Online corrected version: (2006–) "transition element".

[110] Xuefang Wang; Lester Andrews; Sebastian Riedel; Martin Kaupp (2007). "Mercury Is a Transition Metal: The First Experimental Evidence for HgF$_4$". *Angew. Chem. Int. Ed.* **46** (44): 8371–8375. doi:10.1002/anie.200703710. PMID 17899620.

[111] William B. Jensen (2008). "Is Mercury Now a Transition Element?".*J. Chem. Educ.***85**(9): 1182–1183.Bibcode:2008J. doi:10.1021/ed085p1182.

[112] Rayner-Canham, G; Overton, T. *Descriptive inorganic chemistry* (4th ed.). New York: W H Freeman. pp. 484–485. ISBN 0-7167-8963-9.

[113] Thyssen, P.; Binnemanns, K. (2011). "1: Accommodation of the rare earths in the periodic table: A historical analysis". In Gschneidner Jr., K. A.; Büzli, J-C. J.; Pecharsky, V. K. *Handbook on the Physics and Chemistry of Rare Earths* **41**. Amsterdam: Elsevier. pp. 80–81. ISBN 978-0-444-53590-0.

[114] Keeler, J.; Wothers, P. (2014). *Chemical Structure and Reactivity: An Integrated Approach*. Oxford: Oxford University. p. 259. ISBN 978-0-19-9604135.

[115] Scerri, E. (2012). "Mendeleev's Periodic Table Is Finally Completed and What To Do about Group 3?". *Chemistry International* **34** (4).

[116] Castelvecchi, Davide (8 April 2015). "Exotic atom struggles to find its place in the periodic table". *Nature News*. Retrieved 20 Sep 2015.

[117] Emsley, J. (2011). *Nature's Building Blocks* (new ed.). Oxford: Oxford University. p. 651. ISBN 978-0-19-960563-7.

[118] See, for example: "Periodic Table". Royal Society of Chemistry. Retrieved 20 Sep 2015.

[119] See, for example: Brown, T. L.; LeMay Jr., H. E.; Bursten, B. E.; Murphy, C. J. (2009). *Chemistry: The Central Science* (11th ed.). Upper Saddle River, New Jersey: Pearson Education. p. endpapers. ISBN 0-13-235848-4.

[120] Scerri, E (2015). "Five ideas in chemical education that must die - part five". *educationinchemistryblog*. Royal Society of Chemistry. Retrieved Sep 19, 2015. It is high time that the idea of group 3 consisting of Sc, Y, La and Ac is abandoned

[121] Jensen, W. B. (2015). "Some Comments on the Position of Lawrencium in the Periodic Table" (PDF). Retrieved 20 Sep 2015.

[122] "Periodic Table of the Elements". International Union of Pure and Applied Chemistry. Retrieved 3 April 2010.

[123] Owen, S. M. (1991). *A Guide to Modern Inorganic Chemistry*. Harlow, Essex: Longman Scientific & Technical. p. 190. ISBN 0-58-206439-2.

[124] Francl, Michelle (May 2009)."Table manners"(PDF).*Nature Chemistry***1**(2): 97–98.Bibcode:2009NatCh...1...97F.doi:10.1 PMID 21378810.

18.10 Bibliography

- Ball, Philip (2002). *The Ingredients: A Guided Tour of the Elements*. Oxford: Oxford University Press. ISBN 0-19-284100-9.

- Chang, Raymond (2002). *Chemistry* (7th ed.). New York: McGraw-Hill Higher Education. ISBN 978-0-19-284100-1.

- Gray, Theodore (2009). *The Elements: A Visual Exploration of Every Known Atom in the Universe*. New York: Black Dog & Leventhal Publishers. ISBN 978-1-57912-814-2.

- Greenwood, Norman N.; Earnshaw, Alan (1984). *Chemistry of the Elements*. Oxford: Pergamon Press. ISBN 0-08-022057-6.

- Huheey, JE; Keiter, EA; Keiter, RL. *Principles of structure and reactivity* (4th ed.). New York: Harper Collins College Publishers. ISBN 0-06-042995-X.

- Moore, John (2003). *Chemistry For Dummies*. New York: Wiley Publications. p. 111. ISBN 978-0-7645-5430-8. OCLC 51168057.

- Scerri, Eric (2007). *The periodic table: Its story and its significance*. Oxford: Oxford University Press. ISBN 0-19-530573-6.

- Scerri, Eric R. (2011). *The periodic table: A very short introduction*. Oxford: Oxford University Press. ISBN 978-0-19-958249-5.

- Venable, F P (1896). *The Development of the Periodic Law*. Easton PA: Chemical Publishing Company.

18.11   External links

- M. Dayah. "Dynamic Periodic Table". Retrieved 14 May 2012.

- Brady Haran. "The Periodic Table of Videos". University of Nottingham. Retrieved 14 May 2012.

- Mark Winter. "WebElements: the periodic table on the web". University of Sheffield. Retrieved 14 May 2012.

- Mark R. Leach. "The INTERNET Database of Periodic Tables". Retrieved 14 May 2012.

18.12   Text and image sources, contributors, and licenses

18.12.1   Text

• **Matter** *Source:* https://en.wikipedia.org/wiki/Matter?oldid=682761689 *Contributors:* Carey Evans, CYD, The Anome, Tarquin, Ted Longstaffe, X JaM, Roadrunner, Ben-Zin~enwiki, Heron, Camembert, Ryguasu, Isis~enwiki, Stevertigo, Patrick, D, Tim Starling, Gabbe, Menchi, Ixfd64, To mos, Minesweeper, Looxix~enwiki, Ahoerstemeier, Suisui, Angela, Glenn, Cyan, Mxn, Charles Matthews, Reddi, Hyacinth, Rm, Xevi~enwiki, Ga krivas, Jerzy, BenRG, RadicalBender, Rashack~enwiki, Chuunen Baka, Robbot, Kizor, Gandalf61, Ashley Y, Auric, Hadal, Papadopc, Lupo , HaeB, Jan Lapère, Alan Liefting, Enochlau, Giftlite, Djinn112, Art Carlson, FeloniousMonk, Bensaccount, Solipsist, Alexf, Quadell, Antand rus, OverlordQ, Lesgles, Karol Langner, Mikko Paananen, JimWae, DragonflySixtyseven, Kevin B12, Okapi~enwiki, Int19h, Nike, Trevor M acInnis, Brianjd, EugeneZelenko, Discospinster, Rich Farmbrough, Vsmith, Jpk, Autiger, Martpol, Bender235, Kbh3rd, Ghitis, JoeSmack, RJHall, MisterSheik, Mr. Billion, Zegoma beach, Remember, Jashiin, CDN99, Adambro, Bobo192, Army1987, Whosyourjudas, Rrh02, Smallj im, Orbst, .:Ajvol:., Dungodung, Maurreen, I9Q79oL78KiL0QTFHgyc, Giraffedata, PaRaLyZeDHoRSe, Jojit fb, Kjkolb, Nk, Charonn0, MPe rel, Sam Korn, Haham hanuka, Jayakar, Alansohn, Arthena, Atlant, Paleorthid, AzaToth, Kocio, Walkerma, Jaw959, Wd-farmer, Avenue, B art133, Metron4, Snowolf, Wtmitchell, Oking83, Ott, Angr, Woohookitty, Madchester, Canaen, Sandrapalaje, CharlesC, RuM, Graham87, D pr, Sjö, Rjwilmsi, Jake Wartenberg, Strait, MarSch, Quiddity, Tangotango, Nneonneo, Vav11, Yamamoto Ichiro, Exeunt, FayssalF, FlaBot, Lor kki, Naraht, Nihiltres, Crazycomputers, Andy85719, RexNL, TimSE, TheDJ, BabyNuke, TheMightyGrecian, Cloudo, King of Hearts, Chobo t, DaGizza, Sharkface217, Gwernol, The Rambling Man, Wavelength, RobotE, Sceptre, Jimp, Bhny, Juansmith, Stephenb, Wimt, Ugur Ba sak, NawlinWiki, Injinera, Wiki alf, Grafen, Janarius, SCZenz, Retired username, Anetode, Dhollm, Jpbowen, Chichui, Alex43223, Natkee ran, DeadEyeArrow, Dna-webmaster, Wknight94, FF2010, Vadept, Enormousdude, Zzuuzz, Leliathomas, EWing, Sean Whitton, JuJube, A eon1006, JoanneB, Chez37, CWenger, Katieh5584, Junglecat, Banus, Mejor Los Indios, DVD R W, Smack-Bot, Unschool, KnowledgeOfSelf, Hydrogen Iodide, Unyoyega, C.Fred, Bomac, Gilliam, Skizzik, Fogster, Dauto, Andy M. Wang, Grokmoo, Rmosler2100, Master of Puppets, Sc hfiftyThree, Sbharris, Colonies Chris, Hallenrm, Darth Panda, Can't sleep, clown will eat me, OrphanBot, Rrburke, Addshore, Nakon, Jiddisch~e nwiki, John D. Croft, Dreadstar, SpiderJon, DMacks, Ultraexactzz, Kotjze, Kalathalan, Where, Bob-byPeru, Yevgeny Kats, Byelf2007, Dbtfz, LarchOye, Ocanter, Gobonobo, Breno, AstroChemist, Edwy, IronGargoyle, Vidit1, Simonalexan-der2005, Loadmaster, Munita Prasad, Noah Salzman, Hiiiiiiiiiiiiiiiiiiii, Kyoko, Dicklyon, Waggers, Dr.K., RichardF, MHWiki, Supaman89, CrazedEwok, Levineps, Lord Anubis, Joseph So lis in Australia, Newone, Casull, Tony Fox, Esurnir, Courcelles, JRSpriggs, JForget, DSatYVR, Ale jrb, Dycedarg, Van helsing, Vyznev Xnebara, MargyL, Flapping Fish, McVities, MarsRover, Penbat, Gregbard, Fl, Bvcrist, Peterdjones, Gogo Dodo, 01011000, Anonymi, Rracecarr, Stude rby, Dr.Kane, Abtract, RickDC, Epbr123, Barticus88, Mbell, O, TonyTheTiger, Gior-gio51, N5iln, Headbomb, Trevyn, Marek69, Grayshi, Nick Number, MichaelMaggs, Troy392004, Dualactionblend, Escarbot, Oreo Priest, Dantheman531, AntiVandalBot, Majorly, JHFTC, Cpko ndas, Quintote, Jayron32, CHollman82, Jj137, Madbehemoth, Danger, Astavats, Canadian-Bacon, Curlingpro47, Res2216firestar, Ioeth, JAn Dbot, Jimothytrotter, Elias Enoc, The Transhumanist, Sanchom, Blood Red Sand-man, Smiddle, Db099221, Andonic, Tergadare, Kerotan, Ac roterion, Connormah, Bongwarrior, VoABot II, AuburnPilot, JamesBWatson, Kinston eagle, Kajasudhakarababu, Think outside the box, Nytte nd, Cic, Avicennasis, Indon, Animum, Dirac66, Cpl Syx, MindReality, Vssun, DerHexer, Hbent, Arnesh, Tojo940, Greenguy1090, S3000, Fish erQueen, MartinBot, Rettetast, Anaxial, Sm8900, Dan.g, Dogatdog, R'n'B, AlexiusHoratius, Qwertuy, Ash, LedgendGamer, J.delanoy, Trusilv er, Bogey97, Maurice Carbonaro, Ginsengbomb, Eliz81, It Is Me Here, DarkFalls, Ben robbins, Gurchzilla, Vanished User 4517, Tcisco, New EnglandYankee, In Transit, Newtman, Minesweeper.007, Ionescuac, Ju-liancolton, Cometstyles, Tiggerjay, DH85868993, DorganBot, Natl1, S quids and Chips, Funandtrvl, Wikieditor06, Vranak, HamatoKameko, Deor, CWii, Christophenstein, JohnBlackburne, Bry9000, Haade, Philip Trueman, Eastgate, TXiKiBoT, Kww, SCRiBu, Anonymous Dissi-dent, Qxz, Seraphim, Melsaran, DennyColt, Abdullais4u, Noformation, Un itedStatesian, Vgranucci, Whammes2, Shadowlapis, CaptColon, Isis4563, Iluso, Ilkali, Wykypydya, Brainmuncher, Synthebot, Lova Falk, M CTales, Spinningspark, Kchiles, Conostrov, The Strange Kid, Insanity Incarnate, Alcmaeonid, Fireglowe, AlleborgoBot, Logan, TheXenocide , BriEnBest, J. Naven, SieBot, Vijai Singh, Portalian, Nihilnovi, Gerakibot, YourEyesOnly, Nathan, Triwbe, Wing gundam, Gravitan, Flyer22 , Tiptoety, Oda Mari, Prestonmag, Frank.hedlund, Granf, Oxymoron83, Harry~enwiki, Steven Crossin, Iain99, Techman224, The-G-Unit-Boss, Pantuflas, Kudret abi, OKBot, Jonlandrum, Asperal, As-cidian, Denisarona, ClueBot, GorillaWarfare, Snigbrook, Foxj, The Thing That Should N ot Be, Rjd0060, Wwheaton, Drmies, TheOldJacobite, Goshwak, CounterVandalismBot, Classified as matter, Rotational, Superguy342, Puchiko, Excirial, Jusdafax, Afoxtrotn00b, Editorman12342, MorrisRob, Bassoonboy, Rhododendrites, Brews ohare, NuclearWarfare, Cenarium, World, Jotterbot, Mr.24SevenCrashHolly, Tnxman307, Razorflame, SchreiberBike, Oswald07, Saebjorn, Polly, La Pianista, Calor, Taranet, Thingg, A itias, Versus22, Phynicen, MelonBot, SoxBotIII, Bibibita, Goodvac, Hattiel, Nishu pcp, DumZiBoT, Templarion, TimothyRias, Jmanigold, J Keck, AlexGWU, Mattermatters, -: Wik3dPlayful:-, Lord pain377, Rreagan007, Zkunz1, Facts707, WikHead, SilvonenBot, Mifter, Vianello, ZooFari, ElMeBot, Alexcs123, RyanCross, Thatguyflint, HexaChord, Gimie the beat boys, Gatorsalldawa, Addbot, Willking1979, Some jerk on the Internet, DougsTech, Ronhjones, Fieldday-sunday, Laurinavicius, Leszek Jańczuk, Ashanda, Protonk, Brentdeezee, Chamal N, Chzz, X RK, Favonian, 5 albert square, Kisbes-bot, Numbo3-bot, Ehrenkater, Tide rolls, Luckas Blade, Lrrasd, Zorrobot, MuZemike, Grandpsykick, Yobot, Ht686rg90, Senator Palpatine, Legobot II, THEN WHO WAS PHONE?, Brougham96, Azcolvin429, AnomieBOT, Somecrazydude, Joule36e5, Eminem69041, Killion-dude, Galoubet, 9258fahsflkh917fas, Piano non troppo, Icalanise, Ipatrol, Kingpin13, Yachtsman1, Abshir dheere, Ulric1313, Bluerasberry, Materialscientist, Magixdx, ShikyoSays, The High Fin Sperm Whale, Citation bot, Vuerqex, Neurolysis, Ba gumba, Ianwestapleton, Xqbot, Lloydsd, Bihco, YakbutterT, Jsharpminor, Grim23, The Evil IP address, Tricko20, J04n, Paul Sinclair, Amaury, Reflections of Mem-ory, Doulos Christos, Codyfitz8, Tjsnuff, Bigger digger, Nisamayarg, A. di M., Peter470, Griffinofwales, Joel gr over, CES1596, FrescoBot, Singhking97, Alaphent, Sebastiangarth, Machine Elf 1735, Citation bot 1, Redrose64, Pinethicket, I dream of horses, Vicenarian, The Arbiter, Tom.Reding, A8UDI, SpaceFlight89, Merlion444, Heiji hattori is LOVE, Irbisgreif, IVAN3MAN, PhilOak, Gamewi zard71, FoxBot, CalleCool, Trappist the monk, Kimtiger12345, Daniel G J, Vrenator, Darsie42, Reaper Eternal, Wst Nam, Devildude666, Suffusion of Yellow, Tbhotch, Oualidi13, Keegscee, DARTH SIDIOUS 2, Regancy42, Devcas, DASHBot, John of Reading, Ajraddatz, Was hout4, Fotoni, So-larra, Turnurban, Wikipelli, P. S. F. Freitas, Anirudh Emani, Tiniet, Thecheesykid, Jpfairweather, Jmanvball, JSquish, Zér oBot, Anir1uph, Quiqui1, ElvisPresley1, Hazard-SJ, Aeonx, Quondum, L1A1 FAL, Arman Cagle, YvonneM, Inka 888, Damirgraffiti, Carmich ael, Negovori, RockMagnetist, TYelliot, DASHBotAV, Wussification, Morgis, Efiiamagus, BigBarrelRollingHoss, Khestwol, Petrb, Xanchester, C lueBot NG, Mifapetrenko, Satellizer, Piast93, Waroe, Movses-bot, Erichardson2626, Wallawaller, Gamerstuff75, Goober6521, Dream of Nyx, Ca roleHen-son, 336, Be94ware, Widr, Helpful Pixie Bot, Ramaksoud2000, Bibcode Bot, Trunks ishida, BG19bot, Sanjeevgen, Soptx, Vagobot, Moustacheluvr69, Aozf05, MusikAnimal, Thumani Mabwe, Mark Arsten, Lolbading, Soukhoi, Altaïr, Jaloqin, Rs2360, MehFooL, GSMOL, Chiles Malesters, Glacialfox, Cimorcus, Pratyya Ghosh, Mdann52, Khazar2, EuroCarGT, Southwing17, Maxgaineducation, EagerToddler39, Brittany taylorr, Djdjordje12345, CoreyM1232, Lugia2453, Frosty, Graphium, Weather228, Wywin, NievesO, Reatlas, Tuckah parent, Hacker8484, Faizan, Asharia283, Austinben1011, Sosaalhabab, Bumoa, Professor Caramel Revenge, Guymandudeme, Leompersi, Trololsir, Ginsuloft, DavRosen, Jackmcbarn, Frank12123, Demigod mcg1, Victortran2000, JaconaFrere, AspaasBekkelund, Kiran cb, Kaltakus, Gvdjcfhhdfvuxhhrfbj, Havj, Harrydafitz, FourViolas, Fartser, Steven ngsteven, Epic2500, Vicallssssshdggsj, Jayster101, Uwharrie MilSim, NateBillingsley, KasparBot, Kafishabbir and Anonymous: 1142

- **State of matter** *Source:* https://en.wikipedia.org/wiki/State_of_matter?oldid=670655796 *Contributors:* CYD, Zardoz, Olof, Zundark, William Avery, Heron, Ubiquity, Ixfd64, Docu, Julesd, Seani, Reddi, Alexfiles, Robbot, Mayooranathan, Academic Challenger, Pepijn Schmitz, Rursus, Giftlite, Smjg, Art Carlson, Jackol, SoWhy, Rdsmith4, Oneiros, JimWae, DragonflySixtyseven, Glogger, ArthurDenture, Mike Rosoft, Discospinster, Rich Farmbrough, Vsmith, Walden, Rgdboer, Shanes, Smalljim, Zetawoof, Nsaa, Alansohn, Hydriotaphia, Wtmitchell, Mikeo, Vuo, Kazvorpal, Oleg Alexandrov, LOL, Pol098, Ruud Koot, MONGO, Yuriybrisk, Rjwilmsi, DoubleBlue, ElfQrin, Srleffler, Kri, TheSun, Imnotminkus, King of Hearts, Moocha, DVdm, Wavelength, Phantomsteve, Jeffhoy, Arado, Madkayaker, Gaius Cornelius, CambridgeBayWeather, Pseudomonas, NawlinWiki, Dhollm, BOT-Superzerocool, Wknight94, 2over0, Closedmouth, Pb30, Willtron, Yonir, Moomoomoo, JDspeeder1, Cmglee, Knowledgeum, SmackBot, McGeddon, Edgar181, Skizzik, Quinsareth, Miquonranger03, Colonies Chris, Darth Panda, Addshore, RedHillian, Ghiraddje, Valenciano, Bombshell, Keramos, Kipala, A. Parrot, Beetstra, Cratylus3, Mark999, BranStark, Iridescent, Joseph Solis in Australia, Newone, Igoldste, RekishiEJ, Courcelles, Tltltetd, Dycedarg, MarsRover, Cydebot, Mato, Xxanthippe, Bazzargh, Christian75, DumbBOT, Mtpaley, Epbr123, Urdna, N5iln, Headbomb, Kathovo, James086, Yettie0711, PJtP, Escarbot, Mentifisto, Daniels220, Jayron32, Jj137, Gökhan, JAnDbot, D99figge, YK Times, Bongwarrior, VoABot II, JNW, Father Goose, Rich257, Animum, Dirac66, Philg88, Khalid Mahmood, FisherQueen, Anaxial, Sm8900, Jonathan Hall, R'n'B, CommonsDelinker, Wiki Raja, Mausy5043, J.delanoy, Pharaoh of the Wizards, MITBeaverRocks, Uncle Dick, BobEnyart, Victuallers, Myrin1, Gombang, Chemicalrubber, NewEnglandYankee, Kraftlos, Bonadea, Ja 62, CrazyRob926, Useight, Martial75, Funandtrvl, CWii, ABF, Flyingidiot, Jeff G., Indubitably, Barneca, Philip Trueman, TXiKiBoT, Monkey Bounce, Dendodge, Corvus cornix, Cremepuff222, RiverStyx23, Venny85, Madhero88, Blurpeace, Falcon8765, Pageman~enwiki, Brianga, Universaladdress, HPeugeot~enwiki, NHRHS2010, Cryonic07, SieBot, Alessgrimal, Nubiatech, Euryalus, Paradoctor, Winchelsea, Yintan, Happysailor, Flyer22, Socal gal at heart, TrufflesTheLamb, Dillard421, Svick, Mygerardromance, Pinkadelica, Denisarona, Escape Orbit, Kanonkas, ShajiA, Troy 07, Ainlina, Faithlessthewonderboy, Elassint, ClueBot, Wookie501, Rumping, GorillaWarfare, Snigbrook, The Thing That Should Not Be, Ariadacapo, Tizeff, Polyamorph, GoEThe, Dylan620, Mad.martian999, Prognitor~enwiki, Quantumspinhall, Skihatboatbike, Djr32, Excirial, Alexbot, Coralmizu, Ykhwong, Pot, Singhalawap, Dekisugi, The Red, Oswald07, JasonAQuest, Mikhailov Kusserow, Thingg, Horselover Frost, PCHS-NJROTC, Burner0718, SoxBot III, Jmanigold, XLinkBot, Spitfire, PseudoOne, Leftspk, Andypandy2020, Saeed.Veradi, Libcub, Skarebo, SilvonenBot, Noctibus, JinJian, ZooFari, Thatguyflint, HexaChord, IBeAiMpErS0naT0r, Xvijayx, Addbot, Xp54321, Some jerk on the Internet, Jojhutton, PaterMcFly, Ronhjones, TutterMouse, Fieldday-sunday, Bertrc, Icelazer3, Shirtwaist, Vishnava, CanadianLinuxUser, NjardarBot, Download, PranksterTurtle, Glane23, Chzz, FCSundae, Kyle1278, CuteHappyBrute, Eh kia, Tide rolls, Gail, Micki, Cesaar, Angrysockhop, Legobot, Luckas-bot, Yobot, Dr. Footfoe, 2D, Andreasmperu, II MusLiM HyBRiD II, THEN WHO WAS PHONE?, Empireheart, IW.HG, Tempodivalse, Nintendo6, Quangbao, Hairhorn, Daniele Pugliesi, Jim1138, Lewismith3, Piano non troppo, Kingpin13, Dinesh smita, Westonpark, Materialscientist, Spirit469, ImperatorExercitus, 90 Auto, Citation bot, OllieFury, ArthurBot, Xqbot, TinucherianBot II, Beaserbebbeb, Pvkeller, Grim23, GrouchoBot, Charliekarst, Hamhat, Backpackadam, Amaury, Logger9, JediMaster362, Wasteman1066, Lagooncopperorange, FrescoBot, LucienBOT, Lothar von Richthofen, Hielor, Cannolis, Citation bot 1, Athenanoenvoy, Pinethicket, Boulaur, Overthinkingly, Tom.Reding, ContinueWithCaution, Johann137, Jauhienij, Gamewizard71, FoxBot, Yunshui, Zvn, January, Allen4names, Theo10011, Reaper Eternal, Jeffrd10, Tbhotch, Hornlitz, Marie Poise, DARTH SIDIOUS 2, NerdyScienceDude, Chuanenlin123, Jack Schlederer, Kenvancleve, WikitanvirBot, Ajraddatz, DillonLMcCabe, Mordgier, Katherine, Gcastellanos, The Sharminator, Tommy2010, Wikipelli, John Cline, Zane45177, Alpha Quadrant (alt), DidgeGuy, AndrewN, Yamumyadad, Jay-Sebastos, Jesanj, Brandmeister, Donner60, Ashunigam, Orange Suede Sofa, ChuispastonBot, Llightex, ShatteredSpiral, ClueBot NG, Gareth Griffith-Jones, MelbourneStar, UAK32, Rtucker913, Yourmomblah, Theimmaculatechemist, 123Hedgehog456, Whazie, Lfmm11, Mtheodric, Kumamah, ساجد امجد ساجد, Friend20gan, MerlIwBot, Helpful Pixie Bot, Candleabracadabra, Bibcode Bot, Mingmingla, Mark Fole, BZTMPS, Lowercase sigmabot, Yetisyny, Hallows AG, MusikAnimal, Frze, AvocatoBot, Planetary Chaos Redux, Tatchell, Nsda, ChE Fundamentalist, Ugncreative Usergname, Joydeep, Averysamantha, Snow Blizzard, ImhotepBallZ, Mantike, Kydon Shadow, Th4n3r, The Illusive Man, ChrisGualtieri, EuroCarGT, Sidsandyy, Lugia2453, CaSJer, Frosty, SFK2, Jamesx12345, Perfecttwoegan, Reatlas, Passengerpigeon, Epicgenius, I am One of Many, Lsmll, Surfer43, Tentinator, Cebr1979, Amdz96, Backendgaming, Lince celeste, DavidLeighEllis, Ginsuloft, WikiMannWikiMann, Annieminnie21, Rickdutta, Meteor sandwich yum, Muneeb abdelhadi, ShahryarAhmad27900, Nijuthomasgeorge, Bilorv, Saaaaaaaaaa, Ookami-to-Koneko, J73364, S r u fejvd u ub creating u of, Amortias, Natdeso, Adenine2k, Lalalalalala is too great, ElmoLover88545, Darkmatter1435, Chetan082, Alango1998, XxX$pace5Xxx, Cambles13, Johno2000, Jsbdjejnwn, Weegeerunner, Slayeredwarrior, Account50, Blissy2005, Minnie431, KasparBot and Anonymous: 823

- **Solid** *Source:* https://en.wikipedia.org/wiki/Solid?oldid=680590069 *Contributors:* Mav, Olof, Css, Heron, FlorianMarquardt, Luckymama58, Lir, Tim Starling, SebastianHelm, Ahoerstemeier, Suisui, Andrewa, Cgs, Glenn, Andres, Smack, Charles Matthews, Dysprosia, Taxman, Shizhao, Jusjih, Digital-h~enwiki, Donarreiskoffer, Gentgeen, Mayooranathan, Tosha, Giftlite, Bensaccount, Yekrats, Youngoat, Joeblakesley, Icairns, Discospinster, Vsmith, MisterSheik, El C, Joanjoc~enwiki, Bobo192, Smalljim, Eddideigel, Jumbuck, Zachlipton, Danski14, Alansohn, Keenan Pepper, Lectonar, Walkerma, Gene Nygaard, Woohookitty, Mindmatrix, WadeSimMiser, The Wordsmith, Kmg90, Graham87, BD2412, FreplySpang, Rjwilmsi, Vary, Bruce1ee, ElKevbo, FlaBot, Mathbot, Nihiltres, Lmatt, TheSun, Chobot, Sharkface217, DVdm, Stoive, Gwernol, Roboto de Ajvol, YurikBot, Wavelength, RussBot, Salsb, NawlinWiki, Grafen, Anetode, Zwobot, Kooky, Bota47, Wknight94, 2over0, StuRat, Dspradau, Willtron, RunOrDie, Ybbor, Bdve, SmackBot, FocalPoint, C.Fred, Bomac, Jrockley, Delldot, Vincent de Ruijter, Edgar181, Gilliam, Chaojoker, Andy M. Wang, Chris the speller, Bluebot, MalafayaBot, SchfiftyThree, Akanemoto, PureRED, DHN-bot~enwiki, Gracenotes, Jahiegel, PeteShanosky, DMacks, SashatoBot, ArglebargleIV, Kuru, John, Kipala, Gobonobo, Yms, Optakeover, Jose77, Levineps, G1076, NativeForeigner, Amalas, Dycedarg, Nunquam Dormio, Black and White, MarsRover, WeggeBot, Neelix, Equendil, HawkShark, Fl, Xxanthippe, ST47, Pinky sl, Epbr123, MarkBuckles, Headbomb, A3RO, Rhrad, Huigh4444, Mentifisto, AntiVandalBot, Seaphoto, Fatkat61, JAnDbot, Leuko, Wizardboy777, VoABot II, Esmhead, Animum, Cgingold, Dirac66, User A1, Shijualex, JaGa, MartinBot, Anaxial, R'n'B, CommonsDelinker, Tgeairn, J.delanoy, Trusilver, NightFalcon90909, Dbiel, AgainErick, Afluegel, Belovedfreak, Darrendeng, Kaleau, Alan012, Fatkid72, Squids and Chips, CardinalDan, RAult, Youbejoking, VolkovBot, Lear's Fool, Ryan032, Philip

Trueman, TXiKiBoT, Rei-bot, Warlord dehacker, Monkey Bounce, Dilos100, JhsBot, Brandonrush, Synthebot, Falcon8765, AlleborgoBot, Sparkylad, Sfztang, SieBot, Nubiatech, Caltas, Yintan, Toddst1, Lightmouse, Fratrep, Mojoworker, ClueBot, Venske, Snigbrook, The Thing That Should Not Be, Newsjunker, Polyamorph, Blanchardb, Mvp08, Djr32, Excirial, Jusdafax, Anon lynx, PixelBot, California847, Cenarium, SchreiberBike, Unmerklich, Vanished user uih38riiw4hjlsd, Vanished User 1004, DumZiBoT, Steven McTowelie the second, Bodhisattv-aBot, Rror, Addbot, Narayansg, Eric Drexler, Fleetsbridge, Nath1991, X VeNTe, Ballervision, Ilovelakai, EconoPhysicist, AtheWeatherman, Numbo3-bot, Tide rolls, Lrrasd, Quantumobserver, Luckas-bot, Zaereth, Thatgains, A Stop at Willoughby, Kingpin13, Bonusballs, Materialscientist, Spirit469, The High Fin Sperm Whale, Citation bot, OllieFury, Xqbot, Sketchmoose, Capricorn42, Blindgrapefruit2, J04n, Grou-choBot, Amaury, Logger9, Shadowjams, Methcub, FrescoBot, LucienBOT, Fahed ahmed, Fast kartwheels, Tetraedycal, Citation bot 1, Krish Dulal, Pinethicket, I dream of horses, Boulaur, Tom.Reding, BigDwiki, Gruntler, Attack 2400, Jauhienij, ActivExpression, Gamewizard71, FoxBot, TobeBot, Yunshui, Lotje, Javierito92, Vrenator, Theo10011, Ghjthgh, Marie Poise, DARTH SIDIOUS 2, Ripchip Bot, VernoWhit-ney, Solidboy123, DASHBot, EmausBot, WikitanvirBot, Immunize, Super48paul, Hplovecraftpwns, Dewritech, RenamedUser01302013, Solarra, Tommy2010, Wikipelli, K6ka, Lou1986, Xman0444, Hhhippo, JSquish, Gr8xoz, Tolly4bolly, Asiaglass, Ventilate, Donner60, Jan 1922, Lovetinkle, ClueBot NG, Bidpald253, MelbourneStar, This lousy T-shirt, Baseball Watcher, O.Koslowski, Widr, Theopolisme, Lugia77, Pbuffat, Helpful Pixie Bot, Bibcode Bot, Vagobot, MusikAnimal, Writ Keeper, Djhfsdhf, Klilidiplomus, BattyBot, WhiteNebula, Ducknish, Frosty, AgaRed, Reatlas, Faizan, Hareesa, Harlem Baker Hughes, JamesMoose, PhantomTech, EvergreenFir, TCMemoire, Aravinduser, Hisas-hiyarouin, Ukulelemollyjade, Cyberalchemyst, Poopoohead54321, Poopyhead321, Lofle, JHeid2014, JP216, PotatoNinja, Aryaman SInghi, EduacationDino, Stixky, GeneralizationsAreBad, KasparBot, Ethan's biggest fan, Chris1238, Kuantingl, KSFT and Anonymous: 369

- **Liquid** *Source:* https://en.wikipedia.org/wiki/Liquid?oldid=682791534 *Contributors:* Kpjas, Brion VIBBER, Bryan Derksen, Olof, Koyaa-nis Qatsi, William Avery, Peterlin~enwiki, Imran, Heron, Karl Palmen, Luckymama58, Olivier, Lir, Michael Hardy, Ixfd64, SebastianHelm, Ahoerstemeier, Александър, Poor Yorick, Tantalate, Jusjih, PuzzletChung, Fito, Donarreiskoffer, Gentgeen, Robbot, Swestrup, Altenmann, Lowellian, Merovingian, Blainster, Robinh, Jeremiah, Giftlite, Cfp, J heisenberg, BenFrantzDale, Ævar Arnfjörð Bjarmason, Moyogo, Ben-saccount, Yekrats, Sonjaaa, Antandrus, Beland, ClockworkLunch, Vc-wp, H Padleckas, Icairns, Rgrg, Fg2, Clemwang, Frau Holle, Zondor, Mike Rosoft, Moverton, Discospinster, Cacycle, Vsmith, Kbh3rd, El C, Bobo192, Robotje, Smalljim, Mareino, Ranveig, Jumbuck, Danski14, Siim, Alansohn, Gary, Penwhale, Walkerma, Malo, KingTT, Garzo, Shoefly, Vuo, LFaraone, Agquarx, Revived, Woohookitty, Mindmatrix, Georgia guy, TigerShark, MGTom, Esben~enwiki, The Nameless, Graham87, FreplySpang, Sjakkalle, Rjwilmsi, Nneonneo, Tomtheman5, Yamamoto Ichiro, FlaBot, Mishuletz, Gurch, Tedder, Zotel, Physchim62, Chobot, DVdm, YurikBot, Wavelength, RobotE, Arado, Yyy, Nawl-inWiki, Brian Crawford, Moe Epsilon, Bucketsofg, Kooky, BOT-Superzerocool, Everyguy, Mike92591, Wknight94, Tetracube, Tigershrike, Closedmouth, Kevin, ArielGold, Junglecat, Hl540511, SmackBot, FocalPoint, David.Mestel, Unyoyega, Bomac, Jrockley, HalfShadow, Yam-aguchi⁇⁇, Gilliam, Skizzik, Chris the speller, MalafayaBot, Deli nk, Akanemoto, Zachorious, Dethme0w, Jmccabe871, Divna Jaksic, FiveR-ings, Paul Slocum, DRLB, B jonas, -Ozone-, Ryan Roos, DaiTengu, Acdx, Starghost, Bidabadi~enwiki, Bejnar, Chaldean, Nathanael Bar-Aur L., John, Kipala, Mbeychok, A. Parrot, JHunterJ, Jose77, Ginkgo100, CapitalR, Courcelles, Van helsing, Woudloper, MarsRover, Karenjc, Ma-niacalMonkey, Slazenger, Acornembryo, Gogo Dodo, Albert0, Xxanthippe, Difluoroethene, Dancter, Christian75, FastLizard4, JoshHolloway, Omicronpersei8, Thijs!bot, Epbr123, Marek69, Nick Number, Dawnseeker2000, Mentifisto, AntiVandalBot, BokicaK, SummerPhD, MECU, Gökhan, MikeLynch, Bsmithurst, JAnDbot, Bhamv, Xeno, Hut 8.5, Steveprutz, Acroterion, Bongwarrior, VoABot II, AuburnPilot, Redaktor, Animum, Cgingold, Allstarecho, Adacus12, DerHexer, Khalid Mahmood, MartinBot, Mufo, CommonsDelinker, Tgeairn, J.delanoy, Ginsen-gbomb, Mintz l, Hodja Nasreddin, Rvaznyvfgxrvazny, Afluegel, Gurchzilla, Tanaats, Dhaluza, KylieTastic, Cometstyles, Tiggerjay, Useight, Omc, Dan Hickman, Black Kite, Deor, VolkovBot, CWii, Ryan032, Philip Trueman, DoorsAjar, TXiKiBoT, Yupi666, Fcb981, Sean D Mar-tin, Psyche825, Cremepuff222, Wikiisawesome, Meters, Enviroboy, Dogah, SieBot, Mikemoral, Tresiden, Euryalus, Laoris, Winchelsea, Da Joe, Yintan, Yankszack, Toddst1, JohnaG-A, Oxymoron83, Son111, Marquetry28, LidiaFourdraine, Kutera Genesis, Yaluen, Lehasa, Sal-lyForth123, Loren.wilton, Sfan00 IMG, ClueBot, LAX, The Thing That Should Not Be, Remag Kee, DanielDeibler, Polyamorph, Boing! said Zebedee, Djr32, Excirial, Jusdafax, Cenarium, Jotterbot, Iohannes Animosus, Rash, Sdrtirs, Aitias, Egmontaz, Crowsnest, Chhe, Tacos r friends, Matthieumarechal, Avoided, SilvonenBot, WikiDao, ZooFari, Wikiman405, Addbot, Brumski, Eric Drexler, Manuel Trujillo Berges, Some jerk on the Internet, Yoenit, Fgnievinski, Richmond96, Ilovelakai, دمرقندى‎, Debresser, Favonian, Exor674, Jasper Deng, Numbo3-bot, Erutuon, Tide rolls, MuZemike, Mozillaman425, Luckas-bot, Yobot, Zaereth, Fraggle81, Sarrus, Johnsatterfield, WizardOfOz, Democrati-cLuntz, Jim1138, Memphis670, Kingpin13, Bluerasberry, Jrobinjapan, Materialscientist, Spirit469, The High Fin Sperm Whale, Citation bot, MauritsBot, Xqbot, JimVC3, Ryomaandres, GrouchoBot, Nayvik, Logger9, Shadowjams, WaysToEscape, LucienBOT, A little insignificant, Wireless Keyboard, Citation bot 1, Krish Dulal, Pinethicket, I dream of horses, Boulaur, Drerhymez, Hard Sin, Tom.Reding, Calmer Wa-ters, A8UDI, Σ, Abaoabao, Robo Cop, Zorroeatsmaypo, Jauhienij, Gamewizard71, FoxBot, Vrenator, Marie Poise, DARTH SIDIOUS 2, The Utahraptor, Stj6, NerdyScienceDude, 4students, EmausBot, John of Reading, WikitanvirBot, Qurq, Scholar333, Racerx11, Faolin42, RA0808, RenamedUser01302013, Foodeatingperson, K6ka, Savh, Sepguilherme, JSquish, HugeGrayLover, Jman12369874, Hal3y+st3phh, Wayne Slam, Tolly4bolly, Ventilate, TyA, L Kensington, Donner60, EvenGreenerFish, DennisIsMe, Fowen123, Peter Karlsen, Targaryen, DASH-BotAV, Rememberway, ClueBot NG, CocuBot, LOLSmaterThanYou, Satellizer, Bike CharlieCard, Widr, MerlIwBot, Helpful Pixie Bot, Bradynowinstores, Pikapika123, Bibcode Bot, 2001:db8, Mouchumi, Vagobot, 1sgfxn, Nikos 1993, Teamavolition, AwamerT, Mark Arsten, Cadiomals, Insidiae, Poophole10101, D.bolmatov, WhiteNebula, Priyamd, Ameerajanmohamed, Physicsandshiz, Ledgermayne101, Poopy-head88, EuroCarGT, Phillip is the name, Ducknish, Dexbot, TwoTwoHello, ComfyKem, Kevin12xd, Reatlas, PhantomTech, Abc123def455, EvergreenFir, Leoesb1032, Buffbills7701, Blackbombchu, Ugog Nizdast, Aravinduser, JaconaFrere, Carlos Tao, Natty Stott, Samygemayel, Trackteur, CHK101, Cnbr15, Vistardhvaj, KasparBot, Quiesce, Letsdisko, The pro guy, Wikiman900909, Hermionedidallthework and Anony-mous: 455

- **Gas** *Source:* https://en.wikipedia.org/wiki/Gas?oldid=681787714 *Contributors:* Bryan Derksen, Olof, Tarquin, Andre Engels, Youssefsan, William Avery, SimonP, Peterlin~enwiki, Ben-Zin~enwiki, Heron, Luckymama58, Patrick, D, Michael Hardy, Nixdorf, Delirium, Looxix~enw iki,Ahoerstemeier, Stan Shebs, Mac, Jimfbleak, Александър, Glenn, Nikai, Llull, Edmilne, Malbi, Jaimeglz, Wernher, Shizhao, Bcorr, Ju sjih,Johnleemk, Donarreiskoffer, Gentgeen, Robbot, Hankwang, Academic Challenger, Mervyn, LX, Dina, Tobias Bergemann, Giftlite, Tom har-rison, Herbee, Theon~enwiki, Peruvianllama, Everyking, Bensaccount, Gareth Wyn, Alexf, Antandrus, OverlordQ, Scott MacLean, KarolLangner, Maximaximax, Icairns, Zfr, Iantresman, GdB, ⁇⁇, Adashiel, Grstain, D6, N328KF, Jiy, Discospinster, Qutezuce, Vsmith, Cl oseap-ple, Brian0918, RJHall, El C, Hayabusa future, Aude, Bobo192, Fremsley, Giraffedata, Jerryseinfeld, MPerel, Haham hanuka, Jake w, Jum-buck, Danski14, Siim, Gary, Anthony Appleyard, Keenan Pepper, Katefan0, Wtshymanski, Psmither, Yuckfoo, RJFJR, Gene Nygaard,CoolMike, Noz92, Roland2~enwiki, Sylvain Mielot, Woohookitty, BillC, DrAwesome, Pdn~enwiki, Allen3, Behun, BD2412, Phillipedison1891, Rjwilmsi, Panoptical, Bruce1ee, SMC, Krash, Sango123, Yamamoto Ichiro, Gurch, Wars, Drumguy8800, Mogest, Jittat~enwiki, Chobot, Sharkface217, Chachu207, Algebraist, The Rambling Man, YurikBot, Wavelength, RussBot, Jeffhoy, Backburner001, Conscious, SpuriousQ, KevinCuddeback, Stephenb, Jugander, Rsrikanth05, Wimt, NawlinWiki, Grafen, RazorICE, Lepidoptera, Dureo, Irishguy, Dhollm, Zwobot, EEMIV, Kkmurray, Wknight94, Lt-wiki-bot, Closedmouth, Sean Whitton, Willtron, AGToth, RunOrDie, Nsevs, Junglecat, NeilN, GrinBot~enwiki, Bo Jacoby, Amberrock, DVD R W, ChemGardener, SmackBot, FocalPoint, Prodego, KnowledgeOfSelf, Unyoyega, Vald, Bomac, ScaldingHotSoup, Jrockley, Delldot, Ilikeeatingwaffles, RobotJcb, Abbeyvet, Canthusus, Nethency, Fentonrobb, Edgar181, Hmains, Carl.bunderson, Kmarinas86, Chris the speller, TimBentley, SlimJim, Quinsareth, EncMstr, MalafayaBot, Complexica, Stevage, Akanemoto, Pencilcomics, DHN-bot~enwiki, Colonies Chris, Gracenotes, Dethme0w, Can't sleep, clown will eat me, Shalom Yechiel, DéRahier, Sephiroth BCR, Pablo9000, Amazins490, SundarBot, Jmlk17, Zrulli, Nibuod, TedE, RJN, Blake-, Astroview120mm, DMacks, Bidabadi~enwiki, Sadi Carnot, Ged UK, SashatoBot, Lambiam, BrownHairedGirl, Kuru, John, J 1982, Gobonobo, Disavian, Ishmaelblues, JorisvS, Olin, Iron-Gargoyle, SpyMagician, Beetstra, Stizz, Doczilla, Dhp1080, Dcflyer, Ryulong, H, Etafly, Caiaffa, YipYip, Hu12, Ginkgo100, Iridescent, Wwallacee, Blehfu, CP\M, Eassin, Tawkerbot2, Stifynsemons, DangerousPanda, Dgw, Jp-hickson, McVities, MarsRover, WeggeBot, Rak-wiki, Cydebot, SyntaxError55, Rifleman 82, Gogo Dodo, Flowerpotman, Corpx, Nabz~enwiki, DumbBOT, Narayanese, Btharper1221, Gimmetrow, Kablammo, Ucanlookitup, Mojo Hand, Headbomb, Marek69, John254, Nathaniel Zhu, Cool Blue, AgentPeppermint, CharlotteWebb, Escarbot, AntiVandalBot, M84, Majorly, Bigtimepeace, Jj137, Madbehemoth, Dylan Lake, North Shoreman, Waynesewell, Myanw, Dreaded Walrus, JAnDbot, Barek, MER-C, Hut 8.5, PhilKnight, GoodDamon, Yahel Guhan, Bongwarrior, VoABot II, AuburnPilot, Hasek is the best, Mbc362, Kaiserkarl13, Jim Douglas, Kevinmon, Avicennasis, Fabrictramp, Animum, Cgingold, Nposs, Allstarecho, Adacus12, Lukecarpenter169, Spellmaster, PrincessBrat, Glen, DerHexer, Edward321, Greenguy1090, Hdt83, MartinBot, Bullet4troubles, Chaos Wolf, Rettetast, Tholly, R'n'B, AlexiusHoratius, Johnpacklambert, Harrichr, J.delanoy, Trusilver, Sp3000, Jakesdamajorbomb, GeoWriter, Socrgrl3426, Rufous-crowned Sparrow, AtholM, Katalaveno, Abhijitsathe, Hawkmaster9, Ryan Postlethwaite, Afluegel, Mikael Häggström, Sman789, V.V zzzz, Jonodabomb, NewEnglandYankee, SJP, Mufka, Lyctc, Uhai, Darkfrog24, Screwe, Idioma-bot, Lights, X!, VolkovBot, Cireshoe, Thedjatclubrock, TheOtherJesse, Philip Trueman, DoorsAjar, TXiKiBoT, Z.E.R.O., Woodsstock, Qxz, Redmusicjamin, Melsaran, Aycbubbles, Corvus cornix, K193, ^demonBot2, Drappel, Delbert Grady, Mannafredo, Fireman17, Isis4563, Miwanya, Venny85, Wenli, Greswik, Krzysfr, Sbakka, Synthebot, Falcon8765, Emily.xxo, BaByGuRRL19, Brianga, HeirloomGardener, AlleborgoBot, Logan, Resurgent insurgent, NHRHS2010, Comeinayeahaa, Adamboy555, Darting., SieBot, Euryalus, Nicklovesgold, Dennislee272727, Jauerback, Virtual Cowboy, Rystheguy, Gerakibot, Caltas, Cwkmail, Keilana, Flyer22, Tiptoety, Oda Mari, Oxymoron83, Nuttycoconut, Steven Crossin, Lightmouse, Powerofgas, Dcarriker92, Alex.muller, Phoneuser, Torchwoodwho, Coldcreation, Maralia, Hooiwind, Nn123645, Denisarona, Faithlessthewonderboy, Martarius, ClueBot, Drmies, Razimantv, Mild Bill Hiccup, CounterVandalismBot, Hschrage, Toudaiji Neji, Puchiko, Excirial, Granvo, MorrisRob, Sun Creator, 7&6=thirteen, I luv carrots, SchreiberBike, Kpark454, I Has A Username, Katanada, Soulspick, Skier lad, DumZiBoT, Life of Riley, Clausc, Kalin1344, Poopman101, Superman1159, Bailey Pitzer, Ridhwan95, SilvonenBot, Tjmcgowan9, Noctibus, Addbot, Vejvančický, Fyrael, Ashton1983, Leszek Jańczuk, Tedmund, Ilovelakai, Skyezx, CarsracBot, Glass Sword, Mkardous, Chenzw-Bot, Numbo3-bot, Tide rolls, Fryed-peach, VP-bot, Luckas-bot, Yobot, Fraggle81, Azylber, DemocraticLuntz, Bjk343, Materialscientist, Spirit469, The High Fin Sperm Whale, Citation bot, ArthurBot, Obersachsebot, Xqbot, Intelati, Gilo1969, GrouchoBot, IShadowed, GhalyBot, Schekinov Alexey Victorovich, Joaquin008, FrescoBot, LucienBOT, Thayts, Rackmount-guy, DivineAlpha, Cannolis, Krish Dulal, Pinethicket, Red banksy, Adlerbot, Tom.Reding, RedBot, Serols, Σ, IJBall, Jauhienij, Keri, Gamewizard71, FoxBot, TobeBot, Lotje, Vrenator, Begoon, 4, JamAKiska, Jeffrd10, JV Smithy, Tbhotch, DARTH SIDIOUS 2, Mean as custard, Yangosplat222, DASHBot, EmausBot, Gfoley4, 478jjjz, Abcboy151, Hudmaster, Dewritech, Racerx11, GoingBatty, John of Lancaster, Hhhippo, JSquish, ZéroBot, The Nut, AvicAWB, Magasjukur, J1812, Ventilate, Dirtykorean, Donner60, ChuispastonBot, NTox, TYelliot, DASHBotAV, ClueBot NG, Iiii I I I, SusikMkr, Baseball Watcher, Snotbot, Corusant, Prumpa, O.Koslowski, Scottiessoulja, Kasirbot, Widr, Jalenjohnson129, Coiladam, Helpful Pixie Bot, JohnSRoberts99, Bibcode Bot, WNYY98, Lowercase sigmabot, Juro2351, Hallows AG, MusikAnimal, Nikos 1993, Amp71, Piguy101, Mark Arsten, Gorthian, Working Cat, Snow Blizzard, Hamzah145, Shawn Worthington Laser Plasma, BattyBot, RichardMills65, Cyberbot II, CarrieVS, Khazar2, MrNoSignal26, Adam8257, Murshed11, Etcyorz3t789, Saehry, Lugia2453, Frosty, Ashcool1999, Bilalrockstar5, Tyler Daigle, Randykitty, Epicgenius, Hahabacon123, Ugog Nizdast, WikiJuggernaut, Aniapo, DudeWithAFeud, Yolokid1024, Carecat2434, Joeleoj123, BethNaught, New Imam Bukhsh Gas Company, Nibgco, Pipo45, A.Minkowiski, Technick14, NQ, Plop2004, Goolongo325, Ashisbiswas560, Leocatelli09, Luistheguy, Minecrafter123456, Rahulhsp, Arnold22palmer, Megan Mackenzie, Siddhantsaka, Perica85, Spidermanwillruletheworld, Rylee Imhoff, KasparBot, Mynamejeff47, Benjamintylerolmstead2, Lucyiliana, Letsdisko and Anonymous: 589

- **Plasma (physics)** *Source:* https://en.wikipedia.org/wiki/Plasma_(physics)?oldid=682732790 *Contributors:* Trelvis, Vicki Rosenzweig, Mav, Bryan Derksen, Olof, AstroNomer~enwiki, Roadrunner, Secretsaregood, Heron, Stevertigo, Patrick, Michael Hardy, Tim Starling, Tapper of spines, Zeno Gantner, Ellywa, Stevenj, Jebba, Darkwind, Julesd, Glenn, Tantalate, Wikiborg, Reddi, David Latapie, IceKarma, Nv8200pa, Omegatron, Phoebe, Pakaran, Cdupree, April~enwiki, Donarreiskoffer, Robbot, Psychonaut, Moink, Hadal, Papadopc, David Edgar, SoLando, Dbroadwell, Wile E. Heresiarch, Giftlite, Mat-C, Art Carlson, MadmanNova, Monedula, Everyking, Jacob1207, Gracefool, Solipsist, Bobblewik, SarekOfVulcan, Beland, Mako098765, WhiteDragon, Gunnar Larsson, Karol Langner, APH, Icairns, Bk0, Nickptar, Iantresman, Kelson, Joyous!, Fermion, Jh51681, Deglr6328, Kate, PhotoBox, Spiffy sperry, Jiy, Noisy, Discospinster, Brianhe, Rich Farmbrough, Pjacobi, Vsmith, Florian Blaschke, Warpflyght, Bender235, Lou Crazy, Evice, El C, Huntster, Femto, Bobo192, Illuvatar,, Smalljim, Func, Enric Naval, Evgeny, .:Ajvol:., Elipongo, Tmh, Maurreen, I9Q79oL78KiL0QTFHgyc, Sparkgap, Photonique, Boredzo, B0at, Obradovic Goran, Yoweigh, Danski14, Alansohn, Anthony Appleyard, Ungtss, Arthena, Keenan Pepper, Craigy144, ABCD, RoySmith, PAR, Pion, SMesser, Radical Mallard, Knowledge Seeker, Cburnett, RJFJR, Sciurinæ, Mikeo, Vuo, Pauli133, DV8 2XL, Gene Nygaard, NuVanDibe, HenryLi, Pediddle, Umapathy, Brookie, Feezo, Stemonitis, Gmaxwell, Linas, Justinlebar, WadeSimMiser, Mouvement, Jwanders, Fred J, SDC, Palica, Allen3, Mr Anthem, Marudubshinki, Mandarax, Aarghdvaark, Graham87, Teknic, BD2412, RxS, Kissekatt, Grammarbot, Canderson7, Ketiltrout, Rjwilmsi, Mayumashu, Koavf, Vary, Beng341, HappyCamper, Mbutts, Frenchman113, Krash, DoubleBlue, GregAsche, Yamamoto Ichiro, Algebra, Lcolson, MinorEdit, Eyas, The ARK, Old Moonraker, JohnElder, RexNL, Crouchingturbo, Preslethe, Srleffler, Kri, Physchim62, King of Hearts, Chobot, Helios, Tone, Roboto de Ajvol, YurikBot, Borgx, Madhan49, Hairy Dude, RussBot, Arado, Conscious, Hydrargyrum, CambridgeBayWeather, Yyy, Shaddack, Giro720, Anomalocaris, Wiki alf, Bachrach44, Spike Wilbury, Grafen, NickBush24, Jaxl, JDoorjam, Nucleusboy, Brian Crawford, Coderzombie, Zirland, Figaro, DeadEyeArrow, Oliverdl, Kkmurray, Wknight94, Pr1268, Jezzabr, Tetracube, Nfm, Mütze, Tigershrike, Poppy, Phgao, 2over0, Bobryuu, Theda, Closedmouth, Jwissick, Ameyabapat, Modify, Vicarious, Peter, Willtron, Katieh5584, Kungfuadam, RG2, MAROBROS, Dkasak, Mejor Los Indios, DVD R W, Vedant lath, That Guy, From That Show!, Mhardcastle, TravisTX, Lviatour, SmackBot, Superfreaky56, Unschool, Hkhenson, Sonoma-rich, CarbonCopy, Rex the first, Prodego, KnowledgeOfSelf, Wegesrand, Sharaith, KocjoBot~enwiki, Fitch, FRS, Eskimbot, Jab843, Evanreyes, Yamaguchi⁇⁇, Brianski, Richfife, MPD01605,

Bluebot, Thumperward, Miquonranger03, Mr Poo, SchfiftyThree, Complexica, Akanemoto, MIB4u, Baa, DHN-bot~enwiki, William Allen Simpson, Shamiryan, Exaudio, Can't sleep, clown will eat me, Ioscius, Berland, Rrburke, RedHillian, DavidStern, Nibuod, Ian01, John D. Croft, Astroview120mm, EdGl, SpiderJon, DMacks, Whoville, Mion, Bidabadi~enwiki, Tesseran, Sina2, Kuru, John, JorisvS, IronGargoyle, Ben Moore, Applejuicefool, S zillayali, Frokor, Slakr, SCOTT FISHER, Dicklyon, Waggers, Mets501, Dr.K., Arstchnca, P199, MathStuf, Dl2000, Tusenfem, Iridescent, Electrified mocha chinchilla, Laurens-af, Polymerbringer, Joseph Solis in Australia, Tony Fox, Tawkerbot2, Dlohcierekim, Jafet, DanHickstein, Tommysun, JForget, Sakurambo, CmdrObot, Tanthalas39, Sir Vicious, Alexey Feldgendler, Neachili, Van helsing, Jcoetzee, JohnCD, Bill.albing, PRhyu, Dgw, Dashpool, Rigby~enwiki, Gegorg, MarsRover, Shandris, Inferno32, Guitarmankev1, Tjoneslo, Myasuda, Eecon, Cydebot, Lightblade, W.F.Galway, ArgentTurquoise, Corpx, Bazzargh, Quibik, Gerinych, DumbBOT, Thijs!bot, Epbr123, Andyjsmith, Dougsim, Headbomb, Marek69, FST777, Dawnseeker2000, Elert, Escarbot, Ileresolu, AntiVandalBot, Amideg, Widefox, Seaphoto, Blue Tie, Opelio, Zachwoo, Joe Schmedley, Stepan Roucka, JAnDbot, CosineKitty, Plantsurfer, Andonic, East718, Howsthatfordamage25, PhilKnight, Rothorpe, Acroterion, FaerieInGrey, Mg007, Magioladitis, Wizymon, Kilogray, Karlhahn, Pedro, FJM, Jjpratt, Swikid, Bongwarrior, VoABot II, Mgmirkin, Sushant gupta, Mindgame123, باسم, Rich257, Nikevich, ClaudeSB, Indon, Cgingold, JJ Harrison, Efwiz, 793184265, Just James, DerHexer, EightBall1989, Edward321, Khalid Mahmood, Eeera, TheRanger, SquidSK, Zahakiel, MartinBot, Sm8900, Kostisl, CommonsDelinker, Leyo, Ash, Conundrumer, Tgeairn, Manticore, J.delanoy, Xhb, 2012Olympian, Hans Dunkelberg, Rhinestone K, All Is One, Eliz81, AgainErick, Martino3, RIckOzone, Davidprior, Shawn in Montreal, Katalaveno, McSly, TheTrojanHought, Afluegel, Coollettuce, SJP, Killer Chao, KylieTastic, Juliancolton, Bogdan~enwiki, Strig, Remember the dot, Gwen Gale, Nigelloring, Mike V, S, Izno, CardinalDan, Idioma-bot, Spellcast, Beatnik Party, Macedonian, Soliloquial, Philip Trueman, PNG crusade bot, Malinaccier, RagnarokEOTW, Hqb, Andrewrhchen, Ask123, Someguy1221, Corvus cornix, Plasmadyne, Saibod, LeaveSleaves, Opiate5555, Ilyushka88, Guest9999, Maxim, Quindraco, Ilke71, Madhero88, K10wnsta, Nagy, Kharissa, Wagaf-d, Insane Burner, News0969, Hmwith, Noobz1112, Hans-wai, Derfel071, Linarutouzumaki, Qlarsen, RJaguar3, Smsarmad, Yintan, Revent, Stonejag, Djdan4961, YemeniteCamel, Flyer22, Tiptoety, James.Denholm, Sunday8, Dingemansm, LarryRiedel, Huevosgrandes, DevOhm, Hardware Hank, Svick, Reginmund, Vituzzu, C'est moi, Spartan-James, Bharatmittal2007, Bwili00, Mygerardromance, Hamiltondaniel, Dust Filter, DRTllbrg, Denisarona, Velvetron, Atif.t2, Jrw6736, Kunak, Martarius, Jiminezwaldorf, ClueBot, Rumping, Connor.carey, GorillaWarfare, Cigarshaped, The Thing That Should Not Be, Liverpoolian!, Rjd0060, MahriMan, Wysprgr2005, Sterlsilver, Alex Lazar, Drmies, Razimantv, CounterVandalismBot, Neverquick, Sniffles22, LeoFrank, Excirial, John Nevard, Wikiuser96, Megiddo1013, Tyler, Dspark76, Subdolous, Editor510, BOTarate, Aitias, Kruusamägi, ShipFan, AnonyScientist, Crowsnest, Htmlcoderexe, Deathknightlord, 03md, XLinkBot, Vivaviagra, Fastily, Spitfire, Gnowor, Roxy the dog, Aldude14, NatulisKaos, Ost316, Avoided, Skarebo, Petedskier, M2o6n0k2e7y3, Mhsb, Firebat08, Eleven even, JinJian, Raso mk, Earvy, Marian123456789, Addbot, Some jerk on the Internet, Wiki guy752, NVLAUR, Fgnievinski, Devansh.sharma, 15lsoucy, Fieldday-sunday, CanadianLinuxUser, Leszek Jańczuk, Ashanda, MrOllie, Maniac7553, Ld100, MrAnderson7, BobMiller1701, Justpassin, Numbo3-bot, Jdawg34520, Tide rolls, Jan eissfeldt, Ostrichdude7, Gail, Ochib, Yobot, TaBOT-zerem, DisillusionedBitterAndKnackered, Fizyxnrd, Wikipedian2, Tamtamar, Eric-Wester, MacTire02, Vsalimova, AnomieBOT, Eminem69041, 1exec1, Jim1138, IRP, Galoubet, World-Famous K-Mart, Piano non troppo, Dinesh smita, EryZ, Materialscientist, Spirit469, Rtyq2, The High Fin Sperm Whale, Citation bot, Jtamad, LouriePieterse, OllieFury, Maxis ftw, Neurolysis, Hammack, Hatchetman459, Capricorn42, Melifluous One, Gilo1969, Romodahomo, Tomwsulcer, Inferno, Lord of Penguins, WingedSkiCap, Brflorance, Blungee, YukioSanjo, Ruy Pugliesi, Nayvik, Honeyfur, Franco3450, Bellerophon, JediMaster362, Terrierhere, Shadowjams, RDC23, AutoMe, LucienBOT, Originalwana, Paolo Di Febbo, Danzence, Recognizance, Saehrimnir, Hgevs, Kthapelo, Allowrocks2003040957, Drew R. Smith, Citation bot 1, Pinethicket, I dream of horses, Boulaur, Elockid, Hard Sin, Jonesey95, Tom.Reding, V.narsikar, RedBot, KaliHyper, ContinueWithCaution, Jauhienij, IVAN3MAN, Gamewizard71, FoxBot, Iæfai, ItsZippy, Sternenstaub~enwiki, Vrenator, Begoon, Raidon Kane, Reaper Eternal, Stalwart111, Pedalle, Pranavsharma67, Jhenderson777, Adi4094, Earthandmoon, Unbitwise, Pimp3245, Minimac, Angelito7, DARTH SIDIOUS 2, Whisky drinker, NameIsRon, Legofreak666, Beyond My Ken, Cjc38, EmausBot, WikitanvirBot, Optiguy54, Smithu1976, Kavantiger62, Wikipelli, Billy9999, JSquish, Susfele, Claudio M Souza, Fæ, Melmaola, Ш, PMcQuillen, Sivaraman99999, A930913, Wayne Slam, Tolly4bolly, Rassnik, Maxfisch, Donner60, Orange Suede Sofa, ChuispastonBot, Explica, Rocketrod1960, Jungleback, Warharmer, Yamagawa10k, Rudolfensis, ClueBot NG, La bla la, Chocolateoak, SusikMkr, Movses-bot, Iwsh, GoldenGlory84, TheOriginalMexi, Abdossamad Talebpour, Lookb4uleap, Frietjes, Mesoderm, Marechal Ney, Widr, Mtheodric, Pluma, MerllwBot, Helpful Pixie Bot, Ba11zooka, Bibcode Bot, WNYY98, Midashboy, Lowercase sigmabot, BG19bot, Radrac, Northamerica1000, Sol1869, MusikAnimal, Leo Yeei, Juuomaqk, Vincent Liu, Ninney, FutureTrillionaire, PlasmaSoul, Bapi 123, Trevayne08, Sparkie82, Shawn Worthington Laser Plasma, Glacialfox, BattyBot, Joems1324, WhiteNebula, Devin 496, APerson, Dexbot, Codename Lisa, Mogism, Denis Fadeev, Saehry, Lugia2453, Isarra (HG), Plasmafroid, Moodilman, Jamesx12345, AI126 at wiki, Mruthun Rajkumar, Mark viking, Ruby Murray, Paraneoz, AmaryllisGardener, Eyesnore, 2cuteforyou, CarbonLawyer, Three1two, Backendgaming, Qwertyasdfqwertyfdsa, Comp.arch, Wikisekharja, Dhdpla, ReconditeRodent, The Herald, Prokaryotes, Ultimorino, VirusEditor, Anrnusna, Meteor sandwich yum, JaconaFrere, CHEZBALZ, KillerBottox123, Mickedice, Mahusha, SkateTier, Rahup007, Сяра, Adamekjiri, Asdfghjkl1234567890zxcvbnm, Fleivium, Brant9922, Narliu, Stefan.nettesheim, Tetra quark, Isambard Kingdom, Khalood246, Samfart20, KasparBot and Anonymous: 1032

- **Bose–Einstein condensate** *Source:* https://en.wikipedia.org/wiki/Bose%E2%80%93Einstein_condensate?oldid=680011496 *Contributors:* Kpjas, CYD, Archibald Fitzchesterfield, Bryan Derksen, Olof, Tarquin, Gareth Owen, Josh Grosse, Hfastedge, Spiff~enwiki, Michael Hardy, Gabbe, TakuyaMurata, SebastianHelm, Alfio, Ellywa, Cyp, Stevan White, Darkwind, Glenn, Mxn, Schneelocke, Loren Rosen, Feedmecereal, Dino, Wikiborg, The Anomebot, ElusiveByte, BenRG, JorgeGG, Donarreiskoffer, Chris 73, Nurg, Robinh, GreatWhiteNortherner, Dave6, M-Falcon, Matt Gies, Giftlite, Smjg, Inter, Herbee, Dratman, Tom~, Eequor, Balenman, Chrissmith, Mooquackwooftweetmeow, Toytoy, XxPantherNovaXx, Fangz, Piotrus, Karol Langner, Brian Jackson, Spiralhighway, Sam Hocevar, Kramer, Nickptar, Vivacissamamente, Grunt, Eep², NightMonkey, Lone Isle, Noisy, Discospinster, Guanabot, ThomasK, Vsmith, Aardark, Paul August, Bender235, TOR, RJHall, El C, Lycurgus, Ruyn, Laurascudder, Jpgordon, Fuxx, Directorstratton, Slicky, Sasquatch, Haham hanuka, Alansohn, Arthena, Nwinther, PAR, Pion, Kfitzgib, Cjnm, Tom12519, Snowolf, Einstein9073, BRW, KapilTagore, Pauli133, Gene Nygaard, Joriki, OwenX, Woohookitty, Linas, David Haslam, Benbest, Ruud Koot, Jeff3000, Astrophil, BlaiseFEgan, Bugman, Sjö, Rjwilmsi, Amire80, Rillian, BlueMoonlet, Salix alba, Keimzelle, Exeunt, Azure8472, FlaBot, SchuminWeb, The.valiant.paladin, Shade², Pete.Hurd, Srleffler, Erik4, King of Hearts, Chobot, DVdm, Sasoriza, YurikBot, Wavelength, Taurrandir, Rob T Firefly, Hairy Dude, Huw Powell, Flameviper, Michael Slone, JabberWok, David Woodward, Shell Kinney, WulfTheSaxon, Truetyper, Howcheng, Chakazul, Katrielalex, Dogcow, Grafikm fr, Zwobot, Wangi, Wknight94, FF2010, 2over0, Closedmouth, Dr.alf, Stuhacking, Otto ter Haar, Groyolo, Allium, SmackBot, Serg3d2, RossyMiles, Oxford Comma, Olegt1, Nickst, Eskimbot, Dilbert3, Gaff, Kmarinas86, Thumperward, Hichris, DHN-bot~enwiki, Raistuumum, Salmar, Tcb Beany, Karpita, Can't sleep, clown will eat me, Nick Levine, Kelvin Case, Neo139, Onorem, MBlume, Voyajer, Rrburke, Xyzzyplugh, GeorgeMoney, TedE, Bigmantonyd, Rich.lewis, DMacks, Xiutwel, Ligulembot, Mion, Sadi Carnot, Josellis, Tethros, Lucretius~enwiki, Lambiam, Andi47, John, MagnaMopus, Jaganath, SteveG23, Mgiganteus1, Ckatz, BillFlis, Kyoko, Dicklyon, Inquisitus, Phuzion, Brienanni, Iridescent, JMK, Clarityfiend, FelisSchrödingeris, Frank Lofaro Jr., CRGreathouse, ZICO, BeenAroundAWhile, DSachan, Orannis, Myasuda, Leakeyjee, Equendil, Stebbins, Kanags, MC10, Tashafairbairn, Mato, Gogo Dodo, JFreeman, Mattjball, Omicronpersei8, Thijs!bot, Epbr123, Wikid77, Trevheg, Fiction Alchemist, Sam Van Kooten, Headbomb, Second Quantization, Iviney, CharlotteWebb, AntiVandalBot, 17Drew, Gökhan, MSBOT, Boleslaw, Sinnerwiki, Sophosmoros, Magioladitis, WolfmanSF, VoABot II, Sushant gupta, Bakken, Ggorelik, Tonyfaull, BatteryIncluded, Dirac66, David Eppstein, LorenzoB, Talon Artaine, Torsionalmetric, Starryharlequin, N734LQ, Anonymous 57, Sketchjoy, Custos0, J.delanoy, MITBeaverRocks, Jtw11, Bogey97, Maurice Carbonaro, AquamarineOnion, Glaux, AppleMacReporter, AntiSpamBot, Tendays, Enix150, Neil Dodgson, Idioma-bot, Austinmohr, Gnipahellir, A4bot, Qxz, Martin451, Mitchell26, Natural Philosopher, Mazarin07, Akhuettel, Spinningspark, Kapalama, Cryonic07, PaddyLeahy, Biscuittin, Awemond, WereSpielChequers, Cmossol, Matthew Yeager, Deathgleaner, Reuqr, Likebox, JD554, Reinderien, Topher385, Scorpion451, Lightmouse, Jakeng, Coldcreation, Psycherevolt, Melcombe, AllHailZeppelin, Crazz bug 5, Martarius, ClueBot, MonkeyMensch, Snigbrook, The Thing That Should Not Be, EoGuy, Emil70, Zero over zero, Razimantv, Maymay, Thegeneralguy, DrakeUnlimited, NuclearWarfare, Dboiko, Doktor Mephisto, SchreiberBike, Thingg, Jonverve, SoxBot III, Egmontaz, DumZiBoT, Ost316, Rreagan007, SilvonenBot, SkyLined, MaizeAndBlue86, Csingh23592, Addbot, DOI bot, Jojhutton, Miskaton, Friginator, Download, ChenzwBot, AtheWeatherman, 84user, Tide rolls, Lightbot, OlEnglish, Teles, SPat, Megaman en m, Ben Ben, Yobot, Wireader, VectorField, AnomieBOT, TheUfoFiles, Aaagmnr, Materialscientist, Limideen, Citation bot, MetaplecticGroup, Natural RX, Xqbot, BME-physics, Lunaintern, RibotBOT, Verbum Veritas, Nixón, HJ Mitchell, Quantum 235, Citation bot 1, Maan361, Gil987, Gaba p, I dream of horses, Coekon, RedBot, Akalabeth, Keri, Asrrin29, Senra, Canuckian89, JV Smithy, DARTH SIDIOUS 2, Obankston, Hajatvrc, Nkf31, EmausBot, John of Reading, Gfoley4, Physics16, GoingBatty, KHamsun, Solarra, Lent1999, H3llBot, Quondum, Timetraveler3.14, DougEFresh1122, Donner60, Fairskys, Carmichael, Jalexander-WMF, ClueBot NG, Gareth Griffith-Jones, Movses-bot, All Hail Hypnotoad!, Zak.estrada, Widr, Fqr2010, Helpful Pixie Bot, HMSSolent, Jubobroff, Bibcode Bot, BG19bot, Northamerica1000, AvocatoBot, Wowwii, Rm1271, Mr.viktor.stepanov, BattyBot, Mrt3366, ChrisGualtieri, Adwaele, Baileybrooks, FlappyJenkins, Dexbot, Makecat-bot, Baldoc83, Jamesx12345, Stewwie, Avrahamleib, Sakurai23, Marcela louis, Reatlas, Epicgenius, Nonsenseferret, Aarya19991111, Ginsuloft, Aritcle, KillerKira, John Doppler, Aarjun Rampal, Happy Attack Dog, Arnaud Migres, Tusharkashyap2001, AfrikanischePost, Hans8654, Sumandark8600, Shengxingwu, Antsiepantsie, Yohoona, Mysterious Gopher, KasparBot and Anonymous: 496

- **Fermionic condensate** *Source:* https://en.wikipedia.org/wiki/Fermionic_condensate?oldid=678482641 *Contributors:* CYD, SebastianHelm, Glenn, Cimon Avaro, Schneelocke, Charles Matthews, Fuzheado, Phys, Jeffq, Fredrik, Cedars, Xerxes314, Steuard, Gdr, Srbauer, Laurascudder, Prsephone1674, Lysdexia, Camw, Bluemoose, Nightscream, BradBeattie, YurikBot, Hellbus, Długosz, Alain r, That Guy, From That Show!, KnightRider~enwiki, SmackBot, Hmains, Kmarinas86, Kcordina, Ohconfucius, WhiteHatLurker, Usgnus, WeggeBot, Difluoroethene, Thijs!bot, Mbell, Headbomb, AntiVandalBot, Majorly, Maliz, MartinBot, Philip Trueman, ClueBot, The Help Fishy, Vanished user uih38riiw4hjlsd, MystBot, Addbot, Mjamja, Luckas-bot, AnomieBOT, Citation bot, Cowgoesmoo2, ProtectionTaggingBot, Tom.Reding, Full-date unlinking bot, EmausBot, KHamsun, ZéroBot, Alpha Quadrant (alt), Aeonx, Quondum, ClueBot NG, Yen-Tzu and Anonymous: 44

- **Quark–gluon plasma** *Source:* https://en.wikipedia.org/wiki/Quark%E2%80%93gluon_plasma?oldid=681306263 *Contributors:* Taw, Michael Hardy, Cyde, Karada, SebastianHelm, Charles Matthews, David Newton, Grendelkhan, Phys, Dmytro, David Edgar, JerryFriedman, Art Carlson, Herbee, Rick Block, HorsePunchKid, Mako098765, Deglr6328, Squash, MuDavid, Ylai, Bender235, CheekyMonkey, Haxwell, Bradkittenbrink, Enric Naval, Cmdrjameson, Supercrisis, Jag123, Sam Korn, Fwb22, Anthony Appleyard, Axl, Hu, Knowledge Seeker, Cal 1234, Vuo, Joriki, Firsfron, Mpatel, GregorB, SDC, Palica, RichardWeiss, Ashmoo, Yuriybrisk, Maros, Ae77, Bubba73, Nihiltres, Goudzovski, Silversmith, YurikBot, Wavelength, Mushin, Bambaiah, Hairy Dude, Hellbus, Salsb, NawlinWiki, CecilWard, E2mb0t~enwiki, Curpsbot-unicodify, Ilmari Karonen, Ybbor, KasugaHuang, Neier, SmackBot, Stepa, PeterSymonds, Skizzik, Kmarinas86, Chris the speller, Silly rabbit, Jbergquist, Khazar, Dark Formal, Vampus, Fangfufu, JayHenry, Petr Matas, CmdrObot, Foice, Van helsing, Ruslik0, Michael C Price, Thijs!bot, Headbomb, Nick Number, Eb.eric, JAnDbot, Xeno, Yill577, Savant13, 28421u2232nfenfcenc, Ethron, MartinBot, Pagw, CommonsDelinker, J.delanoy, Maurice Carbonaro, Jeepday, DorganBot, 1812ahill, Momo Hemo, Fences and windows, BotKung, Pamputt, Ptrslv72, AlleborgoBot, Logan, SieBot, BotMultichill, Triwbe, Maelgwnbot, ClueBot, Flaming, Thunderhippo, Brews ohare, Mstrickl, Healyhatman, DumZiBoT, XLinkBot, Oldnoah, SkyLined, Truthnlove, Stormcloud51090, Addbot, Mjamja, Qmark42, Tide rolls, Lightbot, OlEnglish, מלמד בין, Luckas-bot, Yobot, Amirobot, 4th-otaku, AnomieBOT, Essin, Citation bot, ArthurBot, ProtectionTaggingBot, False vacuum, Ciceronibus, FrescoBot, Citation bot 1, Naxuesen, Tom.Reding, RedBot, Johann137, IVAN3MAN, Meier99, Puzl bustr, EmausBot, Mnkyman, Naznin farhah, ZéroBot, SalGiandinoto, Arbnos, Yiosie2356, SporkBot, Jesanj, Rangoon11, ClueBot NG, Jack Greenmaven, Raktimabir, Theopolisme, Helpful Pixie Bot, Bibcode Bot, 2001:db8, Shawn Worthington Laser Plasma, BattyBot, Kalmiopsiskid, Chemya, Saehry, Epicgenius, Prokaryotes, Polytope24, Pcharito, Vieque, Sofia Koutsouveli, KH-1, Crystallizedcarbon, Isambard Kingdom, Qulos and Anonymous: 115

- **Atom** *Source:* https://en.wikipedia.org/wiki/Atom?oldid=679943114 *Contributors:* AxelBoldt, Trelvis, Lee Daniel Crocker, Mav, Zundark, The Anome, Tarquin, AstroNomer~enwiki, Stokerm, Andre Engels, Youssefsan, Branden, Ben-Zin~enwiki, Drbug, Heron, Comte0, PeterBohne, Stevertigo, Hfastedge, Lir, Patrick, Infrogmation, D, Michael Hardy, Tim Starling, FrankH, Pit~enwiki, Wapcaplet, Ixfd64, Bcrowell, Dcljr, Shoaler, Delirium, PingPongBoy, Card~enwiki, NuclearWinner, Looxix~enwiki, Mdebets, Ahoerstemeier, LoonBB, Suisui, Angela, Jebba, Darkwind, Александър, Julesd, Glenn, Poor Yorick, Andres, Kaihsu, Evercat, Samw, Rob Hooft, Denny, Schneelocke, Ddoherty, Frieda, Timwi, Wikiborg, Stone, Dysprosia, The Anomebot, Doradus, Hr oskar, Big Bob the Finder, Morwen, VeryVerily, SEWilco, Flockmeal, Ldo, BenRG, Gromlakh, Donarreiskoffer, Gentgeen, Robbot, ChrisO~enwiki, Arkuat, Merovingian, Sverdrup, Der Eberswalder, DHN, Caknuck, Rebrane, Paul G, Hadal, UtherSRG, Wikibot, Roozbeh, Mandel, Anthony, Diberri, Cutler, Dina, Timemutt, Tobias Bergemann, David Gerard, Ancheta Wis, Vonkwink, Giftlite, Christopher Parham, Awolf002, Mikez, Palapala, Yuri koval, Inter, BenFrantzDale, Tom harrison, MSGJ, Xerxes314, Everyking, Curps, Michael Devore, Bensaccount, Frencheigh, Guanaco, Andrea Parri, Jorge Stolfi, SWAdair, Jurema Oliveira, Kandar, Stevietheman, Gadfium, Utcursch, Bact, Pcarbonn, Antandrus, OverlordQ, MisfitToys, Piotrus, Kaldari, Jossi, Karol Langner, JimWae, DragonflySixtyseven, Icairns, Tail, Sam Hocevar, Darksun, Engleman, Deglr6328, Achven, M1ss1ontomars2k4, Adashiel, Trevor MacInnis, Grunt, Bluemask, Freakofnurture, Amxitsa, DanielCD, Discospinster, Rich Farmbrough, Guanabot, Huffers, Hidaspal, Rama, Vsmith, Guanabot2, Wadewitz, MarkS, Jlcooke, ESkog, Kbh3rd, Pmetzger, RJHall, El C, Edward Z. Yang, Shanes, Art LaPella, RoyBoy, Bookofjude, CDN99, Afed, Bobo192, Army1987, Harley peters, Whosyourjudas, Shenme, Viriditas, Brim, Maurreen, Richi, Boxed, Timl, Joe Jarvis, Nk, Deryck Chan, Thewayforward, PiccoloNamek, Obradovic Goran, MPerel, Sam Korn, Pearle, Benbread, Nsaa, Ogress, HasharBot~enwiki, Jumbuck, Storm Rider, JYolkowski, Mennato, Richard Harvey, Thebeginning, Slugmaster, Iothiania, Riana, InShaneee, Spangineer, Malo, Snowolf, Saga City, Dirac1933, Geraldshields11, Henry W. Schmitt, Bsadowski1, BlastOButter42, Com-

Ultraorange260, Vishnava, Mac Dreamstate, WFPM, Apemaster3000, Cst17, Download, LaaknorBot, Chamal N, CarsracBot, Glane23, Tech30, Akazme93, Debresser, AnnaFrance, Favonian, Chateau Brillant, AtheWeatherman, LinkFA-Bot, Sin Aura, Omg123123, IOLJeff, Eas4200c.team0, Numbo3-bot, Psproots, Koliri, Tide rolls, Bfigura's puppy, Luckas Blade, Jarble, Emperor Genius, GiantPea, KarenEdda, Georgieboi123, Angrysockhop, Legobot, Northexit182, Drpickem, XxGOWxx, Luckas-bot, Yobot, Googins, Senator Palpatine, Julia W, Yiplop stick stop, Cepheiden, Wikipedian Penguin, Nerdguy, Milesarise, The Bacon Machine, Jmproductionss, Cowdudemanxboyguymale-senorninoxxxxxxx, Ayrton Prost, Blk48, Yami89~enwiki, Chadisasexybeast, Skin and Bones, Kingpin13, Abshirdheere, Pokehen, Csigabi, Materialscientist, The High Fin Sperm Whale, Citation bot, E2eamon, Xqbot, TinucherianBot II, Intelati, GeometryGirl, Jeffrey Mall, Tolmanator, Tad Lincoln, Grim23, P99am, Almabot, Quixotex, GrouchoBot, Harley2121, Capecodsamm, Omnipaedista, Bahahs, Brutaldeluxe, GhalyBot, MerlLinkBot, N419BH, SchnitzelMannGreek, Sesu Prime, Pauswa, Legobot III, Tobby72, Dogbert66, Astronomyinertia, Cargoking, Machine Elf 1735, Finalius, Saiarcot895, Diremarc, Citation bot 1, Kobrabones, Clevercheetah123, Biker Biker, Pinethicket, MBirkholz, HRoestBot, Tom.Reding, RedBot, Bigad1, Footwarrior, White Shadows, Bsece010, FoxBot, Trappist the monk, Sweet xx, Sheogorath, Mrrstatham, Vrenator, Extra999, EzraNemo123, Burnthefairy56, Theproatlearningstuff, Loki2022, Reach Out to the Truth, Minimac, Marie Poise, DARTH SIDIOUS 2, RjwilmsiBot, TjBot, Ling cao, Edouard.darchimbaud, DASHBot, B20180, EmausBot, Thecreator09, Orphan Wiki, Immunize, Gfoley4, Pete Hobbs, 271196samantha, Smallchief, Hhhippo, Werieth, JSquish, CanonLawJunkie, Arik Islam, Mubasher55, Kbop451, Chaitana, Sambam340, H3llBot, Quondum, DanDao, Ahmetkilit, Lizzy chic38, Kobinks, Wayne Slam, Acdcfan1223, Bender176, Coasterlover1994, Pussvock, Thomasroper, Maschen, Dcgunasekar, Epicstonemason, Lilgas52, Puffin, Happiestmuackz11, ChuispastonBot, RockMagnetist, Oliozzicle, Lo79y123, Benjthedivine, DASHBotAV, ResearchRave, Petrb, ClueBot NG, Ulflund, Hiperfelix, Mitch09876, Sagenfowler, Big-Jason99, Hazhk, Moneya, O.Koslowski, Rezabot, Danim, Helpful Pixie Bot, Art and Muscle, Jubobroff, Bibcode Bot, Branthecan, Leonxlin, Malvel, Mysterytrey, Goodfluff, Xyzmathrules, Begman5, Fuppamaster, Jrobbinz1, Rgbc2000, Typoltion, Dr.Nadon, Connorbishop, IMn00b, Arghya33, DGK1318, Shawn Worthington Laser Plasma, Alarbus, Life421, BattyBot, Jimw338, Hebert Peró, ChrisGualtieri, GoShow, 786b6364, JYBot, Dexbot, Mogism, Falktan, GeeBIGS, MeteMetheus, Jamesx12345, RandomLittleHelper, Reatlas, Anastronomer, Raghav Bhalerao, Cowfdashkli, Pirtert, Homaa, AmericanLemming, Praemonitus, Tedsanders, PianoEngineer18, Mandruss, Anythingcouldhappen, Susan.grayeff, Anrnusna, Stamptrader, Joshdude356, Jazzmusician94, IStoleThePies, Mahusha, Monkbot, IiKkEe, Suntrax south, Mario Castelán Castro, Narky Blert, Y-S.Ko, Tetra quark, Isambard Kingdom, Forscienceonly, CV9933, Nøkkenbuer, KasparBot, White909090lightning, Myaccount7 and Anonymous: 1481

- **Degenerate matter** *Source:* https://en.wikipedia.org/wiki/Degenerate_matter?oldid=680892569 *Contributors:* CYD, Bryan Derksen, Olof, AstroNomer~enwiki, Roadrunner, Michael Hardy, Alan Peakall, HarmonicSphere, Julesd, Doradus, Zoicon5, Donarreiskoffer, Robbot, Donreed, Jheise, DocWatson42, Herbee, Leperous, WhiteDragon, Superborsuk, Bender235, ZeroOne, Lycurgus, Cherlin, ChristopherWillis, Stephan Leeds, VivaEmilyDavies, RJFJR, Dirac1933, Pauli133, Gene Nygaard, Dan East, Killing Vector, Linas, Mindmatrix, Christopher Thomas, Kgbudge, Rnt20, Ashmoo, Eyu100, Wragge, Karch, YurikBot, TSO1D, Conscious, Hede2000, Ytrottier, Hellbus, Anomalocaris, Trovatore, John Newbury, Enormousdude, Modify, Mhenriday, RG2, Tim314, SmackBot, Eskimbot, Lainagier, Ohnoitsjamie, Grokmoo, Kmarinas86, Oni Ookami Alfador, Voyajer, Wikipedia brown, Percommode, Aldaron, ChowRiit, Just plain Bill, Dark Formal, Jaganath, JorisvS, Anescient, Newone, Julian.cancino, Thijs!bot, Headbomb, Yellowdesk, Myanw, JAnDbot, WolfmanSF, Dirac66, Su-no-G, Warren Dew, MartinBot, STBot, Steve98052, Sofar 2, Vranak, Holme053, Kurgus, Lechatjaune, Anna Lincoln, Mardhil, Entropy1963, Anchor Link Bot, Kallog, Denisarona, WurmWoode, PipepBot, Garyzx, SuperHamster, Masterblooregard, Chief buffalo chip, Brews ohare, Coinmanj, Arjayay, 2, Tealwisp, SkyLined, Addbot, Dsmith77, 84user, Phynisha 25, Luckas-bot, Captain Quirk, Hunnjazal, StrontiumDogs, Brithans, Nickkid5, Tomdo08, INick3, Omnipaedista, RibotBOT, Seeleschneider, FrescoBot, Mossmanj, Vhann, Fkmusgrave, Kevinpeck, Fartherred, TobeBot, Erixmix, Hhhippo, HiW-Bot, Liquidmetalrob, Ὁ οἶστρος, Quondum, MajorVariola, ClueBot NG, Bibcode Bot, BG19bot, Trevayne08, Zedshort, Shawn Worthington Laser Plasma, Sschongster, ChrisGualtieri, Bterranova, Zinganthropus, Phleg1, Jjusiopao, Volker Siegel, DoisKoh, ErikNatuurkunde and Anonymous: 107

- **QCD matter** *Source:* https://en.wikipedia.org/wiki/QCD_matter?oldid=650185710 *Contributors:* Ixfd64, Xerxes314, Niteowlneils, WhiteDragon, The Land, FT2, Army1987, Mac Davis, Vuo, Kazvorpal, Ceyockey, Davidfstr, GregorB, SeventyThree, Marudubshinki, Leapfrog314, Nanite, Goudzovski, YurikBot, Wavelength, Bambaiah, Hairy Dude, Archelon, Salsb, SCZenz, Dhollm, Nekura, SmackBot, Mcneile, Colonies Chris, Dark Formal, Mgiganteus1, Xionbox, Dan Gluck, CmdrObot, Headbomb, "", Beasticles, Igodard, WolfmanSF, HiB2Bornot2B, Maurice Carbonaro, LokiClock, Fences and windows, KP-Adhikari, Jmath666, SieBot, Martarius, SteelSoul, Brews ohare, Ordovico, TimothyRias, IngerAlHaosului, Addbot, Rbwolf, OlEnglish, Luckas-bot, Yobot, Citation bot, Eumolpo, Mnmngb, Tom.Reding, 23790AD, Johann137, Trappist the monk, Naviguessor, Gerasime, Suslindisambiguator, Helpful Pixie Bot, Curb Chain, Bibcode Bot, Anwarwaseem, Mogism, RhinoMind and Anonymous: 35

- **Strange matter** *Source:* https://en.wikipedia.org/wiki/Strange_matter?oldid=669591842 *Contributors:* CYD, Bryan Derksen, The Anome, Frecklefoot, Oliver Pereira, SebastianHelm, Schneelocke, Vespristiano, Giftlite, Curps, Beland, EricJamesStone, Pjacobi, Sam Korn, Fwb22, Proteus71, Joriki, Scriberius, Mpatel, Squideshi, Strait, Oalsaker, Eyu100, Margosbot~enwiki, Nihiltres, Spacepotato, Bambaiah, Hellbus, Salsb, NawlinWiki, Kooky, 2over0, JDspeeder1, Algae, Mtffm, MacsBug, SmackBot, Tinz, Xaosflux, Kmarinas86, Marcus Brute, Dark Formal, Yanwen, Zzzzzzzzzzz, Dan Gluck, Siebrand, Michaelbusch, Eassin, OS2Warp, CalebNoble, Orca1 9904, Nilfanion, Zomic13, Chrislk02, Vidale, Mrph, VoABot II, CarlFeynman, Hurax, Idioma-bot, ABF, Fences and windows, Lamro, SieBot, Lethesl, ClueBot, PixelBot, Snookumz, Homocion, SkyLined, Addbot, Micke, Yobot, AnomieBOT, Xqbot, GrouchoBot, ProtectionTaggingBot, Nikto, Johann137, WikitanvirBot, Hhhippo, ZéroBot, Nobelium, Suslindisambiguator, ClueBot NG, Astrocog, Frietjes, Helpful Pixie Bot, Rasheeq1, Cavaliere1, Sebastian5059, Ajomannen, RhinoMind, Erikprantare and Anonymous: 55

- **Phase diagram** *Source:* https://en.wikipedia.org/wiki/Phase_diagram?oldid=669692687 *Contributors:* Uriyan, Tarquin, Gareth Owen, Michael Hardy, Booyabazooka, Kku, GTBacchus, Charles Matthews, Dino, Sbwoodside, Doradus, Phys, Denelson83, Robbot, Lowellian, Diderot, MOiRe, Giftlite, Tom harrison, Beland, H Padleckas, Urhixidur, EugeneZelenko, Discospinster, Vsmith, Ben Standeven, Rgdboer, Femto, Bobo192, Daelanus~enwiki, Mdd, Linuxlad, Arthena, Bantman, Velella, Skatebiker, Oleg Alexandrov, Linas, Karnesky, V8rik, Li-sung, Stardust8212, Ligulem, FlaBot, Alphachimp, CJLL Wright, Borgx, Hairy Dude, Phantomsteve, RussBot, Russell C. Sibley, Sasuke Sarutobi, Yyy, Grafen, Wknight94, Closedmouth, Alexandrov, Cmglee, Mlibby, Itub, KnightRider~enwiki, SmackBot, Terrancommander, Davepape, Hydrogen Iodide, Bigbluefish, BiT, Michbich, Roscelese, Glloq, Vanis314, Vsaishravan, DavidJ710, Rspanton, Reptile209, Khazar, JorisvS, Hemmingsen, Diverman, Arctixfox, EdC~enwiki, Dr.K., Wizard191, JoeBot, LadyofShalott, Lavaka, Meisam.fa, AndyVolykhov, TheTito, Cydebot, OneEyedD0rf, Thijs!bot, Escarbot, JAnDbot, Riceplaytexas, Avicennais, JJ Harrison, Dirac66, Mpa5220, Mythealias, Uvainio, Numbo3, El Belga, Bdodo1992, Afluegel, Rémih, VolkovBot, AlnoktaBOT, Philip Trueman, TXiKiBoT, A4bot, Petergans, Barkeep, SieBot,

Gerakibot, Kopeliovich, Baderimre, Decoratrix, Ken123BOT, ClueBot, PipepBot, Mild Bill Hiccup, Excirial, CohesionBot, CarlosPatiño, Sdrtirs, DumZiBoT, XLinkBot, AngelHerraez, Ngebbett, Matthieumarechal, Addbot, Emok, Ronhjones, Cst17, PentonCourt, Yanksports22, Zorrobot, Arbitrarily0, Luckas-bot, Legobot II, Choij, Piano non troppo, Materialscientist, Ge215, Pvkeller, J04n, Logger9, Alex685, Dr. Walrusgfs, Anterior1, Dheknesn, Duoduoduo, Copistopplayer, TjBot, EmausBot, John of Reading, WikitanvirBot, Bkappius, ClueBot NG, Dltlek, Helpful Pixie Bot, Haminoon, Leslieglasser, Monkbot, BethNaught, KasparBot and Anonymous: 102

• **Antimatter** *Source:*https://en.wikipedia.org/wiki/Antimatter?oldid=682100270*Contributors:*Magnus Manske, MichaelTinkler, Bryan Derk-sen, Robert Merkel, Zundark, Mark Ryan, Andre Engels, XJaM, William Avery, SimonP, Stevertigo, Patrick, Kchishol1970, Michael Hardy,Tim Starling, Tango, Delirium, Andrel, SebastianHelm, Minesweeper, Glenn, BenKovitz, Evercat, Emperorbma, Timwi, Dysprosia, TheAnomebot, Furrykef, Ozuma~enwiki, Morwen, Saltine, Rei, Phys, Omegatron, Frazzydee, Jeffq, Phil Boswell, Vespristiano, Yelyos, Symeon~enwiki,Gandalf61, Academic Challenger, Meelar, DHN, Intangir, Hadal, Papadopc, Wikibot, HaeB, Peter L, Alan Liefting, Matt Gies, Giftlite,DocWatson42, Harp, Pretzelpaws, BenFrantzDale, Geeoharee, Fastfission, Herbee, Xerxes314, Fleminra, Alibaba, Bobblewik, DemonThing, Alexf, Knutux, Antandrus, Mako098765, Piotrus, DRE, Jossi, Karol Langner, DragonflySixtyseven, Elroch, Cructacean, Tmxxine, FoeNyx, Tumbarumba, Ianneub, JFM, Jh51681, Deglr6328, TheObtuseAngleOfDoom, Mh, Eisnel, Mike Rosoft, Ouro, Mormegil, Lone Isle, Discospinster, Rich Farmbrough, Mhowkins, Guanabot, Pak21, Jaedza, FT2, Pjacobi, Vsmith, Jpk, Silence, ArnoldReinhold, Bender235, Kaiser-shatner, RJHall, Livajo, José Gnudista, Bletch, Parklandspanaway, Chairboy, Shanes, C1k3, RoyBoy, Bootedcat, CDN99, Bobo192, Chefox, Dreish, I9Q79oL78KiL0QTFHgyc, Nk, B0at, Zelda~enwiki, Apostrophe, Krellis, Karlheg, Jumbuck, Alansohn, Gary, Anthony App-leyard, 119, Arthena, Scottib, Keenan Pepper, Ronline, Iothiania, Ciaran H, Sligocki, Pion, Idont Havaname, Bart133, Cookiemobsta, Wt-mitchell, Rebroad, Keepsleeping, Knowledge Seeker, Cal 1234, Bananaclaw, Count Iblis, Bsadowski1, DV8 2XL, Gene Nygaard, Deroravi, Oleg Alexandrov, Itinerant, Mindmatrix, Daniel Case, StradivariusTV, Uncle G, Robert K S, MONGO, Fett0001, GregorB, CharlesC,Doc Ruby, Nleseul, Christopher Thomas, Palica, Sneakums, Miroku Sanna, Sparkit, Magister Mathematicae, Qwertyus, Edison, Sjö, Rjwilmsi,Nigh tscream, Strait, Marasama, Bill37212, Eyu100, HappyCamper, Ligulem, Bubba73, Matt Deres, Lotu, Yamamoto Ichiro, Titoxd, FlaBot,Ian P itchford, El Cid, Alhutch, JiFish, Hiding, Nivix, Fragglet, RexNL, Ayla, Goudzovski, Alphachimp, Cannywizard, Srleffler, Erik4, Kingof He arts, Chobot, Stephen Compall, Amaurea, YurikBot, Bambaiah, Jimp, Brandmeister (old), Ohwilleke, Arado, Jtkiefer, JabberWok, Fab-ricatio nary, Stephenb, Centurion328, Gaius Cornelius, CambridgeBayWeather, Ihope127, Ugur Basak, MidnightWolf, Shanel, Zhaladshar,Wiki a lf, Mipadi, Bachrach44, BGManofID, Veledan, BryanJones, Unmake, E2mb0t~enwiki, JPMcGrath, Dbfirs, EEMIV, Sabariajay, Dark-fred, E mpty2005, Gzabers, Javajunkiewa, Wknight94, Ms2ger, ReCover, Super Rad!, Gtdp, Chase me ladies, I'm the Cavalry, Arthur Rubin,Esprit1 5d, Shawnc, Jdmitr2nr, Rocketrye12, JLaTondre, Grifter tm, Nimbex, Archer7, Karlwilbur, Kungfuadam, RG2, JDspeeder1, Syedgj,Mussno on, Luk, Sycthos, SmackBot, Amcbride, MattieTK, Terrancommander, Hydrogen Iodide, Od Mishehu, Prototime, Jrockley, Frymaster,AnOdd Name, Edgar181, Dyslexic agnostic, Gilliam, Ccw412, Dauto, Jcarroll, Ottawakismet, Tree Biting Conspiracy, Sbharris, Justintime32,DTR, C an't sleep, clown will eat me, Vladislav, Fiziker, Onorem, Skydiver, Rrburke, VMS Mosaic, LouScheffer, Dbdb, Thrane, Wen DHouse, F lyguy649, Jumping cheese, Massen~enwiki, Nakon, Savidan, Rajrajmarley, Dreadstar, Hteen, Richard001, Wirbelwind, Iridescence,Ligulemb ot, Starghost, Vina-iwbot~enwiki, Bejnar, Pilotguy, Hmoul, SashatoBot, ArglebargleIV, Rory096, DA3N, Titus III, Rodri316, Ken-Fehling, Dhesi, JorisvS, Kingdom heartless, Scetoaux, Lumpio-, Mikieminnow, Hypnosifl, Dammit, Doczilla, Kvng, Shoo tsukino, FredilYupigo, Newone, UncleDouggie, Amakuru, Domitori, MyNikko, MottyGlix, Courcelles, Lynch82, Tawkerbot2, Dlohcierekim, DKqwerty,Harold f, Doceddi, Mellery, Kentaru z, CmdrObot, Tobes00, Mattbr, Van helsing, Blackserenity, GHe, Whatty123, Zbell4, Eternalmonkey, BillPapa, Gog o Dodo, Corpx, Chasingsol, ANTIcarrot, Kyutieboi, DumbBOT, Optimist on the run, SuperGerbil, Daniel Olsen, UberScienceNerd,Cfslattery1 , EnglishEfternamn, DJBullfish, Epbr123, Carlitosrosario, Carloschavez, Orchestral, MrXow, Nonagonal Spider, Headbomb, Moul-der, Norwe gianBlue, Davidhorman, Emstidor, Rhysis, Escarbot, Paleo1, AntiVandalBot, Yonatan, Tyco.skinner, Ostertag~enwiki, Mister Mac-beth, Erfa, RobJ1981, Danger, MECU, Ingolfson, Skomorokh, CosineKitty, Instinct, Db099221, Maalyex, Noimnotokay, S Quinn, Bongwar-rior, VoAB ot II, JamesBWatson, Trugster, Froid, Tonyfaull, Mother.earth, ValleyOfMegiddo, 28421u2232nfenfcenc, Nelon123, JoergenB,DerHexer, J aGa, WLU, Pigman3211, Video game fan11, Stephenchou0722, AussieBoy, EyeSerene, Verkle, Mermaid from the Baltic Sea,Mont95, Tge airn, J.delanoy, Mrtangent, Paulamicela, Yonidebot, RoryFilberg1102, Siryendor, Extransit, Foober, Acalamari, Charmed4ever,McSly, Alph achrome, SteveChervitzTrutane, AntiSpamBot, Linuxmatt, Plasticup, Spoxjox, Cooldude7273, NewEnglandYankee, 83d40m,Student7, Ta naats, MetroCentral, Equazcion, Spiesr, Lighted Match, KuroTatsuijin, Cs302b, Halmstad, Tttecumseh, Thecinimod, Cardi-nalDan, Spell cast, Eipakten, ScarletSpiderfan, Antimattersquared, VolkovBot, Nburden, AlnoktaBOT, Philip Trueman, TXiKiBoT, Osh-wah, Pjstewart , Wikibot.1, JohnBoyTheGreat, Wolfire9, WazzaMan, Fizzackerly, Martin451, LeaveSleaves, Mannafredo, Dljazz, Therma-ctor, SwordSm urf, Enigmaman, Pepve, ImmortalKnight, Enviroboy, Vector Potential, Spinningspark, Bill of rights1-10, Munkay, Neparis,Beckett25, Tid dly Tom, Sakkura, Winchelsea, Alexbook, Agesworth, Crimson Enk, AlbertHall, Edc343, Android Mouse, RadicalOne, Oxy-moron83, The-G-Unit-Boss, Dsmith7707, DeathRidesAhorse, Evanissowrong, Anchor Link Bot, Hamiltondaniel, Simnia, BashM, Martarius,ClueBot, Child arsonist, Hustvedt, The Thing That Should Not Be, Jan1nad, Aguyforyou, Drmies, Ilithi Dragon, Vipanvig, Meph0, Coun-terVandalismBo t, U5K0, Bennirubber, Switchcraft, ChandlerMapBot, Excirial, Crywalt, Jnate19, Eeekster, Beamjockey, Leonard^Bloom,Fdodelap~enwik i, Brews ohare, Executor Tassadar, NuclearWarfare, Cenarium, C628, Thingg, Agentbonbon, Dana boomer, Teacherbrock,DJ Sturm, SoxB ot III, Sparkygravity, Oore, DumZiBoT, AlexGWU, Vexmutz, H0dges, Klotus7, Ost316, Rreagan007, Facts707, WikHead,Mhsb, SkyLined, NCDane, Skeletor 0, Dethdeeler, Mortense, Roentgenium111, Some jerk on the Internet, DOI bot, JosieSmosie, Landon1980,Captain-tucker, F riginator, RHugh-baa, Hah tea bag, Cl8936, AnOddShaman, D0762, Xen64~enwiki, Looie496, Jim10701, Proxima Cen-tauri, LaaknorBot, Bassbonerocks, Austin123457, Sissysue, 84user, Numbo3-bot, VASANTH S.N., Tide rolls, Lightbot, OlEnglish, TenthPlague, ScAvenger, Legobot, आशीष भटनागर, Luckas-bot, Yobot, Nuclear power plank, Fraggle81, Zafhore, Legobot II, Rsquire3, Alexan-der336, Naipicnirp, Tempodivalse, 0rangotang0, Synchronism, AnomieBOT, 1exec1, IRP, Piano non troppo, Ham12343, SJ71, Powerzilla,Citation bot, Maxis f tw, Buda55x, Gawdl3y, Xqbot, Whitley777, Ryantrudeau, Cscseccot, GrouchoBot, ProtectionTaggingBot, Marzs, Mac-bookair3140, Amaury , Doulos Christos, JediMaster362, Tjsnuff, Shadowjams, A. di M., Anti-Chronon, FrescoBot, LucienBOT, Originalwana,Ds1420, DivineAlpha, HamburgerRadio, Citation bot 1, Arctic Night, Jonesey95, El estremeñu, Tom.Reding, Jsjunkie, Meaghan, Random-StringOfCharacters, Sh adeofTime09, Nick b 89, Trappist the monk, Jonkerz, Vrenator, Seahorseruler, 564dude, Rayray59063, Dsfsdfsdf,Brambleclawx, Chinybo y123, DARTH SIDIOUS 2, Sandraobianwuzia, Fcy, Becritical, Vishwarun.kannan, Kerrick Staley, Chibby0ne, DASH-Bot, EmausBot, John of Reading, EbaumsHellYeah, Chuckie.awesome, Octaazacubane, Boundarylayer, RenamedUser01302013, Slightsmile,Wikipelli, Hhhippo, Wer ieth, JSquish, Bryce Carmony, Cobaltcigs, Quondum, Chris81w, L Kensington, Alexbowyer, WaterCrane, Chris857,Irie212, Herk1955, Speci al Cases, Modern Democritus, ClueBot NG, Michaelmas1957, Thiboded, 1300cjs, RaptorHunter, JfLowell, Simon-stone, TheArchman, Oak montowls, 1068411lck, Dhardtke, Szarek, GlassLadyBug, Grouchyowl, Jorgenev, Helpful Pixie Bot, Midlandman,TORNELLcello, Bobttla, Bibcode Bot, DBigXray, Slaughter182, BG19bot, Hoult66, Arionstone, MrBill3, Klilidiplomus, BattyBot, Chris-

Gualtieri, Gdrg22, Dexbot, Aj8uppal, Mr. Guye, Inayity, Yourmotherismyfriend, CuriousMind01, TheDuckLair, Illuusio, Tomasedasonno3, Thermoroach, Reatlas, Anrnusna, Kylejaylee, Monkbot, Shabab Rahman, HMSLavender, Aech34, Epigogue, Thomashly, IGoPro HD, Gaurav6897, KasparBot, Delikeva, Garrettparks, Thewinner01, Bridgetistheboss and Anonymous: 885

- **Dark matter** *Source:* https://en.wikipedia.org/wiki/Dark_matter?oldid=682534510 *Contributors:* AxelBoldt, Chenyu, Derek Ross, CYD, BF, Bryan Derksen, The Anome, Tarquin, Taw, XJaM, Arvindn, William Avery, Roadrunner, Mintguy, Bth, Stevertigo, Edward, Nealmcb, Boud, FrankH, Cprompt, DopefishJustin, Bobby D. Bryant, Ixfd64, SebastianHelm, Alfio, CesarB, Looxix~enwiki, Mkweise, William M. Connolley, JWSchmidt, Glenn, Mxn, Charles Matthews, Timwi, Fuzheado, Rednblu, Haukurth, DW40, Dragons flight, Furrykef, Saltine, Dogface, Populus, Jusjih, Finlay McWalter, Bearcat, Robbot, Zandperl, Korath, Nurg, Naddy, Arkuat, Gandalf61, Pingveno, Rursus, Rtfisher, Wereon, Diberri, Adam78, Aasim75, Marc Venot, Ancheta Wis, Giftlite, Graeme Bartlett, Laudaka, Barbara Shack, Herbee, Fropuff, Xerxes314, Dratman, Curps, Joconnor, Jdavidb, Unconcerned, Eequor, Bobblewik, Andycjp, Alexf, Geni, Antandrus, HorsePunchKid, Melikamp, PDH, Rdsmith4, Bosmon, Bbbl67, Icairns, Sam Hocevar, Cynical, Lumidek, Iantresman, Burschik, Joyous!, Adashiel, Urvabara, Discospinster, Rich Farmbrough, Oliver Lineham, Vsmith, Jpk, ArnoldReinhold, Murtasa, D-Notice, JPX7, KaiSeun, SpookyMulder, Bender235, Kjoonlee, Kaisershatner, Pk2000, PsychoDave, RJHall, Mr. Billion, El C, Bletch, PhilHibbs, Shanes, Frankenschulz, Art LaPella, RoyBoy, Themusicgod1, Bobo192, Smalljim, Shenme, Cmdrjameson, Reuben, Kmaguire, I9Q79oL78KiL0QTFHgyc, Zelda~enwiki, Mr. Brownstone, E is for Ian, Jumbuck, Storm Rider, Alansohn, Gary, Anthony Appleyard, Guy Harris, Eric Kvaalen, Arthena, Keenan Pepper, Kocio, Bart133, RPellessier, Benna, ClockworkSoul, Cal 1234, Count Iblis, Guthrie, H2g2bob, Bsadowski1, GabrielF, Pauli133, Leondz, DV8 2XL, Gene Nygaard, Feline1, Oleg Alexandrov, Brookie, Natalya, Flying fish, WilliamKF, Yeastbeast, Mindmatrix, RHaworth, Plek, BillC, JPFlip, Benbest, ^demon, WadeSimMiser, Gxojo, MONGO, Jwanders, Torqueing, 🯄🯄🯄🯄🯄, Joke137, Wisq, Christopher Thomas, Palica, Mandarax, RedBLACKandBURN, Aarghdvaark, RichardWeiss, Ashmoo, Graham87, Malangthon, Mamling, Jclemens, Drbogdan, Loris Bennett, Rjwilmsi, Lars T., Strait, Patrick Gill, Tangotango, Tawker, Smithfarm, Stevenscollege, Mike Peel, HappyCamper, SeanMack, ScottJ, Krash, Dermeister, Rangek, Madcat87, FlaBot, Ian Pitchford, PlatypeanArchcow, A scientist, Margosbot~enwiki, Gark, Nivix, Gparker, Pathoschild, Gurch, Stevenfruitsmaak, Goudzovski, Tomer Ish Shalom, Smithbrenon, Chobot, Moocha, DVdm, Gwernol, The Rambling Man, YurikBot, Wavelength, RobotE, Koveras, Hairy Dude, Huw Powell, Phmer, Hillman, RussBot, Michael Slone, Ohwilleke, Bhny, JabberWok, GLaDOS, DanMS, Zelmerszoetrop, Eleassar, Merick, Big Brother 1984, NawlinWiki, Alpertron, Długosz, Schlafly, FFLaguna, BlackAndy, Dbmag9, SCZenz, Haoie, Raven4x4x, Ospalh, Durval, Bota47, Supspirit, Pegship, Noosfractal, Charlie Wiederhold, WAS 4.250, Smoggyrob, Reyk, Tvaughan, Joedixon, Eric TF Bat, Emc2, Ilmari Karonen, Allens, Bernd in Japan, InsayneWrapper, Bclayabt, Attilios, SmackBot, Cubs Fan, Ashill, IddoGenuth, Tomer yaffe, Stellea, InverseHypercube, KnowledgeOfSelf, Clpo13, Nickst, RedSpruce, Nightbat, Doc Strange, Herbm, Edgar181, HalfShadow, Flux.books, Dheerajkakar, Yamaguchi🯄🯄, Richmeister, Gilliam, Folajimi, The Gnome, Oscarthecat, Skizzik, Kmarinas86, Chris the speller, SuperBuuBuu, Quinsareth, Persian Poet Gal, Sirex98, MalafayaBot, Silly rabbit, Sangrolu, Villarinho, DHNbot~enwiki, Sbharris, Hongooi, Jdthood, CheerLeone, Gtkysor, Can't sleep, clown will eat me, Nick Levine, Tamfang, Kelvin Case, V1adis1av, Vanished User 0001, Rrburke, Jgoulden, Auvii, Krich, Wen D House, Radagast83, Engwar, Nakon, VegaDark, John D. Croft, Alexander110, KimO, Adrigon, SpiderJon, Ultraexactzz, Zadignose, Tesseran, Byelf2007, L337p4wn, K7lim, SashatoBot, Mchavez, Swatjester, Leftydan6, Minaker, John, Ashoat, Scientizzle, Acitrano, Linnell, JoshuaZ, James.S, JorisvS, Coredesat, Goodnightmush, ICBB, Plunge, JHunterJ, Hypnosifl, Silverthorn, Descubes, Freederick, Dr.K., Vanished user, Iridescent, Darkerprojects, Astrobayes, Newone, MOBle, Igoldste, CapitalR, AGK, Courcelles, Tawkerbot2, Dlohcierekim, Chetvorno, Hammer Raccoon, Owen214, Eastlaw, Peledre, Pukkie, Anakata, Runningonbrains, DKOH, NickW557, Gregbard, MikeWren, Vttoth, Necessary Evil, Ryan, Viciouspiggy, Gogo Dodo, Anonymi, Xxanthippe, A Softer Answer, Odie5533, Tawkerbot4, DumbBOT, Robertinventor, Kozuch, Mtpaley, Philza85, Starship Trooper, UberScienceNerd, Crum375, Thijs!bot, Epbr123, Astroceltica, Passaggio, Barbarina, Mbell, Eugenespeed, N5iln, Mojo Hand, Carlif, Headbomb, Tonyle, Marek69, Lars Lindberg Christensen, OtterSmith, SusanLesch, Mmortal03, Hmrox, Hires an editor, AntiVandalBot, Seaphoto, Orionus, Opelio, Shirt58, Rehnn83, Joehodge, AaronY, Jj137, TTN, Dylan Lake, Chill doubt, Spencer, Yellowdesk, Sniktaw, CPitt76, Gökhan, Jcarter1, Res2216firestar, JAnDbot, Leuko, Husond, MER-C, CosineKitty, Plantsurfer, Mcorazao, Therealintellectual, Folkform, Balbers, 100110100, Autotheist, Wasell, Magioladitis, Bongwarrior, VoABot II, Timothy McVeigh, Charlesrkiss, AuburnPilot, Krkaiser, Mbarbier, Kaivosukeltaja, Foroa, Swpb, Stigmj, T a y l o s, Ekantik, Brusegadi, Bubba hotep, Fabrictramp, Catgut, Lilian.Kaufmann, Zhanghia, Acornwithwings, Vssun, LtHija, Whisky5, DerHexer, Prisca6023, PeteSF, Rickard Vogelberg, NatureA16, DancingPenguin, MartinBot, Schmloof, STBot, Pagw, Fs644, Nikpapag, Anaxial, CommonsDelinker, Jean-Pierre Petit~enwiki, PrestonH, WelshMatt, Chrishy man, Tgeairn, J.delanoy, Pharaoh of the Wizards, Trusilver, Adavidb, Kudpung, Rod57, Arion 3x3, PedEye1, McSly, Tarotcards, Davy p, HiLo48, NewEnglandYankee, Ohms law, Jorfer, Blckavnger, Potatoswatter, KylieTastic, Joshua Issac, Infiniteglitch, Remember the dot, Pitpif, Vanished user 39948282, Neekap, Natl1, Ldebain, BernardZ, SoCalSuperEagle, Squids and Chips, CardinalDan, Idioma-bot, Sheliak, Funandtrvl, Lights, VolkovBot, Craigheinke, Itsfullofstars, ColdCase, Jeff G., JohnBlackburne, Mocirne, AlnoktaBOT, Scikid, Grammarmonger, Leojohns, Larry R. Holmgren, Philip Trueman, TXiKiBoT, Oshwah, Docanton, Authorized User, Theophilus reed, Drestros power, Strichek, MarekMahut, Monkey Bounce, Lradrama, Sintaku, Carillonatreides, Martin451, Broadbot, Wikiisawesome, Mazarin07, Inductiveload, Knightshield, Telecineguy, Spiral5800, Kurowoofwoof111, Greswik, RobertFritzius, SwordSmurf, Falcon8765, Hellothere17, Enviroboy, Littlehollah, Wanchung Hu, Illumini85, SonOfMog Worf, Jazzman123, PGWG, 19merlin69, NHRHS2010, Neparis, Bfpage, S-n-ushakov, SieBot, Calliopejen1, Tresiden, Wibubba48, Tachyonics, Pallab1234, Paradoctor, KGyST, Bentogoa, Jimlester51, Battlepace, Oda Mari, Aaarnooo, Suomichris, Crowstar, PromX1, Lightmouse, Tombomp, Cyberplasm, Diego Grez-Cañete, Spartan-James, Thinghy, Mygerardromance, Hamiltondaniel, Superbeecat, Denisarona, JL-Bot, Escape Orbit, Starcluster, Troy 07, Atif.t2, ArepoEn, Ak47gforce, Ratemonth, Sfan00 IMG, ClueBot, Phoenix-wiki, GorillaWarfare, The Thing That Should Not Be, ArdClose, Rodhullandemu, Cptmurdok, Drmies, Uncle Milty, Iuhkjhk87y678, Niceguyedc, MrBosnia, Bhaskarns, Andwor, Ktr101, Excirial, Dombom12, Cromescythe, Barbarinaz, FOARP, Brews ohare, Jotterbot, Iohannes Animosus, R.Andrae, Kentgen1, Ordovico, Mastertek, Rgoogin, Thehelpfulone, 1ForTheMoney, Versus22, Palmer666palmer, PCHS-NJROTC, Burner0718, Pillar of Babel, SoxBot III, Erodium, Vanished user uih38riiw4hjlsd, 1ofhissheep, TimothyRias, Arianewiki1, XLinkBot, DCCougar, Oldnoah, Rror, Gwark, Ost316, Avoided, Webmaster369, Gthomson, Tugrul irmak, Noctibus, Ploversegg, ZooFari, Parejkoj, Tayste, Addbot, Xp54321, Grayfell, Experimental Hobo Infiltration Droid, Willking1979, Some jerk on the Internet, Uruk2008, 04aeverington, DOI bot, Tcncv, Nohomers48, CharlesChandler, Gmeyerowitz, Haasfelix, Download, Proxima Centauri, Ashirgo, RTG, Redheylin, Glane23, Darkmatter654, SamatBot, Nanzilla, Lzkelley, Clone 209, Tassedethe, Numbo3-bot, Peridon, Chinchinthehun, Evildeathmath, Tide rolls, Lightbot, OlEnglish, Qemist, Gail, North Polaris, Legobot, Artichoke-Boy, Luckas-bot, Yobot, WikiDan61, Cosoce, Dov Henis, Aldebaran66, KillYourLove, CzechFalcon, Amble, Mmxx, CinchBug, Perusnarpk, IW.HG, Einstein vs Dark energys, Eric-Wester, Tempodivalse, Synchronism, AnomieBOT, Letuño, Girl Scout cookie, IRP, JackieBot, RBM 72, AdjustShift, Nicolaas Vroom, Henrykandrup, Iluziat, Materialscientist, Dendlai, ImperatorExercitus, The High Fin Sperm Whale, Citation bot, Ternity0127, Maxis ftw, Frankenpuppy, Quebec99, LilHelpa, Aksel89, Xqbot, Stlwebs, Random astronomer, Sionus, Cureden, Jradis1337, Capricorn42, Wperdue, Deleance, Raspw, Tomwsulcer, Magicxcian, AbigailAbernathy, Srich32977, NOrbeck, Artemis6234, Almabot, Abell 1367, Feldhaus, False vacuum, RibotBOT, Waleswatcher, Mikedr, Kongkokhaw, Rvnieuwe, Shadowjams, MeDrewNotYou, A. di M., Peter470, Sageman7, ☐☐, Luminique, Captain-n00dle, Imyfujita, FrescoBot, Andyradke0, Ag allstar, Paine Ellsworth, Originalwana, Styxpaint, Mark Renier, VS6507, PhysicsExplorer, Dbirkhofer, Steve Quinn, Nestlefolife, Adrian Akau, 1414rwbt, SF88, Citation bot 1, Redrose64, DUUJEEGWEEM, Tyler6298, Pinethicket, I dream of horses, Grammarspellchecker, Danlof, 10metreh, Jonesey95, Tom.Reding, Pmokeefe, A8UDI, For.a.limited.time.only, Elentirno, TedderBot, Aknochel, SkyMachine, IVAN3MAN, Kgrad, Nieuwenh, Trappist the monk, Puzl bustr, Fama Clamosa, Domeinthebumhole, Michael9422, UrukHaiLoR, Allen4names, JLincoln, Jeffrd10, Lovemybluetooth, Diannaa, Fastilysock, Innotata, DrCrisp, Whisky drinker, Onel5969, RjwilmsiBot, 5mgoblue5, Blakelewis122, Þorri, Mathewsyriac, Leandro.lelas, Mserard313, Mdznr, Sbugnon, Ultima821, EmausBot, Francophile124, Grrow, Super48paul, GoingBatty, RA0808, Gimmetoo, Solarra, Jmencisom, Slightsmile, Tommy2010, Winner 42, SusanaMultidark, Gocows2, Wikipelli, Serketan, Krifferjel, Zurich Astro, Hhhippo, Mhatthei, Svolin, Micahqgecko, JSquish, Josve05a, Trojanmice, MithrandirAgain, Edwinkaren, Devilaza, Arbnos, Oraclan, Suslindisambiguator, AlbertusmagnusOP, Tolly4bolly, L1A1 FAL, Ancient Anomaly, L Kensington, Maj den, Corabilek, Donner60, Aldnonymous, Ihardlythinkso, RockMagnetist, Terra Novus, TYelliot, DASHBotAV, Kroupap, D Phoesheezey, Travies10, Jxraynor, TheTimesAreAChanging, ClueBot NG, Rich Smith, Afjvanraan, Crystal7878, Catinthehat93, Bped1985, Infinifold, Wiggit002, Jj1236, PapaMike, MonEyshOt42069210, Muon, Esdacosta, Asukite, Masssly, Ph.d Carl edenburgh, Widr, Gavin.perch, Helpful Pixie Bot, Curb Chain, Calabe1992, Bibcode Bot, BG19bot, Dualus, Kishanparekh, Stevenwilkins, NacowY, Cheeseray1, Cyberguy5, Darkmatter adam, Yomomma8102, Hza a 9, Rarelight, Cyberpower678, Cosmologist77, தென்காசி சுப்பிரமணியன், Dahliamtl, Dodshe, Mark Arsten, Darkmatterotheruniverses, Cadiomals, Zedshort, Achowat, Rolandwilliamson, A2Die, Clint55555, Mgka79, NotEither, BattyBot, Ronin712, Babymushrooms, Davidmexican, Drphilmarshall, Dilaton, Quin71901, U-95, ChrisGualtieri, Npmay, Kvark92, Lukasz.astrus, Ducknish, JYBot, Davidlwinkler, Astrohap, Hunterf12, Dexbot, Caroline1981, Gravityking100, Junavia, Fredrikdn, CuriousMind01, Lugia2453, Wjs64, Andwor42, Frosty, Honneydewp243, Junjunone, DrHowzer73, JustAMuggle, WadiElNatrun, Reatlas, Rfassbind, Acetotyce, I am One of Many, DirkXcal, Melonkelon, Ybidzian, Gig9876, M.ashrafinia, Trolololman12, Ilikedeletingstufffromhere, DavidLeighEllis, Onecreation, Zenibus, Jernahthern, Hipposaregrey, Frinthruit, Stamptrader, Cyberalchemyst, Aaronknowsitall, FelixRosch, Darkmer, Doubleknockout, Monkbot, Wardinstrument, Leegrc, Vikas Rauniyar, Apipia, Upsalla, Jkvaternik, Lol kaptyn troll, HMSLavender, Mohammedshukoor, Callum92, Stefania.deluca, Ashweigh, Oldstone James, Astezar, 39Debangshu, YoYoDude012, Anunaki truth, Pyrotle, Tetra quark, Carazmatic, God of matterrr, Silversparkcontributions, Isambard Kingdom, Rizi0909, Absolutelypuremilk, Anand2202, Kbap2002, Kb2002, DN-boards1, Yohoona, KasparBot, I love trains sooo much, Id6040, Boowiebear, Stephane Le Corre, Mustachman71, Abdelrahmam shawky and Anonymous: 1228

- **Exotic matter** *Source:* https://en.wikipedia.org/wiki/Exotic_matter?oldid=681113111 *Contributors:* Bryan Derksen, Alan Peakall, Aragorn2, David Latapie, Omegatron, Phil Boswell, Korath, Bkell, Intangir, Geeoharee, Lethe, Mboverload, ConradPino, Beland, WhiteDragon, Icairns, Erik Garrison, Tzarius, EricJamesStone, Pjacobi, Paul August, JYolkowski, Mac Davis, Woohookitty, LOL, Meneth, CharlesC, Christopher Thomas, Driftwoodzebulin, Teknic, Quale, Matjlav, FireCrack, Ejurgaite, Krackpipe, Slant, Nimur, YurikBot, Salsb, Astral, ErkDemon, Xiroth, Dbfirs, Geoffrey.landis, JDspeeder1, Itub, SmackBot, Hmains, Kmarinas86, MovGP0, Vladis1av, Nakon, Nathanael Bar-Aur L., Vampus, JorisvS, Hypnosifl, RMHED, Inquisitus, Michaelbusch, TwistOfCain, Happy-melon, Göörgë Büsh, P0LARIS, Alexnye, Keraunos, Headbomb, Escarbot, Lfstevens, Ca7ch, BatteryIncluded, Cpl Syx, R'n'B, Hans Dunkelberg, OneDoubleO, Aliento, Red Act, Lamro, Antixt, Universaladdress, AlleborgoBot, PseudoAdmin^*(3d), WereSpielChequers, SteveThePhysicist, PedantryIsMyMiddleName, Gigo300, ClueBot, Superwj5, The Thing That Should Not Be, John.D.Ward, SeaRisk, Dean Wormer, Alexbot, DumZiBoT, Thatguyflint, Kbdankbot, Addbot, CanadianLinuxUser, Tassedethe, Luckas-bot, NotARusski, AnomieBOT, 9258fahsflkh917fas, Unara, Aaagmnr, Citation bot, RibotBOT, Locobot, Samwb123, Pinethicket, Vuyane, Unbitwise, Slightsmile, Eternalemp, ClueBot NG, Bibcode Bot, Cadiomals, Dexbot, Darth Sitges, Capitolcity, DavidLeighEllis, Krowizard113, Usernameonwiki, Lxplot and Anonymous: 120

- **Periodic table** *Source:* https://en.wikipedia.org/wiki/Periodic_table?oldid=682400345 *Contributors:* AxelBoldt, Dreamyshade, Chuck Smith, Lee Daniel Crocker, Mav, Bryan Derksen, Timo Honkasalo, The Anome, Tarquin, DanKeshet, Rjstott, Andre Engels, XJaM, Christian List, PierreAbbat, Heron, Fonzy, Youandme, Olivier, Someone else, Bob Jonkman, Patrick, Infrogmation, Michael Hardy, Erik Zachte, TMC, Kwertii, Dan Koehl, Shellreef, Taras, Wapcaplet, Ixfd64, Dcljr, Tomi, Eric119, Kosebamse, Egil, Mdebets, Ahoerstemeier, Stan Shebs, Ronz, Jpatokal, Theresa knott, Snoyes, Suisui, Den fjättrade ankan~enwiki, Kragen, Salsa Shark, Cyan, Stefan-S, Poor Yorick, Kwekubo, Jiang, Eirik (usurped), Mxn, BRG, Smack, Schneelocke, Jengod, Okome~enwiki, Emperorbma, EL Willy, Eszett, Adam Bishop, Reddi, Stone, Piolinfax, Dtgm, Selket, Tpbradbury, Rarb, Maximus Rex, Nv8200pa, Tempshill, Bevo, Traroth, Shizhao, Stormie, Dpbsmith, Bcorr, Secretlondon, Jusjih, Just another user 2, Darthchaos, Jeffq, Lumos3, Denelson83, Jni, Nofutureuk, Gromlakh, Gentgeen, Robbot, Phisite, Juve82, Fredrik, Chris 73, WormRunner, Altenmann, Romanm, Naddy, Lowellian, WebElements, Yosri, Rfc1394, Texture, Hippietrail, Caknuck, Bkell, David Edgar, Borislav, Eliashedberg, Radagast, David Gerard, Giftlite, DocWatson42, Haeleth, Ævar Arnfjörð Bjarmason, Tom harrison, Lupin, Everyking, Bkonrad, No Guru, NeoJustin, Bensaccount, Zaphod Beeblebrox, AJim, Avsa, Yekrats, Dmmaus, Archenzo, Brockert, Darrien, SWAdair, Bobblewik, Deus Ex, Edcolins, Lucky 6.9, Peter Ellis, Gadfium, Zed0, Ran, Antandrus, Ctachme, PDH, Jossi, Exigentsky, Kesac, Vbs, Icairns, Sam Hocevar, Clemwang, Karl Dickman, Adashiel, Iwilcox, EagleOne, Mike Rosoft, Alkivar, D6, Andrew11, Poccil, Zarxos, EugeneZelenko, Felix Wan, A-giau, Noisy, Discospinster, Rich Farmbrough, KarlaQat, Cacycle, Inkypaws, Vsmith, Samboy, Joeclark, SpookyMulder, Bender235, TerraFrost, Sunborn, Klenje, RJHall, El C, Kwamikagami, Shanes, Briséis~enwiki, RoyBoy, Femto, Semper discens, Grick, Bobo192, AlHalawi, Whosyourjudas, Nyenyec, Reinyday, Clawson, Cwolfsheep, Dbchip, Giraffedata, SpeedyGonsales, Jojit fb, Nk, Eddideigel, Conget~enwiki, Jhd, Conny, Stephen G. Brown, Danski14, Honeycake, Orzetto, Alansohn, Mo0, Atlant, Keenan Pepper, Plumbago, Sl, Damnreds, AzaToth, Mac Davis, Caesura, Blobglob, Wtmitchell, ClockworkSoul, Unconventional, Helixblue, Stephan Leeds, Harej, RJFJR, Skatebiker, Computerjoe, GabrielF, Ghirlandajo, HGB, Feline1, Weyes, Lucent, Philthecow, Cimex, TigerShark, Benbest, Mpatel, Schzmo, U10ajf, Bluemoose, CharlesC, Waldir, SeventyThree, EarthmatriX, MarcoTolo, Cataclysm, V8rik, Qwertyus, Kbdank71, FreplySpang, DePiep, Dwaipayanc, Canderson7, Drbogdan, Saperaud~enwiki, Angusmclellan, Joe Decker, Koavf, Oblivious, SeanMack, Shalmanese, Sango123, Ptdecker, Yamamoto Ichiro, RobertG, Pumeleon, Nivix, Pathoschild, RexNL, Gurch, Kolbasz, Brendan Moody, Scerri, Alphachimp, Kri, Dalta~enwiki, Glenn L, Physchim62, Imnotminkus, Chobot, Visor, Jared Preston, DVdm, Bgwhite, Gwernol, EamonnPKeane, Roboto de Ajvol, Mercury McKinnon, YurikBot, Wavelength, Hairy Dude, Deeptrivia, Phantomsteve, RussBot, Vlad4599, Fabartus, SpuriousQ, IanManka, Stephenb, Rintrah, Alvinrune, Schoen, Rsrikanth05, Bovineone, Wimt, Stassats, Anomalocaris, EngineerScotty, NawlinWiki, Wiki alf, E123, Test-tools~enwiki, Jaxl, Terfili, Yahya Abdal-Aziz, Mkouklis, Nick, Ragesoss, Dhollm, Cholmes75, Dmoss, Matticus78, RUL3R, AdiJapan, Ryanminier, Juanpdp, Hv, Misza13, Beanyk, Aaron Schulz, Bota47, CorbieVreccan, Derek.cashman, DRosenbach, Elkman, Phaedrus86, Smaines, Wknight94, Tetracube, FF2010, Ageekgal, Closedmouth, Jwissick, Ketsuekigata, Sean Whitton, Petri Krohn, DGaw, CWenger, Smurrayinchester, Kungfuadam, Junglecat, RG2, NeilN, DVD R W, Itub, Thecroman, SmackBot, Android 93, Bobet, Reedy, InverseHypercube, KnowledgeOfSelf, TestPilot, Melchoir, Unyoyega, KocjoBot~enwiki, Davewild, Thunderboltz, Milesnfowler, Anastrophe, Delldot, J0lt C0la, Knowhow, Elk Salmon, Edgar181, HalfShadow, Eupedia, Srnec, Gilliam, Ohnoitsjamie, Skizzik, Carbon-16, JRSP, Chris the speller, Keegan, Iskander32, RDBrown, Thumperward, Fuzzform, Lollerskates, EncMstr, MalafayaBot, OrangeDog, Roscelese, Bonaparte, Xoyorkie13, Metacomet, Dustimagic, DHN-bot~enwiki, DNAmaster, Darth Panda, Suicidalhamster, Can't sleep, clown will eat me, Onorem, Clorox, Konczewski, Squadoosh, Andy120290, Ddon, DR04, UU, Grover cleveland, Jachapo, PiMaster3, TotalSpaceshipGuy3, Savidan, Dreadstar, Pwjb, Aco47, Peterwhy, Jklin, DMacks, BrotherFlounder, Suidafrikaan, Sadi Carnot, The undertow, SashatoBot, Mchavez, Nishkid64, Tarantola, LtPowers, Archimerged, Khazar, Vitall, Scientizzle, Gobonobo, Btg2290, Anoop.m, Olin, ManiF, JohnWittle, Moop stick, Jaywubba1887, Ckatz, Dale101usa, Chrisch, Garudabd, Digger3000, Slakr, Rainwarrior, Beetstra, Mr Stephen, AxG, Arkrishna, Mets501, Ambuj.Saxena, Ryulong, RichardF, Jose77, DGtal, WOWGeek, Sifaka, Asyndeton, Ramuman, Dead3y3, Michaelbusch, Walton One, Tabfugnic, David Little, J Di, CapitalR, DavidOaks, Supertigerman, Pearson3372, Az1568, Courcelles, Túrelio, Ziusudra, Dpeters11, Tawkerbot2, Bobby131313, VinceB, Cryptic C62, Kaischwartz, Lincmad, TranClan, JForget, Betaeleven, Deon, Eli84, Van helsing, NullAshton, CBM, Rawling, DSachan, GHe, Fork me, Egmonster, Black and White, FlyingToaster, Wikiman7~enwiki, WeggeBot, Logical2u, Pi Guy 31415, Johnlogic, MrFish, Bill Sayre, Dmsc893, Rudjek, Nebular110, Reywas92, Grahamec, MC10, Rasmus vendelboe, Vanished user vjhsduheuiui4t5hjri, Rifleman 82, Corpx, GeorgeTopouria, Islander, Mycroft.Holmes, Methyl~enwiki, Fifo, Christian75, DumbBOT, Chrislk02, Shrikethestalker, Taylor4452, Ndufour, Memorymike, JodyB, Rowlaj01, Calvero JP, Satori Son, Casliber, Thijs!bot, Full On, Barticus88, ShayneRyan, Opabinia regalis, Kiwi137, Corsair18, Dagrimdialer619, Headbomb, Sobreira, Marek69, Ydoommas, John254, Racantrell, Dmitri Lytov, Philippe, Nezzington, Nemti, Escarbot, Mentifisto, Lani123, Tom dl, AntiVandalBot, Luna Santin, Michael phan, Bigtimepeace, Random user 8384993, Chill doubt, LegitimateAndEvenCompelling, Myanw, Figma, Tomertomer, JAnDbot, Barek, MER-C, Zerotjon, Sanchom, Hut 8.5, Kirrages, Kerotan, Maurakt, Magioladitis, Canjth, Bongwarrior, VoABot II, JNW, Kinston eagle, Redaktor, Aa35te, SparrowsWing, Avicennasis, Superworms, Ahecht, Nposs, Wikiak, Dirac66, Adrian J. Hunter, Allstarecho, ChrisSmol, StuFifeScotland, User A1, Musicloudball, Cpl Syx, Vssun, Just James, DerHexer, Khalid Mahmood, TheRanger, DancingPenguin, FisherQueen, Hdt83, MartinBot, HLewis, PostScript, Rettetast, Roastytoast, 1993 lol, Glrx, Kateshortforbob, CommonsDelinker, Supia, PrestonH, Thomasrive, Exodecai101, Slash, J.delanoy, Pharaoh of the Wizards, Ilovestars89, ChickenMarengo, Hans Dunkelberg, Psycho Kirby, Smartweb, I2yu, Sergeibernstein, Tempnegro, Extransit, WarthogDemon, Munkimunki, Richard777, Thom.fynn, Tdadamemd, Yvonr, Adamsbriand, EH74DK, Rescorbic, McSly, Aonrotar, Ryan Postlethwaite, Ephebi, Notapotato, Gurchzilla, Wasitgood69, Wasitgood, Paulbkirk, Monkeybutt5423, AntiSpamBot, Gffootball58, Coin945, Bigsnake 19, NewEnglandYankee, Creator58, Joka1991, SJP, C0RNF1AK35, Shoessss, Bob, Vanished user 39948282, BrianScanlan, Nat682, Darklama, Hyuuganeji0123, Arjun Rana, AnjuX, Suuperturtle, Idioma-bot, Johnnieblue, Deor, 28bytes, VolkovBot, Thedjatclubrock, Iosef, Jmocenigo, Christophenstein, Jeff G., JohnBlackburne, TheOtherJesse, RemoteCar, Barneca, Philip Trueman, Af648, Drunkenmonkey, Sweetness46, TXiKiBoT, TheVault, A4bot, Quilbert, Caster23, GDonato, Miranda, Chrisk12, Sankalpdravid, Qxz, Littlealien182, Anna Lincoln, Corvus cornix, Martin451, Jackfork, LeaveSleaves, Andrewrost3241981, DBragagnolo, Luuva, Quindraco, Gona.eu, RadiantRay, Madhero88, Jinglesmells999, Finngall, Aciddoll, Superjustinbros., Deeryh01, Synthebot, Enviroboy, Chengyq19942007, Insanity Incarnate, Everybody's Got One, Why Not A Duck, Brianga, Jaybo007, HiDrNick, LuigiManiac, Petergans, ConnTorrodon, NHRHS2010, SieBot, Coffee, Mikemoral, TJRC, Rihanij, PlanetStar, Jmwwiki, Borgdylan, Gprince007, Tiddly Tom, Scarian, WereSpielChequers, Jauerback, Jack Merridew, Gerakibot, Dawn Bard, Viskonsas, Caltas, Kragenz, The way, the truth, and the light, Michfg, Sat84, Whiteghost.ink, Til Eulenspiegel, Purbo T, Tiptoety, Exert, Elcobbola, Nopetro, Hamilton hogs, Hiddenfromview, Segalsegal, SquirrelMonkeySpiderFace, Oxymoron83, Nuttycoconut, Lightmouse, Hwn tls, Hak-kâ-ngìn, BenoniBot~enwiki, Jack the Stripper, Sunrise, Dillard421, Werldwayd, Pappapasd, Maelgwnbot, DixonD, Nergaal, Precious Roy, Escape Orbit, Into The Fray, Jimmy Slade, Kanonkas, Georgedriver, Mr. Granger, Twinsday, ClueBot, PipepBot, Dinamik, The Thing That Should Not Be, Apastrophe, Techdawg667, Gawaxay, Syhon, Thompsontm, Drmies, Polyamorph, Elsweyn, Ryoutou, CounterVandalismBot, Ansh666, Blanchardb, LizardJr8, Dylan620, Timex987, Wifiless, Puchiko, Natasha.fielding, Dlorang, Robert Skyhawk, Excirial, Alexbot, BirgerH, Omgosh2, Jazjaz92, NuclearWarfare, Pearrari, Kaeso Dio, Jotterbot, LarryMorseDCOhio, Psinu, Realm up, DeltaQuad, Kaiba, Dekisugi, Pwntskater, NolanRichard, Thehelpfulone, La Pianista, Another Believer, Kiran the great, Viper275, Aitias, Blargblarg89, Versus22, Sch00l3r, SoxBot III, MairAW, JDT1991, BalkanFever, Vanished user uih38riiw4hjlsd, Indopug, TimothyRias, Jean-claude perez, Neuralwarp, XLinkBot, Shpakovich, Gonzonoir, Nsimya, TheSickBehemoth, Ryuken14, Nsim, WikHead, Noctibus, JinJian, ZooFari, MystBot, RyanCross, Thatguyflint, Roentgenium111, Tomilee0001, Yoenit, OmgItsTheSmartGuy, NicholasSThompson, JenR32, Vchorozopoulos, WFPM, Cst17, Download, SoSaysChappy, EconoPhysicist, Mathmarker, Glane23, Plutonium55, Debresser, LinkFA-Bot, Drova, LiveAgain, PoliteCarbide, Numbo3-bot, Tide rolls, BrianKnez, Luckas Blade, Greyhood, Arbitrarily0, Angrysockhop, Jack who built the house, Luckas-bot, Yobot, Essam Sharaf, Azylber, KamikazeBot, Jobroluver98, EnDaLeCoMpLeX, MacTire02, Tempodivalse, AnomieBOT, Zhieaanm, Helixer, Rubinbot, Jake Fuersturm, Degg444, Daniele Pugliesi, Piano non troppo, Collieuk, AdjustShift, Kingpin13, Ulric1313, Materialscientist, RadioBroadcast, The High Fin Sperm Whale, Citation bot, ArthurBot, Carturo222, Xqbot, Ziaix, Timir2, Sionus, Gopal81, Vidshow, Tfts, YBG, Grim23, Srich32977, WingedSkiCap, Pirateer, GrouchoBot, RibotBOT, Ian Fraser at Temple Newsam House, Spesh531, FaTony, AlimanRuna, A. di M., Nolimits5017, Dougofborg, Thehelpfulbot, R8R Gtrs, FrescoBot, Ylime715, Dogposter, StaticVision, Michael93555, Car132, Zach112233, Weetoddid, Strongbadmanofme, Maxus96, Grandiose, Xhaoz, Dellacomp, Citation bot 1, Pezzells, Pshent, Redrose64, ArnaudContet, Pinethicket, I dream of horses, M.pois, Hamtechperson, BTolli, Gemmi3, RedBot, Fishekad, NarSakSasLee, Kangxi emperor6868, Hardwigg, Abc518, Tjlafave, Lightlowemon, Gamewizard71, FoxBot, Double sharp, Adult Swim Addict, TobeBot, Graniggo, Jaeger Lotno, Pitcroft, Sillyboy67, Hughbert512369, Dinamik-bot, BZRatfink, Fyandcena, Kelvin35, Mattvirajrenaudbrandon, Turn off 2, Imawsome 09, Mass09, Tbhotch, Pldx1, Naughtysriram, DARTH SIDIOUS 2, Luhar1997, Twonernator, Pickweed, Dancojocari, EmausBot, Sir Arthur Williams, WikitanvirBot, GA bot, Franjklogos, Tf1321, Tommy2010, P. S. F. Freitas, Kitrkatr, 陈桂连, Zainiadragon10000, JSquish, Jakers69, StringTheory11, Dffgd, Cobaltcigs, Шугун, Joshlepaknpsa, Hzb pangus, Elly4web, Arman Cagle, KotVa, Sven nestle2, Hh73wiki, Ego White Tray, Negovori, ChuispastonBot, GermanJoe, Sunshine4921, Rmashhadi, Chaotic iak, Whoop whoop pull up, Heidslovesearl, Xanchester, ClueBot NG, Ihakeycakeyabreak, Wd930, Ozkithar Salas, Manubot, CocuBot, Lanthanum-138, Frietjes, Hazhk, Moneya, Parcly Taxel, Metaknowledge, Willzuk, Helpful Pixie Bot, RobertGustafson, ಉಪ್ಪು ಕಾಯಿ್ನ, Curb Chain, Bibcode Bot, BG19bot, MKar, Vagobot, Nagasturg, Sandbh, Mark Arsten, IraChesterfield, Soerfm, Zedshort, FeralOink, Razzat99, SoylentPurple, BattyBot, Justincheng12345-bot, Judiakok1985, Ziggypowe, Ushau97, Maxronnersjo, ChrisGualtieri, JYBot, Dexbot, LightandDark2000, Aditya Mahar, Mogism, Carpelogos, Jtrevor99, Leitoxx, Kazim5294, AmericanLemming, WikiEditor2563, Michel Djerzinski, The Herald, Shearflyer, Quenhitran, Asadwarraich, Kind Tennis Fan, Monkbot, HiYahhFriend, Deepak harshal nagle, Trackteur, Encyclopedia Lu, IiKkEe, Shane Stachwick, Mogie Bear, R. Portela F., Forscienceonly, KasparBot, Alistairgray42, Equinox, Lexi sioz and Anonymous: 1391

18.12.2 Images

- **File:080998_Universe_Content_240_after_Planck.jpg** *Source:* https://upload.wikimedia.org/wikipedia/commons/b/b6/080998_Universe_Content_240_after_Planck.jpg *License:* Public domain *Contributors:* http://map.gsfc.nasa.gov/media/080998/index.html updated data from http://www.nasa.gov/mission_pages/planck/news/planck20130321.html *Original artist:* NASA, Modified by User:▢▢

- **File:1D_normal_modes_(280_kB).gif** *Source:* https://upload.wikimedia.org/wikipedia/commons/9/9b/1D_normal_modes_%28280_kB%29.gif *License:* CC-BY-SA-3.0 *Contributors:* This is a compressed version of the Image:1D normal modes.gif phonon animation on Wikipedia Commons that was originally created by Régis Lachaume and freely licensed. The original was 6,039,343 bytes and required long-duration downloads for any article which included it. This version is 4.7% the size of the original and loads *much* faster. This version also has an interframe delay of 40 ms (v.s. the original's 100 ms). Including processing time for each frame, this version runs at a frame rate of about 20–22.5 Hz on a typical computer, which yields a more fluid motion. Greg L 00:41, 4 October 2006 (UTC). (from http://en.wikipedia.org/wiki/Image:1D_normal_modes_%28280_kB%29.gif) *Original artist:* Original Uploader was Greg L (talk) at 00:41, 4 October 2006.

- **File:1e0657_scale.jpg** *Source:* https://upload.wikimedia.org/wikipedia/commons/a/a8/1e0657_scale.jpg *License:* Public domain *Contributors:* Chandra X-Ray Observatory: 1E 0657-56 *Original artist:* NASA/CXC/M. Weiss

- **File:2006-01-14_Surface_waves.jpg** *Source:* https://upload.wikimedia.org/wikipedia/commons/4/43/2006-01-14_Surface_waves.jpg *License:* CC-BY-SA-3.0 *Contributors:* picture taken by Roger McLassus (improved by DemonDeLuxe, Sep 2006) *Original artist:* Roger McLassus

- **File:3D_model_hydrogen_bonds_in_water.svg** *Source:* https://upload.wikimedia.org/wikipedia/commons/c/c6/3D_model_hydrogen_bonds_in_water.svg *License:* CC BY-SA 3.0 *Contributors:* File:3D model hydrogen bonds in water.jpg *Original artist:* User Qwerter at Czechwikipedia:Qwerter. Transferred fromcs.wikipedia; Transfer was stated to be made byUser:sevela.p. Translated to english by by MichalMaňas (User:snek01). Vectorized by Magasjukur2

- **File:Acap.svg** *Source:* https://upload.wikimedia.org/wikipedia/commons/5/52/Acap.svg *License:* Public domain *Contributors:* Own work *Original artist:* F l a n k e r

- **File:Ambox_current_red.svg** *Source:* https://upload.wikimedia.org/wikipedia/commons/9/98/Ambox_current_red.svg *License:* CC0 *Contributors:* self-made, inspired by Gnome globe current event.svg, using Information icon3.svg and Earth clip art.svg *Original artist:* Vipersnake151, penubag, Tkgd2007 (clock)

- **File:Ambox_important.svg** *Source:* https://upload.wikimedia.org/wikipedia/commons/b/b4/Ambox_important.svg *License:* Public domain *Contributors:* Own work, based off of Image:Ambox scales.svg *Original artist:* Dsmurat (talk · contribs)

- **File:Antimatter_Explosions.ogv** *Source:* https://upload.wikimedia.org/wikipedia/commons/2/2c/Antimatter_Explosions.ogv *License:* Public domain *Contributors:* Goddard Multimedia *Original artist:* NASA/Goddard Space Flight Center

- **File:Antimatter_Explosions_2.ogv** *Source:* https://upload.wikimedia.org/wikipedia/commons/a/a8/Antimatter_Explosions_2.ogv *License:* Public domain *Contributors:* Goddard Multimedia *Original artist:* NASA/Goddard Space Flight Center

- **File:Artist's_impression_of_the_expected_dark_matter_distribution_around_the_Milky_Way.ogv** *Source:* https://upload.wikimedia.org/wikipedia/commons/0/03/Artist%E2%80%99s_impression_of_the_expected_dark_matter_distribution_around_the_Milky_Way.ogv *License:* CC BY 4.0 *Contributors:* ESO *Original artist:* ESO/L. Calçada

- **File:Asterisks_one.svg** *Source:* https://upload.wikimedia.org/wikipedia/commons/4/49/Asterisks_one.svg *License:* CC BY-SA 3.0 *Contributors:* Own work *Original artist:* DePiep

- **File:Asterisks_two.svg** *Source:* https://upload.wikimedia.org/wikipedia/commons/3/3f/Asterisks_two.svg *License:* CC BY-SA 3.0 *Contributors:* Own work *Original artist:* DePiep

- **File:Atomic_orbital_energy_levels.svg** *Source:* https://upload.wikimedia.org/wikipedia/commons/9/98/Atomic_orbital_energy_levels.svg *License:* CC BY-SA 3.0 *Contributors:* File:High School Chemistry.pdf, page 301 *Original artist:* Richard Parsons (raster), Adrignola (vector)

- **File:Atomic_orbitals_and_periodic_table_construction.ogv** *Source:* https://upload.wikimedia.org/wikipedia/commons/6/61/Atomic_and_periodic_table_construction.ogv *License:* CC BY-SA 3.0 *Contributors:* Own work *Original artist:* Jubobroff

- **File:Atomic_resolution_Au100.JPG** *Source:* https://upload.wikimedia.org/wikipedia/commons/e/ec/Atomic_resolution_Au100.JPG *License:* Public domain *Contributors:* ? *Original artist:* ?

- **File:Binary_Boiling_Point_Diagram_new.svg** *Source:* https://upload.wikimedia.org/wikipedia/commons/6/60/Binary_Boiling_Point_new.svg *License:* CC BY-SA 2.5 *Contributors:*

- Binary_Boiling_Point_Diagram.PNG *Original artist:* Binary_Boiling_Point_Diagram.PNG: H Padleckas

- **File:Binding_energy_curve_-_common_isotopes.svg** *Source:* https://upload.wikimedia.org/wikipedia/commons/5/53/Binding_energy_-_common_isotopes.svg *License:* Public domain *Contributors:* ? *Original artist:* ?

- **File:Bohr_atom_animation_2.gif** *Source:* https://upload.wikimedia.org/wikipedia/commons/1/17/Bohr_atom_animation_2.gif *License:* CC BY-SA 3.0 *Contributors:* Own work *Original artist:* Kurzon

- **File:Bose-Einstein_Condensation.ogv** *Source:* https://upload.wikimedia.org/wikipedia/commons/d/d9/Bose-Einstein_Condensation.ogv *License:* CC BY-SA 3.0 *Contributors:* Own work *Original artist:* Jubobroff Jubobroff J.Bobroff and full list in credits

- **File:Bose_Einstein_condensate.png** *Source:* https://upload.wikimedia.org/wikipedia/commons/a/af/Bose_Einstein_condensate.png *License:* Public domain *Contributors:* NIST Image *Original artist:* NIST/JILA/CU-Boulder

- **File:Boyle_air_pump.jpg** *Source:* https://upload.wikimedia.org/wikipedia/commons/3/31/Boyle_air_pump.jpg *License:* Public domain *Contributors:* New Experiments ... Touching the Spring of the Air ... *Original artist:* Robert Boyle

- **File:Brosen_ironcarbon.svg** *Source:* https://upload.wikimedia.org/wikipedia/commons/3/3a/Brosen_ironcarbon.svg *License:* CC BY 2.5 *Contributors:* Own work *Original artist:* Brosen

- **File:Liquid_helium_Rollin_film.jpg***Source:*https://upload.wikimedia.org/wikipedia/commons/f/f8/Liquid_helium_Rollin_film.jpg*License:*Public domain*Contributors:*Own work*Original artist:*I, AlfredLeitner, took this photograph as part of my movie "Liquid Helium,Superfluid
"

- **File:Magnetic_rope.svg** *Source:* https://upload.wikimedia.org/wikipedia/commons/1/12/Magnetic_rope.svg *License:* Public domain *Contributors:* http://history.nasa.gov/SP-345/ch15.htm#250 *Original artist:* NASA; edited by Jaybear

- **File:Matter_Distribution.JPG** *Source:* https://upload.wikimedia.org/wikipedia/commons/5/50/Matter_Distribution.JPG *License:* CC BY-SA 3.0 *Contributors:* Own work *Original artist:* Brews ohare

- **File:Medeleeff_by_repin.jpg** *Source:* https://upload.wikimedia.org/wikipedia/commons/b/b3/Medeleeff_by_repin.jpg *License:* Public domain *Contributors:* http://www.picture.art-catalog.ru/picture.php?id_picture=4318 *Original artist:* Ilya Repin

- **File:Mergefrom.svg** *Source:* https://upload.wikimedia.org/wikipedia/commons/0/0f/Mergefrom.svg *License:* Public domain *Contributors:* ? *Original artist:* ?

- **File:Mollier_enthalpy_entropy_chart_for_steam_-_US_units.svg***Source:*https://upload.wikimedia.org/wikipedia/commons/f/fa/Moll_enthalpy_entropy_chart_for_steam_-_US_units.svg *License:* CC BY-SA 3.0 *Contributors:* Transferred from en.wikipedia *Original artist:* emok (talk). Original uploader was Emok at en.wikipedia

- **File:MountRedoubtEruption.jpg** *Source:* https://upload.wikimedia.org/wikipedia/commons/1/17/MountRedoubtEruption.jpg *License:* Public domain *Contributors:* http://web.archive.org/web/20051111095409/http://wrgis.wr.usgs.gov/dds/dds-39/album.html *Original artist:* R. Clucas

- **File:NO2-N2O4.jpg** *Source:* https://upload.wikimedia.org/wikipedia/commons/8/8b/NO2-N2O4.jpg *License:* Public domain *Contributors:* en:Image:N02-N2O4.jpg *Original artist:* en:User:Greenhorn1

- **File:Nano_Si_640x480.jpg** *Source:* https://upload.wikimedia.org/wikipedia/commons/9/9d/Nano_Si_640x480.jpg *License:* Public domain *Contributors:* ? *Original artist:* ?

- **File:NeTube.jpg** *Source:* https://upload.wikimedia.org/wikipedia/commons/8/88/NeTube.jpg *License:* CC BY-SA 2.5 *Contributors:* usermade *Original artist:* User:Pslawinski

- **File:Newlands_periodiska_system_1866.png***Source:*https://upload.wikimedia.org/wikipedia/commons/e/e5/Newlands_periodiska_syste_1866.png*License:*Public domain*Contributors:*John Alexander Reina Newlands(1838–1898)*Original artist:*John Alexander Reina Newlands

- **File:Nitrogen.ogg** *Source:* https://upload.wikimedia.org/wikipedia/commons/b/b6/Nitrogen.ogg *License:* CC BY 2.0 *Contributors:* http://www.flickr.com/photos/expictura/2449887718/ *Original artist:* Ryan Poling aka expictura on Flickr.

- **File:Nuvola_apps_edu_science.svg** *Source:* https://upload.wikimedia.org/wikipedia/commons/5/59/Nuvola_apps_edu_science.svg *License:* LGPL *Contributors:* http://ftp.gnome.org/pub/GNOME/sources/gnome-themes-extras/0.9/gnome-themes-extras-0.9.0.tar.gz *Original artist:* David Vignoni / ICON KING

- **File:Office-book.svg** *Source:* https://upload.wikimedia.org/wikipedia/commons/a/a8/Office-book.svg *License:* Public domain *Contributors:* This and myself. *Original artist:* Chris Down/Tango project

- **File:PVT_3D_diagram.png** *Source:* https://upload.wikimedia.org/wikipedia/commons/f/fc/PVT_3D_diagram.png *License:* CC BY-SA 3.0 *Contributors:* Thin Film Deposition // Principle & Practice //Redrawing by Kyu Ho Lee (revised by H Padleckas) *Original artist:* Donald L. Smith//

- **File:PaperAutofluorescence.jpg** *Source:* https://upload.wikimedia.org/wikipedia/commons/d/d5/PaperAutofluorescence.jpg *License:* CC BY-SA 3.0 *Contributors:* Own work *Original artist:* Richard Wheeler (Zephyris)

- **File:People_icon.svg** *Source:* https://upload.wikimedia.org/wikipedia/commons/3/37/People_icon.svg *License:* CC0 *Contributors:* OpenClipart *Original artist:* OpenClipart

- **File:Periodic_Table_overview_(standard).svg** *Source:* https://upload.wikimedia.org/wikipedia/commons/d/d4/Periodic_Table_overview_%28standard%29.svg *License:* CC BY-SA 3.0 *Contributors:* Own work *Original artist:* DePiep

- **File:Periodic_Table_overview_(wide).svg***Source:*https://upload.wikimedia.org/wikipedia/commons/f/f3/Periodic_Table_overview_29.svg *License:* CC BY-SA 3.0 *Contributors:* Own work *Original artist:* DePiep

- **File:Periodic_table_(metals–metalloids–nonmetals,_32_columns).png** *Source:* https://upload.wikimedia.org/wikipedia/commons/7/78/Periodic_table_%28metals%E2%80%93metalloids%E2%80%93nonmetals%2C_32_columns%29.png *License:* CC BY-SA 4.0 *Contributors:* Own work *Original artist:* DePiep

- **File:Periodic_table_(polyatomic).svg** *Source:* https://upload.wikimedia.org/wikipedia/commons/9/98/Periodic_table_%28polyatomic%29.svg *License:* CC BY-SA 3.0 *Contributors:* Own work -- Actually "inspired by"/forked from earlier free versions on Wikipedia/Commons like this, but there is no option to note this in Upload. *Original artist:* DePiep

- **File:Periodic_table_blocks_spdf_(32_column).svg** *Source:* https://upload.wikimedia.org/wikipedia/commons/f/f2/Periodic_table_blocks_spdf_%2832_column%29.svg *License:* CC BY-SA 3.0 *Contributors:* https://commons.wikimedia.org/wiki/File:Periodic_Table_2.svg *Original artist:* User:DePiep

- **File:Periodic_table_by_Mendeleev,_1869.svg***Source:*https://upload.wikimedia.org/wikipedia/commons/c/ce/Periodic_table_by_2C_1869.svg *License:* Public domain *Contributors:* Own work *Original artist:* NikNaks

- **File:Periodic_table_by_Mendeleev,_1871.svg***Source:*https://upload.wikimedia.org/wikipedia/commons/a/aa/Periodic_table_by_Men 2C_1871.svg *License:* Public domain *Contributors:* Own work *Original artist:* NikNaks

- **File:Periodic_trends.svg** *Source:* https://upload.wikimedia.org/wikipedia/commons/f/fe/Periodic_trends.svg *License:* CC0 *Contributors:* Own work *Original artist:* Mirek2

- **File:Periodic_variation_of_Pauling_electronegativities.png***Source:*https://upload.wikimedia.org/wikipedia/commons/b/b4/Periodic_v_of_Pauling_electronegativities.png *License:* CC-BY-SA-3.0 *Contributors:* Own work *Original artist:* Physchim62

- **File:Phase-diag2.svg** *Source:* https://upload.wikimedia.org/wikipedia/commons/3/34/Phase-diag2.svg *License:* CC-BY-SA-3.0 *Contributors:* SVG conversion from raster image Image:Phase-diag.png; some additions from Image:Phase diagram.png *Original artist:* me

- **File:Phase_change_-_en.svg** *Source:* https://upload.wikimedia.org/wikipedia/commons/0/0b/Phase_change_-_en.svg *License:* Public domain *Contributors:* Own work *Original artist:* F l a n k e r, penubag

- **File:Phase_diagram_for_pure_substance.JPG** *Source:* https://upload.wikimedia.org/wikipedia/commons/e/e6/Phase_diagram_for_pure_substance.JPG *License:* CC BY-SA 3.0 *Contributors:* Own work *Original artist:* Brews ohare

- **File:Phase_diagram_of_water.svg** *Source:* https://upload.wikimedia.org/wikipedia/commons/0/08/Phase_diagram_of_water.svg *License:* CC BY-SA 3.0 *Contributors:* Own work *Original artist:* Cmglee

- **File:Physics_matter_state_transition_1_en.svg** *Source:* https://upload.wikimedia.org/wikipedia/commons/8/8f/Physics_matter_state_1_en.svg *License:* GFDL *Contributors:* Own work *Original artist:* ElfQrin

- **File:Plasma-lamp_2.jpg** *Source:* https://upload.wikimedia.org/wikipedia/commons/2/26/Plasma-lamp_2.jpg *License:* CC-BY-SA-3.0 *Contributors:*

- own work www.lucnix.be *Original artist:* Luc Viatour

- **File:Plasma_fountain.gif** *Source:* https://upload.wikimedia.org/wikipedia/commons/b/b9/Plasma_fountain.gif *License:* Public domain *Contributors:* ? *Original artist:* ?

- **File:Plasma_jacobs_ladder.jpg** *Source:* https://upload.wikimedia.org/wikipedia/commons/4/40/Plasma_jacobs_ladder.jpg *License:* CC BY-SA 3.0 *Contributors:* Own work *Original artist:* Chocolateoak

- **File:Plasma_scaling.svg** *Source:* https://upload.wikimedia.org/wikipedia/commons/4/4c/Plasma_scaling.svg *License:* Public domain *Contributors:* Transferred from en.wikipedia to Commons. *Original artist:* Jafet.vixle at English Wikipedia

- **File:Plastic_household_items.jpg** *Source:* https://upload.wikimedia.org/wikipedia/commons/b/b2/Plastic_household_items.jpg *License:* CC BY-SA 3.0 *Contributors:* Own work *Original artist:* ImGz

- **File:Portal-puzzle.svg** *Source:* https://upload.wikimedia.org/wikipedia/en/f/fd/Portal-puzzle.svg *License:* Public domain *Contributors:* ? *Original artist:* ?

- **File:PositronDiscovery.jpg** *Source:* https://upload.wikimedia.org/wikipedia/commons/6/69/PositronDiscovery.jpg *License:* Public domain *Contributors:* Anderson, Carl D. (1933). "The Positive Electron". *Physical Review* **43** (6): 491–494. DOI:10.1103/PhysRev.43.491. *Original artist:* Carl D. Anderson (1905–1991)

- **File:Potential_energy_well.svg** *Source:* https://upload.wikimedia.org/wikipedia/commons/c/c5/Potential_energy_well.svg *License:* Public domain *Contributors:* Based upon Image:Potential well.png, created by User:Koantum. This version created by bdesham in Inkscape. *Original artist:* Benjamin D. Esham (bdesham)

- **File:Pressure-enthalpy_chart_for_steam,_in_US_units.svg** *Source:* https://upload.wikimedia.org/wikipedia/commons/6/64/Pressure-chart_for_steam%2C_in_US_units.svg *License:* CC BY-SA 3.0 *Contributors:* Own work *Original artist:* Emok

- **File:Purplesmoke.jpg** *Source:* https://upload.wikimedia.org/wikipedia/commons/f/f0/Purplesmoke.jpg *License:* Public domain *Contributors:* Own work *Original artist:* Macluskie

- **File:QCD_phase_diagram.png** *Source:* https://upload.wikimedia.org/wikipedia/commons/8/8f/QCD_phase_diagram.png *License:* Public domain *Contributors:* Transferred from en.wikipedia; transferred to Commons by User:Quadell using CommonsHelper. *Original artist:* Original uploader was Dark Formal at en.wikipedia

- **File:Quark_structure_proton.svg** *Source:* https://upload.wikimedia.org/wikipedia/commons/9/92/Quark_structure_proton.svg *License:* CC BY-SA 2.5 *Contributors:* Own work *Original artist:* Arpad Horvath

- **File:Quartz_oisan.jpg** *Source:* https://upload.wikimedia.org/wikipedia/commons/4/4a/Quartz_oisan.jpg *License:* CC BY-SA 4.0 *Contributors:* Own work *Original artist:* Didier Descouens

- **File:Question_book-new.svg** *Source:* https://upload.wikimedia.org/wikipedia/en/9/99/Question_book-new.svg *License:* Cc-by-sa-3.0 *Contributors:*
Created from scratch in Adobe Illustrator. Based on Image:Question book.png created by User:Equazcion *Original artist:* Tkgd2007

- **File:Rotation_curve_(Milky_Way).JPG** *Source:* https://upload.wikimedia.org/wikipedia/commons/7/77/Rotation_curve_%28Milky 29.JPG *License:* CC BY-SA 3.0 *Contributors:* Own work *Original artist:* Brews ohare

- **File:S-p-Orbitals.svg** *Source:* https://upload.wikimedia.org/wikipedia/commons/0/0f/S-p-Orbitals.svg *License:* CC-BY-SA-3.0 *Contributors:* selfmade from [1] *Original artist:* This file was made by **User:Sven**

- **File:Sbs_block_copolymer.jpg** *Source:* https://upload.wikimedia.org/wikipedia/en/9/9a/Sbs_block_copolymer.jpg *License:* PD *Contributors:* ? *Original artist:* ?

- **File:Schlierenfoto_Mach_17_Delta_-_NASA.jpg** *Source:* https://upload.wikimedia.org/wikipedia/commons/a/ac/Schlierenfoto_Mach_17_Delta_-_NASA.jpg *License:* Public domain *Contributors:* ? *Original artist:* NASA

- **File:Selfassembly_Organic_Semiconductor_Trixler_LMU.jpg** *Source:* https://upload.wikimedia.org/wikipedia/commons/8/82/Selfassembly_Organic_Semiconductor_Trixler_LMU.jpg *License:* CC BY-SA 3.0 *Contributors:* Own work *Original artist:* Frank Trixler; adapted from LMU/CeNS: Organic Semiconductor Nanostructures

- **File:ShortPT20b.png** *Source:* https://upload.wikimedia.org/wikipedia/commons/c/c9/ShortPT20b.png *License:* CC BY-SA 3.0 *Contributors:* Own work *Original artist:* Sandbh

- **File:Wikiquote-logo.svg** *Source:* https://upload.wikimedia.org/wikipedia/commons/f/fa/Wikiquote-logo.svg *License:* Public domain *Contributors:* ? *Original artist:* ?
- **File:Wikisource-logo.svg** *Source:* https://upload.wikimedia.org/wikipedia/commons/4/4c/Wikisource-logo.svg *License:* CC BY-SA 3.0 *Contributors:* Rei-artur *Original artist:* Nicholas Moreau
- **File:Wikiversity-logo-Snorky.svg** *Source:* https://upload.wikimedia.org/wikipedia/commons/1/1b/Wikiversity-logo-en.svg *License:* CC BY-SA 3.0 *Contributors:* Own work *Original artist:* Snorky
- **File:Wiktionary-logo-en.svg** *Source:* https://upload.wikimedia.org/wikipedia/commons/f/f8/Wiktionary-logo-en.svg *License:* Public domain *Contributors:* Vector version of Image:Wiktionary-logo-en.png. *Original artist:* Vectorized by Fvasconcellos (talk · contribs), based on original logo tossed together by Brion Vibber
- **File:Woven_bone_matrix.jpg** *Source:* https://upload.wikimedia.org/wikipedia/commons/e/ec/Woven_bone_matrix.jpg *License:* Public domain *Contributors:* Transferred from en.wikipedia; transferred to Commons by User:Magnus Manske using CommonsHelper. *Original artist:* Robert M. Hunt. Original uploader was Robert M. Hunt at en.wikipedia
- **File:Wpdms_physics_proton_proton_chain_1.svg** *Source:* https://upload.wikimedia.org/wikipedia/commons/7/74/Wpdms_physics_proton_chain_1.svg *License:* Public domain *Contributors:* Own work *Original artist:* see below

18.12.3 Content license

- Creative Commons Attribution-Share Alike 3.0

www.ingramcontent.com/pod-product-compliance
Lightning Source LLC
Chambersburg PA
CBHW080801180526
45168CB00006B/2287